quiring Editor: Jerry Papke
oject Editor: Al Starkweather
over design: Dawn Boyd

Library of Congress Cataloging-in-Publication Data

Terplan, Kornel
    Telecom operations management solutions with NetExpert / Kornel Terplan
         p. cm.
      Includes bibliographical references and index.
      ISBN 0-8493-3224-9 (alk. paper)
      1. Telecommunictions systems--Management. 2. NetExpert. 3. Computer networks--Manage-
    ment I. Title.
    TK5102.5.T459 1998
    621.382′1—dc21
                                           98-3925
                                          CIP

No claim to original U.S. Government works
International Standard Book Number 0-8493-3224-9
Library of Congress Card Number 98-3925
Printed in the United States of America 1 2 3 4 5 6 7 8 9 0
Printed on acid-free paper

TELECOM OPERA
MANAGEMENT SOLUT
With NetExpert™

Kornel Terplan

CRC Press

Boca Raton   Boston   London   New York   Washington, D.C.

Author

Dr. Kornel Terplan is a telecommunications expert with more than 30 years of highly successful multinational consulting experience. His book, Communication Networks Management, published by Prentice Hall (now in second edition), and his book on Effective Management of Local Area Networks, published by McGraw-Hill, are viewed as the state-of-the-art compendium throughout the community of international corporate users.

He has provided consulting, training and product development services to more than 75 national and international corporations on four continents, following a scholarly career that to date has combined 140 articles, 16 books, and 115 papers with editorial board services.

Over the last 20 years he has designed five network management related seminars and has made 80 seminar presentations in 15 countries. He received his doctoral degree at the University of Dresden and completed advanced studies, researched and lectured at Berkeley, Stanford University, University of California at Los Angeles, and Rensselaer Polytechnic Institute in Troy, NY.

His consulting work concentrates on network management products and services, outsourcing, central administration of LANs, network management centers, strategy of network management integration, implementation of network design and planning guidelines, product comparisons, and benchmarking network management solutions.

His most important clients include AT&T, GTE, Walt Disney World, Boole & Babbage, BMW, Kaiser Permanente, Siemens, France Telecom, German Telekom, Commerzbank, Union Bank of Switzerland, Creditanstalt Austria, Swiss Credit, State of Washington, Georgia Pacific Corp., Objective Systems Integrators, Unisource, and the Hungarian Telecommunications Co.

He is Industry Professor at Brooklyn Polytechnic University and at Stevens Institute of Technology.

Acknowledgments

The author would like to thank these contributors, listed in alphabetical order:

Amy Butts

Camille Campanale, Carolyn Covington, Ed Craft, and Randy Custeau

Scott Dahlstrom

David Friedman

A. J. Germek

Matt Izzo

Jerry Johnson and Tom Johnson

Lucy Kataoka and Ed Kurzenski

Fillipo Lodo

Jean Murphy

Mo Nikain

Ed Reeder

Tim Sebring and Chris Simon

Dick Vento and Mike Vitella

The author also would like to thank Mary Cate O'Malley for her excellent coordination work with subject matter experts from Objective Systems Integrators. Without her help, the author would not have been able to submit his manuscript on time.

Special thanks are due to Jerry Papke, Al Starkweather, and Susanne Lassandro of CRC Press who were extremely helpful in every phase of this publication.

The book was developed in association with Objective Systems Integrators (OSI) of Folsom, CA.

Preface

The telecommunications industry is in continuous change. Due to innovations in the technology, the life cycle of support systems is getting shorter and shorter. Incumbent and new operators, including RBOCs, LECs, CLECs, CAPs, IXCs, PTTs, and ISPs, try to highlight differentiators in their service offerings. Existing and future support systems address various layers of the telecommunications management networks. Besides operating element managers and network management systems, present and future emphasis is on service management, including service definition, service creation, provisioning, assurance, service usage, and billing. Based on data collected, processed, and consolidated in these layers, business support systems of the future concentrate on usage trends, customer behaviors, consolidated statistics, data warehousing, data mining, and on financial applications. To better match each client's needs and to help differentiate their services, carriers provide service level agreements.

The objective of this book is to show how all the emerged telecommunication technologies, such as voice networks, dedicated lines, ISDN, and packet switching, and emerging technologies, such as frame relay, FDDI, SMDS, SONET/SDH, ATM, xDSL, and wireless can be managed.

Management applications and products are based on the NetExpert framework. NetExpert is not the only OSS solution in the industry, but it combines all the features that network personnel rely on to manage complete, multivendor networks; it offers management from remote locations by implementing web-based technology; it enables service providers with different business models to build comprehensive, customized OSS solutions unique to their needs; and it is the fastest growing OSS-framework in the wireless industry.

Competition to NetExpert comes from software platform providers, system integrators, outsourcers, hardware vendors, and network equipment manufacturers. In many cases, NetExpert is working with management products from other providers. Comparing NetExpert with other products is beyond the scope of this publication.

The book consists of seven parts:

Part I focuses on requirements and standards for operations support systems. Special emphasis is placed on the Telecommunications Management Network (TMN) standards, COBRA, and on the NetExpert management framework.

Part II deals with the management of voice-related services, including service creation, provisioning, assurance, and billing. Two application rulesets, switchMASTER and trafficMASTER, based on the NetExpert framework, demonstrate successful implementation examples.

Part III addresses the needs of the wireless industry. Various wireless technologies are under consideration. Criteria and prerequisites of successful management are detailed and demonstrated by the mobileMASTER ruleset package.

Part IV deals with the management of frame relay, SONET/SDH, and ATM networks. After covering the basics, transportMASTER is shown in its role of managing high-capacity transmission systems. This part also includes a detailed analysis of the North Carolina Information Highway, including the role of NetExpert in the OSS of its principal builders.

Part V concentrates on how to manage broadband services, xDSL and HFC. From the management perspective, everything is new, including consumer expectations, transport protocols, business processes, and supplier alliances. The loopMASTER application ruleset helps to understand the complexity of managing this heterogeneous environment.

Part VI is devoted to helping Internet Service Providers (ISPs) manage their services. Time-to-market with new services is critical because the ISP market is so crowded. Innovative management strategies are absolutely necessary. Examples are given for customer network management, management of SNMP-based components, and managing corporate intranets. Three case studies demonstrate the applicability of the NetExpert framework, and its application toolsets.

Part VII addresses the cost/benefit ratio with management frameworks. This part supports the buy/build-decision for operations support systems. Also, tangible and intangible benefits with a management framework are listed.

Kornel Terplan
Hackensack, NJ

Subject Matter Experts
from Objective Systems Integrators
Folsom, CA

October 1997

Contents

Part I

Requirements and Standards

This part of the book first addresses basic management processes of telecommunications providers. These processes consist in most cases of service creation, service delivery, service assurance and of allocating and distributing service charges. For both incumbent and new providers, customer care is the number one priority. *Chapter 1* details management processes following the recommendations of the Network Management Forum. It also details the requirements for innovative operations support systems. A short market survey tries to quantify the market size for OSSs. The closing segment introduces the NetExpert framework with its principal core and extended modules and rulesets. The rulesets address various levels of managed objects.

Chapter 2 deals with emerged and emerging technologies. After a short introduction into the basics of the technology, the management capabilities and solutions are outlined. Information models, conformance to standards, management protocols and management products are referenced. Most of these technologies are targeted again to show how the NetExpert framework is capable of managing networks and systems based on these technologies.

Chapter 3 gives an overview of TMN (Telecommunication Networks Management). TMN will play a key role in future management architectures. Both the TMN models, and TMN layers, such as network elements, element management, network management, service management and business management are described in detail.

Other industry standards, such as CMIP, SNMP, RMON, CORBA, Java, and WBEM are addressed in *Chapter 4*, which prepares the reader for managing heterogeneous technologies with one umbrella manager.

Management frameworks play a key role in managing networks and systems. *Chapter 5* gives a detailed overview of the NetExpert management framework. It concentrates on gateways between the management server and the managed objects, on the management server with its core expert system, called IDEAS, and the management clients, called workstations. Both core elements and enablers of the framework technology are addressed in some depth. The pre-packaged rulesets focus on different layers of the TMN architecture. Usually they address devices, network elements and services by point, domain and enterprise rulesets. Management enablers concentrate on reporting, package administration, data archiving, and deploying Web capabilities into the NetExpert framework.

Finally, the last segment of this part, *Chapter 6* demonstrates how NetExpert is supporting the TMN architecture and COBRA.

1 New Requirements to the Management Framework

CONTENTS

FIGURES

The communication environment is changing rapidly. The barriers of traditional phone and data technologies are going to break down. The user can expect a true multimedia environment with existing services transferred and brand new services implemented. Completely new suppliers, such as cable companies, will compete with interexchange carriers, RBOCs and local phone companies for the market share. The differentiator is the price/performance ratio and time-to-market delivery of the service under consideration. Today's migrated and new services lack powerful management solutions. NetExpert is able to address all the challenges the innovative telecommunication industry is generating towards management solutions.

1.1   NETWORK MANAGEMENT DEFINITION AND PRINCIPAL FUNCTIONS

Network management means deploying and coordinating resources in order to design, plan, administer, analyze, operate, and expand communication networks to meet service-level objectives at all times, at a reasonable cost, and with optimal combination of resources.

The practical business model from the perspective of a telecommunications service provider shows multiple layers, such as (NMF95):

- Customer care processes
- Service development and maintenance processes
- Network systems management processes.

These specific processes are supported by common processes and databases of the management framework. Figure 1.1 shows a possible breakdown of these three principal layers.

In addition to the overall model, more detailed descriptions of each process also can be created. Each entry represents a process on its own. But, these processes are heavily interrelated. For instance, the service configuration process can be further detailed by defining its principal functions:

- design solution
- assign capacity
- configure network
- configure customer premises equipment
- update customer configuration record
- initiate orders to other service providers
- initiate installation work
- activate service
- report completion

There are also other business models that address the network and systems management processes and functions. Figure 1.2 shows this model with more detailed processes in comparison to Figure 1.1.

The functional groups of this model are described as follows (TERP96):

The entry point into the model is the client. Clients represent internal or external customers or any other users of management services. Clients may report problems, request changes, order equipment or facilities, or just want to get information. This interface is the Client Contact Point, implemented as the Single Point of Contact to handle all client-related problems, changes, orders and inquiries. Principal activities include:

- Receive problem reports
- Handling calls
- Inquiry handling

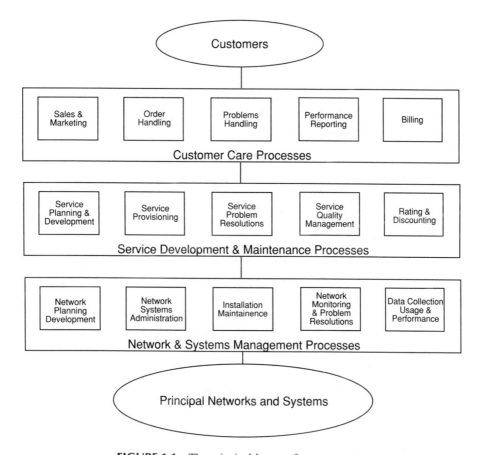

FIGURE 1.1 The principal layers of management.

- Receive change requests
- Handling orders
- Make service requests
- Opening and referring trouble tickets
- Closing trouble tickets

Operations support receives trouble tickets from the Client Contact Point. Major activities of Operations Support include:

- Problem determination by handling trouble tickets
- Problem diagnosis
- Taking corrective actions
- Repair and replacement
- Referring to third parties
- Backup and reconfiguration
- Recovery
- Logging events and corrective actions

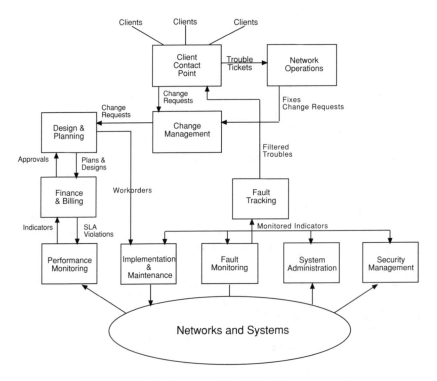

FIGURE 1.2   Business model of network management.

This business area may be further subdivided into second and third level support, support by vendors, and by third-party maintainers. As a result of their troubleshooting work, change requests for fixes are sent to Change Control. Not only clients, but also monitors may report problems to Operations Support. In this case, trouble-tickets are opened in Fault Monitoring, and forwarded to Operations Support via Fault Tracking. Principal functions of Fault Tracking concentrate on:

- Tracking manually reported or monitored faults
- Tracking the progress and escalating problems, if necessary
- Information distribution
- Referral

This is a very central activity playing a key role in supervising and correcting service-quality-related problems. Change Control is dealing with:

- Managing, processing and tracking service orders
- Routing service orders
- Supervising the handling of changes

The result of these activities is validated change requests sent to Planning and Design. This business area is supporting design and planning functions, such as:

- Needs analysis
- Projecting application load
- Sizing resources
- Authorizing and tracking changes
- Raising purchase orders
- Producing implementation plans
- Establishing company standards
- Quality assurance

The results of the work of this group is distributed to Finance and Billing, and to Implementation and Maintenance. Implementation and Maintenance implements changes and work orders sent by Planning and Design, and by Change Control. In addition, this area is in charge of

- Implementing change requests and work orders
- Maintaining resources
- Inspecting
- Maintaining configuration database
- Provisioning

By continuously monitoring systems and networks, status and performance information is collected. Fault Monitoring is proactively detecting problems, opening and referring trouble tickets.

Performance Monitoring deals with:

- Monitoring and reporting the frequency, duration, and severity of error conditions on systems and networks
- Monitoring Service Level Agreement compliance
- Monitoring third party and vendor performance with regard to problem alleviation
- Reporting on performance statistics and trends to management and to users

Performance Monitoring is informing Finance and Billing with status of service quality.

Security Management is responsible for ensuring secure communication and for protecting the management system. In more detail, the following functions are supported:

- Threat analysis
- Administration (access control, partitioning, authentication)
- Detection (evaluating services and solutions)
- Recovery (evaluating services and solutions)
- Protecting the management systems

Systems Administration is responsible to administer the whole distributed processing environment including:

- Software version control
- Software distribution
- Systems management (upgrades, disk space management, job control)
- Administering the user definable tables (user profiles, router tables, security servers)
- Local and remote configuring resources
- Names and address management
- Applications management

In addition to the traditional lists, a number of new management functions have been defined and implemented. Most examples are originating with telecommunications service providers, and most are service oriented. In particular, the following are in the middle of attention. Most of them are defined in the middle part of Figure 1.1 under the umbrella of service development and maintenance:

- **service creation**   design, prototype, test and launch new services
- **service activation**   refers to activities associated with the establishment and/or modification of customer service
- **service provisioning**   refers to a chain of activities based on service orders until all physical and logical resources are turned on in the network
- **connection management**   refers to the management of all physical resources that constitutes connections
- **service management**   refers to a set of activities required to maintain the network and associated services on an ongoing basis
- **workflow management**   optimal and coordinated way of organizing tasks for service creation, activation and provisioning. Usually, it is considered as part of configuration management.
- **workforce management**   optimal dispatch of human resources to provisioning, maintenance and troubleshooting work orders and
- **customer network management**   lets incorporated users of communication services view and alter their segments of the provider's network

TMN will unify and simplify these processes by clearly defining four layers, such as:

1. Business Management Layer (BML)
   - total enterprise responsibility
   - goal achievement
   - executive action
   - supported by billing/rating management, financial systems, sales/marketing applications, trending and statistics, and by data warehousing.
2. Service Management Layer (SML)
   - interaction with BML

- interfacing with customers and other administrators
- interaction with service providers
- maintaining service level agreements
- maintaining statistical data
- interaction between services
- interaction with NML
- supported by service level agreements, performance logs, alarm logs, resource usage distribution, workforce management, and reporting tools.

3. Network Management Layer (NML)
   - interaction with SML on performance, resource usage, availability
   - provision, cessation, or modification of network capabilities for the support of service to customers, control and coordination of the network view of all network elements within its scope or domain
   - supported by management frameworks, correlation applications, planning tools, switch and bandwidth provisioning systems, traffic analysis and testing tools.

4. Element Management Layer (EML)
   - provision of a gateway (mediation) function for the NML to interact with network elements
   - maintaining statistical, log, and other data about elements
   - control and coordination of a subset of network elements supported by point products running on various platforms.

Most processes and functions referenced previously can be ported into the new management architecture represented by TMN.

The practical implementation of processes and functions is accomplished by means of frameworks. Frameworks are sophisticated and powerful software structures to support basic and extended management attributes. They are flexible in terms of underlying operating systems and hardware. The majority of frameworks are based on UNIX, NT or on proprietary systems. The integrated entity of management software, operating system and hardware is defined as Operations Support System from the perspective of a telecommunications service provider. In the commercial business environment, the term *management platform* is used widely. Management platforms do not contain telco-specific applications. In other areas, however, such as operating systems and hardware, similarities cannot be denied.

1.2 EXPECTATIONS FROM USERS AND CHALLENGES TO OPERATIONS

The expectations for the new management framework on behalf of users are very high. These expectations may be summarized as follows:

- Capability of real-time status surveillance on various components of the networks
- Capability of processing a very large number of management related messages, alarms, and notifications within a short period of time
- Support of policy-based management
- Support of real-time alarm correlation to highlight the cause of problems
- Use of case-based-reasoning to utilize experiences from identical or similar problems
- Support of a powerful data repository where static and dynamic attributes of managed objects can be maintained
- Capabilities to distribute databases and management functions
- Access to accounting information in real time or near real time
- Support of client/server-structures with great flexibility of task distribution
- Support of web-based management for certain management functions, where security and performance are not critical
- Real time accessibility of performance information
- Very flexible configuration capabilities by a relational or object oriented database
- Providing accurate data for billing and accounting
- Consistent management of equivalent network elements from different suppliers
- Element management for a diversity of transport and switching technologies, including:
  — traditional telephony (POTS) and ISDN switches
  — wireless (cellular and PCS)
  — digital cross-connect systems
  — intelligent network equipment
  — hybrid fiber/coax (HFC)
  — fiber in the loop (FITL)
  — synchronous optical network (SONET)
  — synchronous data hierarchy (SDH)
  — asynchronous transfer mode (ATM)
  — fast packet switching (frame and cell relay)
  — switched multimegabit data service (SMDS)
  — connectionless data services
  — xDSL equipment and systems
- Common service-level management across different network architectures and technologies
- Common access to corporate data regardless of format and location
- Support for both standard and proprietary operations interfaces
- Integration with third-party, or existing/legacy Operations Support systems using standard and proprietary interfaces

- Rapid customization of OSS management applications and interfaces
- Scalable, reliable support for large networks

1.3　ORIENTATION OF THE NEW OPERATIONS SUPPORT SYSTEMS

The term Operations Support System (OSS) is being used frequently in this book. OSS is — in general — the whole discipline. It is the total operations infrastructure, and that includes things that operate the network and others that care for the customers. In other words, it includes everything that runs or monitors the network, but is not actually the network.

A new perspective for the management environment must begin by recognizing that the network fabric is amorphous. Topology, configuration, and element components will be ever changing. The model must also recognize that each task that the network will be performing places its participating elements in a state of multidimensional virtual identity.

On one dimension, bandwidth spectrum over the same connection will be constantly changing. On another dimension, the view of that bandwidth and related service will change on a user-by-user, and provider-by-provider basis.

The management and support environment model must maintain a view of the total network fabric and its ever-changing state. It must, therefore, be all encompassing, multidimensional, and able to change its relationship with each network element on a virtual basis. Important attributes of the new management model are:

Object Orientation　　Object-based technology is specifically geared to allow logic to be built in an "organic" fashion and forms the basis of a new network management environment. By combining the behavior of a network element with its function, a series of attributes is established for its operation. These attributes, or objects, form the initial building blocks of a dynamic model.

Groups of objects with similar operating methods or processes can then be organized into classes. A series of switches, for instance, may have independent characteristics for which an object (attributes) would be defined. They all, however, switch traffic. The methods and variables of the switching procedures can be defined as a class. Classes and subclasses can be cut and pasted into other classes, allowing for a rapid, yet connected definition of operation. When a new switch is added, its attributes would be declared in a new object along with reference to its class, thus allowing it to be added to the management environment without disruption. Similarly, objects or elements can be removed from the definition of fabric content without having to "update" every other element defined to the fabric.

Interobject, and interclass behaviors are defined as relationships to one another. By using objects and relationships, a high degree of autonomy and independence is maintained, yet the entire network fabric can be modeled.

By defining relationships separately from the object attributes themselves, a multidimensional structure is put into place that when combined into a repository, forms a complete Management Information Base (MIB). The organic nature of the object-oriented approach allows for the modeling of the network management and support environment as it actually is vs. hard coding it for what one may think it might do. In this fashion, network elements are free to "behave" in whatever manner they may be called upon to do . By using object-oriented technology as a network management and support environment fundamental, the unpredictable aspects of the network are not arbitrarily bounded, yet the action of the elements can be discretely monitored and tracked.

Expert Analysis　　With an object-oriented model definition in place, the boundaries inherent in traditional systems have been removed. The next fundamental that must be put into place is a method to handle the enormous amounts of informational data that must flow through the network management and support environment.

A data flow methodology must be implemented that will not only manage the data, but will make automated decisions on the data. Human interfaces, no matter the number, will not be able to keep pace with the torrent of information. Critical data will need to be prioritized. Noncritical data will need to be filtered. Alien data will require translation, and corollaries will be needed to transform status information into actionable events.

To provide the automated decision making capacities that broadband network information traffic flow will demand requires the use of expert systems. With the aid of knowledge-based rules, automation can be applied to the interpretation of network events. These rules are dynamically modifiable, and contain the decision logic for event thresholding and correlation, alert generation, severity definition, and other message states.

With the coupling of expert systems rules interpretation to an object-oriented, relationship driven MIB, a dynamic environment is created that is definable, actionable and responsive in real time.

Standards and Open Systems　　With a framework for a dynamic, real-time management and support system laid out, care must be taken to ensure that the framework is accessible and flexible to a wide variety of networks, network elements, computing systems, and applications.

Fundamental to the achievement of flexibility and accessibility will be the adherence to standards, and the use of Open Systems. Incorporation of open operating environments, database environments, and systems interfaces will allow network and system management environments to interconnect to, and interoperate with, existing systems and network components.

Likewise, standards will allow providers to take advantage of market available platforms, databases, and applications as well as new developments in device technology and the ongoing "commoditization" of computing technology.

Standards compliance and incorporation are also vital to the network management framework. With the unpredictability in the depth to which the network provider's fabric must reach into that of its commercial customers, the provider must be prepared to deliver network interfaces on a variety of levels.

The incorporation of such protocol standards as SNMP and CMIP will allow the provider's fabric to interface to the customer's fabric without concern for the physical topology. Standards, too, will play an important role in the interface to residential network service elements. Compliance with popular database and user interface standards will allow direct support of content provider server technology. CORBA provides a communications standard between OSS components provided by various suppliers.

Beyond the support for device interface and interoperation, the use of standards and open systems provide a path to distribute the network management and support intelligence directly onto the elements themselves, thus allowing for both stand-alone and network fabric-based system management.

Rapid Deployment   The next component to a successful management environment model is to place the power of the system directly into the hands of those network knowledge experts who will be responsible for operating and managing the environment.

Moving the system definition and customization directly to the business procedure, experts will be mandatory to ensure rapid deployment of new service and device support. The dynamic and real-time nature of networks will not be able to function if time windows must be carved out in order to relate functional requirements to programmers, who then must develop and test code.

The delivery of directions to the system therefore must be immediate and free of the intervening step of code development. The system interfaces must be graphical, intuitive, and highly manipulatable. Displays must be customizable to meet the viewing needs of the varying skill sets of the business process experts. This will require support of point and click, pop-up menus and other standard commercial interfaces, plus the ability to use existing PCs, workstations and terminals.

Likewise, the dialogue with the system must be straightforward. Business experts and operators must be able to manipulate object definitions, relationships and rules quickly and easily.

The display of network components, topology and status must be zoomable from its highest level to its lowest, with the capacity to balloon various components and segments on commands. Operators must be able to simultaneously view the relationship between the graphical representation of the network and its elements, and the alert and conditional information associated with those representations. The human interface to the system then must be as sophisticated, yet as simple as the underlying framework model.

Domain Management   The basic principles for distributing operations in telecommunications networks include the following:

- Distribution of management functions specific to subnetworks or network elements to lower layers
- Consolidation of common network-wide management functions at upper layers
- Management of end-to-end services independent of the management of network infrastructures.

One important benefit of following these principles is to partition management functionality. This involves moving network elements and subnetwork-specific operations intelligence close to the devices and facilities being managed. Common network-wide operations and service management then can be performed for the network as a whole. TMN supports this partitioning of operations.

The one-to-one *Network Element–Element Management System* reflects current industry conditions whereby individual vendors often supply proprietary Element Management Layer systems with their own network elements. In a hybrid network, this can yield a mix of proprietary Element Management Layer systems, each with a different set of operations and interface capabilities. Providing a common, cohesive operational support over all services and equipment in this environment can become complex, requiring custom functionality at the Network Management Layer to balance out the varying inconsistencies between element management systems, as well as providing operations functions over the network as a whole. This unduly increases the complexity needed for OSSs at the Network Management Layer. As the network grows, and the diversity of services and equipment increases, the complexity of OSSs needed to provide common network and service level management becomes unwieldy, and the number of interfaces which must be added to these OSSs grows. This growing complexity clearly is not the intention of the TMN standard.

An equally valid, but more valuable implementation of the TMN model, is to provide integrated management over all network elements that make up a single common subnetwork, or domain. A single domain can span common geographical or functional boundaries, such as a city, market region, or operations center. Integrated management of all devices in a domain is needed regardless of the network element supplier. This is done by shielding the individual Element Management Layer operations for network elements in a domain with a common set of domain-wide functions including alarm correlation, performance monitoring and provisioning, and presenting a single interface to upper layer OSSs. OSSs providing integrated domain management are referred to as *Domain Managers.*

The Domain Manager can incorporate Element Managers by using tight integration. Instead of using more complex Element Managers it may be better to use a Mediation Device as part of the Domain Manager. The Mediation Device communicates with the Element Manager that is outside of the boundary of the domain.

Domain Managers serve to hide network elements and subnetwork-specific operations functions and interfaces from upper layer OSSs. The upper layer OSSs may then become more resilient to supplier, technology, and architecture changes in the network infrastructure. This follows the basic guidelines for distributing operations functionality described above and takes better advantage of the full potential of the TMN architecture.

Common Access to Operations Data   Many networks include legacy aspects that present unique challenges. One of these is the relative inaccessibility of management data, which may be the most inflexible aspect of a legacy OSS. The basic architecture of many legacy systems consists of a large internal database, based on a proprietary format, with a layer of operations functionality on top. In legacy, and even many current generation OSSs, this data is tightly integrated with the operations functionality. These systems act as owners of the data, and share it through limited proprietary interfaces. Any other system requiring access to that data must meet the unique interface.

For large or growing networks, this can introduce a great deal of complexity in the operations architecture. As the number of OSSs requiring access to common data grows, the number of interfaces needed between systems and the complexity of operations-data flows between systems increases. Maintaining the reliability, concurrence, and synchronization of this data becomes increasingly complex as well. The ultimate result is a reduced flexibility in the operations network, making it costly and time consuming to add OSS support for new services.

For the successful operations of modern integrated-service networks in a competitive industry, it is important to increase the accessibility of corporate management data stored in the OSS network. This includes allowing data to be stored in different formats, and supporting common access from other systems. Ultimately, this leads to a separation of operations data from operations functionality.

The goal of managing corporate data independently of OSS functionality is often referred to as "data warehousing" and is a stated goal of many telecommunications carriers. But, this does not imply a consolidation or centralization of data storage. Increased accessibility to operations data, and the ultimate decoupling of operations data from OSSs can be achieved independently of the location and format of the data.

Increased Availability Increased availability is a critical need for advanced operations networks and systems. Often the networks and services they manage require 99.99 percent availability. Availability can be improved by several mechanisms:

- Management system/application redundancy
- Interface redundancy
- Data and event replication
- Application backup and recovery

In addition to the mechanisms listed above, the NetExpert High Availability option also provides resynchronization after recovery. Some of these mechanisms are illustrated in Figure 1.3. For example, interface and system redundancy can allow one domain manager to take over if another system fails. Data and event replication between systems will ensure proper synchronization. At the Network Management Layer, a single management system supports both subnetworks. However, a synchronized backup NMS can take over if the primary system fails. Each system can include backup and recovery capabilities to reduce downtime in the event of a system failure.

Scalable Deployment Functional scalability can be achieved by adding to or extending the capabilities of prepackaged OSSs. For example, new interfaces and management applications can be added to a domain manager to support a new network element. Further capabilities can be added via a rapid development tools set, thereby creating functional capacity through rapid customization.

Additional capacity and performance can be achieved by distributing OSS functionality. In addition to the ability to distribute functional modules of the platform, OSS functionality can be partitioned across multiple systems, communicating with internal messaging systems. This allows operations functions to reside on different computing platforms, while sharing alarm notifications, object/attribute, and other events.

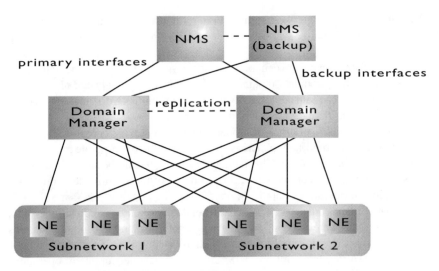

FIGURE 1.3   Increasing availability of management processes.

1.4   MARKET SURVEY ON OPERATIONS SUPPORT SYSTEMS

In the telecommunications industry, business operations, not just technology innovations, will drive progress. New products and new services are the differentiators
for overall market success. The challenge is to bring improved, advanced technology
to the marketplace in an efficient and revenue-generating manner. Time-to-market
is critical.

Customers are using measurable indicators to evaluate the performance of service providers, including service quality, meantime between failures, time-to-repair,
and others. The industry will experience nontraditional alliances, combining com-
plementary businesses, to bring technological expertise and marketing power to the
customer.

Not everything can and will be standardized. Time is again a critical factor.
Standards development, such as TMN, CORBA, TINA and others remain a time-
consuming process. Telecommunications suppliers cannot wait with their solutions
to a business problem for an emerging standard. In terms of new technologies, the
following trends can be observed:

- The Internet and Intranet will continue to drive the communications market in ways which will benefit both users of the Internet, those establishing
  corporate intranets, and the users of everyday communications technology.
- Voice over the Internet or intranets or Internet telephony, xDSL technology, network security, and electronic commerce issues will be resolved
  by technological innovation driven by the demands of Internet and intranet users.
- Many technology advances prompted by other drivers, such as the continued rise of ATM, SONET, SDH, and other broadband technologies, the

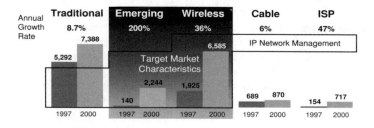

FIGURE 1.4 Growth rates in various market segments

use of video and document conferencing, expansion of the role of the traditional call center, the solidification of ISDN and xDSL in the market, advanced call center operations, wireless technology, EDI and data storage.

Telecommunications suppliers are expected to manage all these emerged and emerging technologies. At the same time, they must bring to the market products and services based on those technologies . The delivery vehicle is the Operational software, or in other words, the Operations Support System.

Telecommunications suppliers can be grouped as follows:

- Traditional or incumbent suppliers representing IXCs and LECs
- Emerging or competitive IXCs and LECs
- Wireless, cellular and PCS services providers
- Cable companies able to provide voice and data services over cable
- Internet Service Providers (ISP)

The annual growth rates are very different in those five groups. Figure 1.4 represents the actual revenues and the growth rates in each of the five groups.

Suppliers of management frameworks should actually target all areas: the traditional group for reengineering legacy OSSs, the wireless group to expand OSS-capabilities, the emerging group for its very high growth rates to supply them with management capabilities, ISPs for the great popularity of Internet/Intranet-applications, and finally the cable companies to provide them with basic management capabilities.

Objective Systems Integrators with the NetExpert management framework has been providing solutions for all five groups. The following chapters will introduce solution examples for each of the five groups.

1.5   NetExpert — OVERVIEW OF OPERATIONAL PRINCIPLES

NetExpert from Objective Systems Integrators consists of a series of coordinated modules that fall into three general groups — external network element and non-NetExpert subsystem gateways, object persistence and behavior servers, and user/operator workstation/Web interfaces. NetExpert is a robust, scalable and distributable architecture that supports a high degree of configuration flexibility while maintaining individual component independence. Easy to use, easy to modify, easy to initiate, NetExpert is quick to roll out and integrate with existing platforms or systems.

Figure 1.5 provides a high-level description of how NetExpert receives from and sends messages to external network elements and Operations Support Systems.

The NetExpert framework is a set of modules covering the basic functions that a distributed application needs, including gateways to the system, a way to send messages or "events," the intelligence to act on those events, and a consistent operator interface. A customer can distribute these modules across a network, gaining the foundation required to monitor continuous and large volumes of events and traffic.

The framework is controlled by rules that replace complex programming languages and enable network analysts to model desired system behaviors. Rules are written with the product's implementation tools. Existing rulesets, called application components, eliminate the cumbersome traditional development process, which entails writing requirements and building a complete solution from scratch. Rule writing is estimated to be up to 10 times more productive than traditional development methods.

The basis of supporting all of these functions is the NetExpert framework.

1.5.1   NetExpert Framework: Functional Components

The NetExpert framework consists of editors, operators, support functions, and other framework enablers. These are represented in Figure 1.6.

OSI technology gives service providers a strategic advantage because it shifts the distribution of development time and effort. When companies attempt to develop systems from the ground up, one hundred percent of their effort goes into requirements planning and customer programming. The development platform environment focuses eighty percent of the effort on system drivers and application program interface levels, while directing twenty percent of the programming effort to application. Conversely, the NetExpert framework and its rule editors target eighty percent of effort at the functional-application component layer, because the framework already contains the basic functions that underlie any application. In addition, writing application components with expert rules is short and simple compared to programming code.

OSI focuses the time-and-effort equation on the most productive component of the system solution. Customers avoid targeting precious resources to extensive software development. They give full attention to the application component layer, where applying intellectual dominance to core business areas is crucial to remaining competitive.

1.5.2   NetExpert Application Rule Packages

NetExpert's modifiable application packages provide a comprehensive subset of functions. These can be further tailored to individual customer environments. This is how the framework accommodates configuration-specific solutions and the

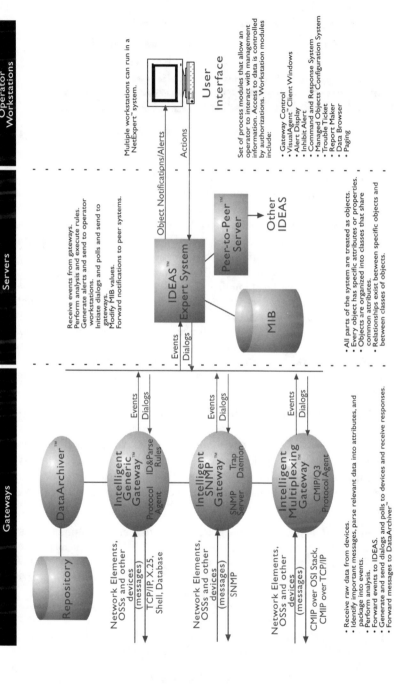

FIGURE 1.5 NetExpert framework operational overview.

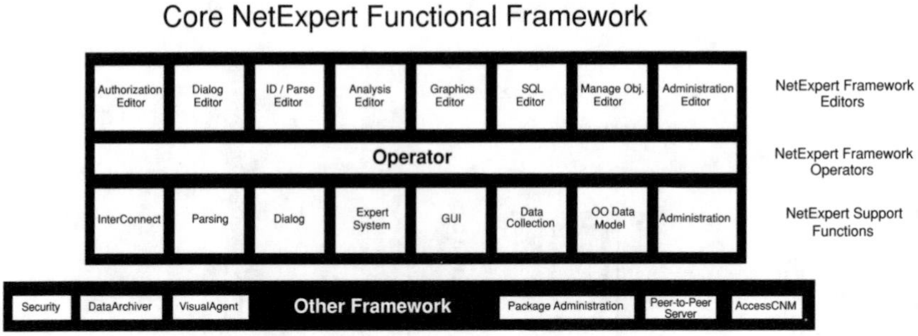

Core NetExpert Functional Framework

Authorization Editor	Dialog Editor	ID / Parse Editor	Analysis Editor	Graphics Editor	SQL Editor	Manage Obj. Editor	Administration Editor	NetExpert Framework Editors
Operator								NetExpert Framework Operators
InterConnect	Parsing	Dialog	Expert System	GUI	Data Collection	OO Data Model	Administration	NetExpert Support Functions

Security	DataArchiver	VisualAgent	**Other Framework**	Package Administration	Peer-to-Peer Server	AccessCNM

FIGURE 1.6   NetExpert functional framework.

demands of the customer's business model. Because they are object-oriented, these rule packages can potentially deliver myriad types of services; manage any number of tangible elements, such as switches or routers; and model intangible elements, such as the knowledge of expert technicians. Rules make it possible for the same NetExpert framework to manage such diverse networks as digital cellular, traditional telephony, high-speed data, or hybrid fiber/coax.

Application rules ride on top of the NetExpert framework. They are categorized as point, domain or corporate level application packages. The differences between each is the business focus they are designed to address. Point applications define the native messages required by a network element during, for example, the provisioning process. Domain applications group higher level commands into those associated with, for example, all switch or transport network devices constituting a service provider's network. Corporate applications perform, manage, and control functions associated with the domain- and point-level applications. The layering of corporate, domain, and point applications are illustrated in Figure 1.7.

Service Deployment, Integration Rules   Driven by the service provider's marketing organization, these types of rules are primarily service creation, change, and destroy oriented. A new service defines a workflow of activities that automatically provision/activate, monitor, and deactivate subscriber services based on service orders. Customization is moderate to heavy as this is where the service provider differentiates himself from his competition. Huge economies of scale are realizable as common service components are reused.

Point, Domain, and Corporate Business Rules   NetExpert-based point, domain, and corporate integration rules roughly align with the TMN element management layer, network management layer, and service management layer. More specifically, each of these types of rulesets offers significant advantages to a service provider who builds its OSS network on TMN principles.

Differences between each ruleset layer are as shown below:

- **Point Integration Rules**   Distinct solutions for managing specific device, standards-based interfaces, or legacy interfaces. Usually point integration rules are vendor-specific and dependent. Customization level is minimal. It corresponds to the element management layer in TMN.

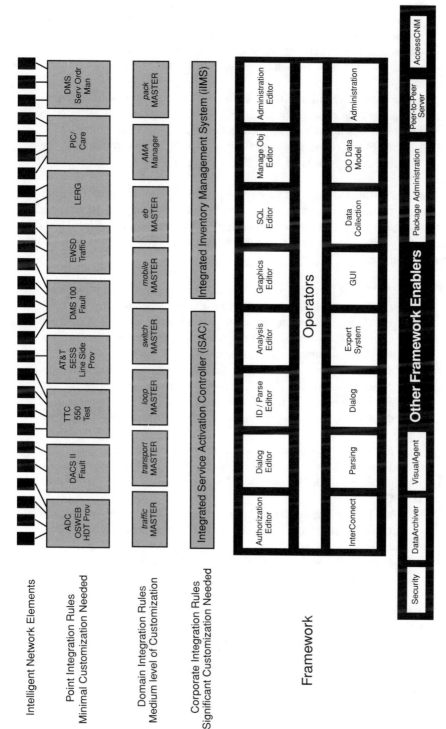

FIGURE 1.7 NetExpert rules based domain applications.

- **Domain Integration Rules**　Provide comprehensive solutions across vendor and process boundaries. These types of rules are primarily function-oriented and dependent on characteristics of the managed operation. Customization level is medium. It corresponds to the network management layer in TMN.
- **Corporate Integration Rules**　Allow for the corporate-wide functions to manage both data and processes. These rules are business and service process oriented and highly dependent on the organization, products, and services of the client. Customization level is high. It corresponds to the service management layer in TMN.

Application Rule Packages are the best combination of delivery time, price, selection, and customization for comprehensive end-to-end management systems. Outlined next are descriptions of each of the domain application rule packages offered by OSI.

Currently available application components for specific network domains or functions include:

- trafficMASTER for real-time performance management and for planning, analyzing, and managing traffic over switched telephony networks
- transportMASTER for managing transport networks containing Synchronous Optical NETwork (SONET), synchronous digital hierarchy (SDH), and digital speed number (DSN) elements
- loopMASTER for broadband access networks that include such elements as host digital terminals (HDTs), network interface units (NIUs), video servers, set-top boxes, and cable modems
- switchMASTER for managing wireline switches and intelligent network equipment
- mobileMASTER for wireless switched networks using technologies including time division multiple access (TDMA), call (or code) division multiple access (CDMA), and Global System for Mobile Communication (GSM)
- ebMASTER for the secure electronic sharing of network information among service providers whose customers' leased lines cross provider boundaries
- amaMANAGER and amaGATEWAY integrated process for the transfer, processing, and management of AMA data generated from network elements
- packetMASTER for packet data switches and ATM related network equipment
- integrated Service Activation Controller (iSAC) to manage and fully automate network service activation

Each of these domain applications reside on top of the NetExpert framework. Additional "point integration" rule packages within a functional domain are also offered to more quickly "jump start" an implementation of NetExpert for a customer's specific business purpose. Various functions including fault, configuration, and performance management are supported.

NetExpert has been used to provide flexible rule-based network management solutions since Objective Systems Integrators (OSI) was formed in 1989. Customers have applied the product to various requirements including fault management, flow-through provisioning, system mediation, electronic bonding, banking machine management, network inventory, AMA/OM data collection, and many more.

Over the last few years, some customers have used the tool to create multifunctional, integrated solutions — to use the NetExpert framework as a foundation product in their network management strategy. This approach has allowed NetExpert customers to utilize the full potential of a rule-based, object-oriented, distributed system. The advantages include:

- Accelerated design procedures
- Flexible integration of new network elements and OSSs
- Cross-vendor, function, and domain correlation
- Common maintenance and operation procedures
- Integrated problem resolution capability
- Substantial reductions in software costs

The framework, running in concert with NetExpert applications, enables users to generate revenue by quickly delivering new services. However, getting to market first is not enough. With NetExpert, users protect past investments, increase the life span of aging equipment, incorporate new elements, and integrate disparate management systems and software ruleset packages. Another advantage OSI users have is their ability to deploy network management systems and OSSs in formerly uncharted niches and integrate these with existing infrastructures. Users are closer than ever to automating their business models because OSI delivers the tools they need to translate key processes across systems that support entire operations.

1.6 SUMMARY

It is obvious that telecommunications suppliers do not start from scratch in terms of implementing operations support systems. The pressing needs of creating, provisioning and maintaining services require flexible and easily customizable management solutions. The needs of incumbent and new service providers differ somewhat, but not in terms of using open standards, rapid prototyping, partitioning by management domains, object orientation, and the use of COBRA to integrate OSSs from different suppliers..

The enrichment of management frameworks by specific applications makes them more powerful. Figure 1.8 illustrates this evolutionary process.

Today, applications and application interfaces dominate; by 2000 and beyond, the framework will incorporate element management, domain management, network and service management applications, as well.

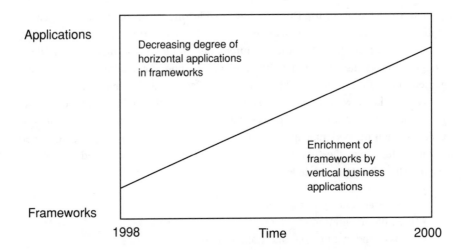

Applications

Decreasing degree of
horizontal applications
in frameworks

Enrichment of
frameworks by
vertical business
applications

Frameworks

1998   Time   2000

1998   Framework attributes provide little functionality
2000   Frameworks incorporate integrated management functions

FIGURE 1.8   Evolving management frameworks.

The following chapters concentrate on implementing NetExpert to address man-
agement challenges in the telecommunications industry.

2 Technology Overview

CONTENTS

FIGURES

This chapter summarizes all the emerged and emerging technologies that are expected to be managed by various management systems. The segmentation of networks into CPE, LEC, and IEX is followed by principal foundation concepts for networking technologies. Connection-oriented and connectionless communication, physical and virtual circuits, switching technologies, routing alternatives, multiplexing techniques, addressing schemes and control strategies are explained to the novice reader. Emerged technologies concentrate on T/E carrier systems, voice networks, SS7, X.25-based packet switching networks and ISDN. Also their conformance to OSI layers and their management solutions are included. Emerging technologies challange network managers by providing state-of-the-art technology without strong management solutions.

Emerging technologies include frame relay, FDDI, SMDS, ATM, SONET/SDH, and mobile and wireless communications. The capabilities of these technologies are presented using the same evaluation criteria. Other emerging IP network technologies, such Digital Subscriber Line, CTI, data and voice over cable, are just briefly addressed. The conformance to OSI layers and early management solutions are addressed, as well.

Managed objects in communication networks are grouped according to the geographical territory of their locations (Figure 2.1).

- CPE (Customer Premises Equipment) defining the location within the customers' territory.
- LEC (Local Exchange Carrier) or CLEC (Competitive Local Exchange Carrier) defining the location within the territory of local exchange.
- IXC (Inter-Exchange Carrier) or PTT (Postal Telephone and Telegraph) defining the wide area, connecting LECs, CLECs, and CPEs following the interexchange rules of the country or of the carriers.

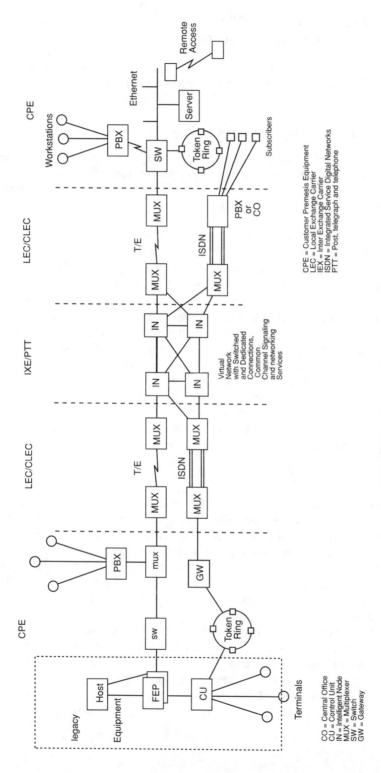

FIGURE 2.1 Segmentation of networks.

2.1 FOUNDATION CONCEPTS
FOR NETWORKING TECHNOLOGIES

The majority of emerged and emerging technologies have a few basic foundation principles. These foundation principles will be addressed in this segment. The basics for this segment can be found in more detail in (BLAC94).

2.1.1 CONNECTION-ORIENTED AND CONNECTIONLESS COMMUNICATIONS

Communication systems that employ the concepts of circuits and virtual circuits are said to be connection-oriented. Such systems maintain information about the users, such as their addresses and their ongoing QOS (Quality of Service) needs. Often, this type of system uses state tables that contain rules governing the manner in which the user interacts with the network. While these state tables clarify the procedures between the user and the communication network, they do add overhead to the communication process.

In contrast, communication systems that do not employ circuits and virtual circuits are said to be connectionless systems. They are also known as datagram networks and are widely used throughout the industry. The principal difference between connection-oriented and connectionless operation is that connectionless protocols do not establish a virtual circuit for the end user communication process. Instead, traffic is presented to the service provider in a somewhat ad hoc fashion. Handshaking arrangements, mutual confirmations are minimal and perhaps nonexistent. The network service points and the network switches maintain no ongoing knowledge about the traffic betwen the two end users. State tables as seen with connection-oriented solutions are not maintained. Therefore, datagram services provide no *a priori* knowledge of user traffic and they provide no ongoing, current knowledge of the user traffic. But, they introduce less overhead.

2.1.2 PHYSICAL AND VIRTUAL CIRCUITS

End users operating terminals, computers and client equipment communicate with each other through a communication channel called the physical circuit. These physical circuits are also known by other names, such as channels, links, lines, and trunks. Physical circuits can be configured wherein two users communicate directly with each other through one circuit, and no one uses this circuit except these two users. They can operate this circuit in half-duplex or full-duplex. This circuit is dedicated to the users. This concept is still widely used in simple networks without serious bandwidth limitations.

In more complex systems, such as networks, circuits are shared with more than one user-pair. Within a network, the physical circuits are terminated at intermediate points at machines that provide relay services on another circuit. These machines are known by such names as switches, routers, bridges, gateways, etc. They are responsible for relaying the traffic between the communicating users. Since many communication channels have the capacity to support more than one user session, the network device, such as the switch, router or multiplexer is responsible for sending and receiving multiple user traffic to/from a circuit.

In an ideal arrangement, a user is not aware that the physical circuits are being shared by other users. Indeed, the circuit provider attempts to make this sharing operating transparent to all users. Moreover, in this ideal situation, the user thinks that the circuit directly connects only the two communicating parties. However, it is likely that the physical circuit is being shared by other users.

The term "virtual circuit" is used to describe a shared circuit wherein the sharing is not known to the circuit users. The term was derived from computer architectures in which an end user perceives that a computer has more memory than actually exists. This additional virtual memory is actually stored on an external storage device.

There are three types of virtual circuits:

1. **Permanent virtual circuits (PVC)**   A virtual circuit may be provisioned to the user on a continuous basis. In this case, the user has the service of the network any time. A PVC is established by creating entries in tables in the network nodes' databases. These entries contain a unique identifier of the user payload which is known by various names, such as a logical channel number (LCN), virtual channel identifier (VCI), or virtual path identifier (VPI).

   Network features such as throughput, delay, security and performance indicators also are provisioned before the user starts with operations. If different types of services are desired, and if different destination end-points must be reached, then the user must submit a different PVC identifier with the appropriate user payload to the network. This PVC is provisioned to the different endpoint, and perhaps with different services.

2. **Switched virtual circuits (SVC)**   A switched virtual circuit (SVC) is not preprovisioned. When a user wishes to obtain network services to communicate with another user,the user must submit a connection request packet to the network. The packet must provide the address of the receiver, and it must also contain the virtual circuit number that is to be used during the session.

   SVCs entail some delay during the setup phase, but they are flexible in that they allow the user to select dynamically the receiving party and the negotiation of networking parameters on a call-by-call basis.

3. **Semi-permanent virtual circuits (SPVC)**   With this approach, a user is preprovisioned, as in a regular PVC. Like a PVC, the network node contains information about the communicating parties and the type of services desired. But, these types of virtual circuits do not guarantee that the users will obtain their level of requested service. In case of congested networks, users could be denied the service.

   In a more likely scenario, the continuation of a service is denied because the user has violated some rules of the communications. Examples are higher bandwidth demand and higher data rates than as agreed to with the supplier.

2.1.3 SWITCHING TECHNOLOGIES

Voice, video, and data signals are relayed in a network from one user to another through switches. This section provides an overview on prevalent switching technologies.

Figure 2.2 compares and contrasts three types of switching technologies.

Circuit switching provides a direct connection between two networking components. Thus, the communicating partners can utilize the facility as they see fit — within bandwidth and tariff limitations. Many telephone networks use circuit switching systems. Circuit switching provides clear channels, error checking, session establishment, frame flow control, frame formatting, and selection of codes. The protocols are the responsibility of the users.

Today, the traffic between communicating parties is usually stored in fast queues in the switch and switched to an appropriate output line with TDM techniques. This technique is known as circuit emulation switching (CES), which is summarized by:

- Direct connection end-to-end
- No intermediate storage unless CES used
- Few value-added functions
- Modern systems use TDM to emulate circuit switching

Message switching was the dominating switching technology during the last two decades. The technology is still widely used in certain applications, such as electronic mail, but it is not employed in a backbone network. The switch is usually a specialized computer. It is responsible for accepting traffic from attached terminals and computers. It examines the address in the header of the message and switches the traffic to the receiving station. Because of the low number of switching computers, this technology suffers under backup problems, performance bottlenecks and lost messages due to congestion. It can be summarized by:

- Use of store-end-forward technology
- Disk serves as buffers
- Extensive value-added functions
- Star topology because of expense of switches

Packet switching relays small pieces of user information to the destination nodes. Packet switching has become the prevalent switching technology of data communications networks. It is used in such diverse systems as private branch exchanges (PBXs), LANs and even with multiplexers. Each packet only occupies a transmission line for the duration of the transmission; the lines are usually fully shared with other applications. This is an ideal technology for bursty traffic. Modern packet switching systems are designed to carry continuous, high-volume traffic as well as asynchronous, low-volume traffic, and each user is given an adequate bandwidth to meet service level expectations.

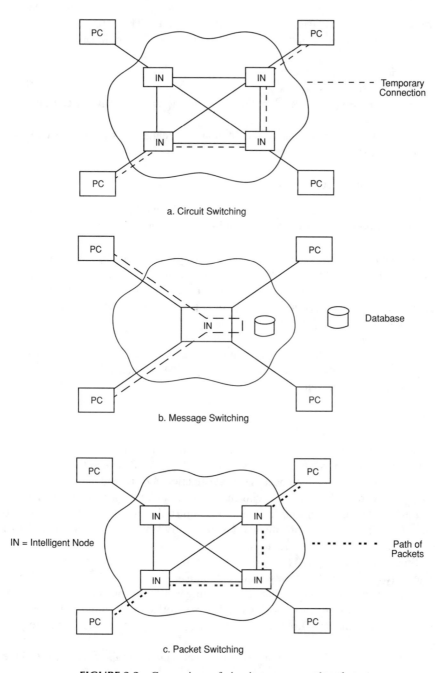

a. Circuit Switching

b. Message Switching

c. Packet Switching

FIGURE 2.2   Comparison of circuit, message and packet.

The concept of packet and cell switching is similar; each attempts to process the traffic in memory as quickly as possible. But cell switching is using much smaller Protocol Data Units (PDUs) relative to packet switching. The PDU-size is fixed with cell switching. The PDU size may vary with packet switching. In summary:

- Hold-and-forward technology
- RAM serves as buffers
- Extensive value-added-functions for packet, but not many for cells.

Switching will remain one of the dominating technologies in the telecommunications industry.

2.1.4 ROUTING TECHNOLOGIES

There are two techniques to route traffic within and between networks: source routing and non-source routing. The majority of emerging technologies use non-source routing.

Source routing derives its name from the fact that the transmitting device — the source — dictates the route of the PDU through a network or networks. The source places the addresses of the "hops" in the PDU. The hops are actually routers representing the internetworking units. Such an approach means that the internetworking units need not perform address maintenance, but they simply use an address in the PDU to determine the destination of the PDU.

In contrast, non-source routing requires that the interconnecting devices make decisions about the route. They don't rely on the PDU to contain information about the route. Non-source routing is usually associated with bridges and is quite prevalent in LANs. Most of the emerging new technologies implement this approach with the use of a virtual circuit identifier (VCI). This label is used by the network nodes to determine where to route the traffic.

The manner in which a network stores its routing information varies. Typically, routing information is stored in a software table, called a directory. This table contains a list of destination nodes. These destination nodes are identifiers with some type of network address. Along with the network address (or some type of label, such as a virtual circuit identifier) there is an entry describing how the router is to relay the traffic. In most implementations, this entry simply lists the next node that is to receive the traffic in order to relay it to its destination.

Small networks typically provide a full routing table at each routing node. For large networks, full directories require too many entries, and are expensive to maintain. In addition, the exchange of routing table information can impact the available bandwidth for user payload. These networks are usually subdivided into areas, called domains. Directories of routing information is kept seperately in domains.

Broadcast networks contain no routing directories. Their approach is to send the traffic to all destinations.

Network routing control is usually categorized as centralized or distributed. The centralized uses a network control center to determine the routing of the packets. The packet switches are limited in their functions. Central control is vulnerable; a backup is absolutely necessary. On the other hand, it increases the operating expense. Distributed control requires more intelligent switches, but provides a more resilient solution. Each router makes its own routing decisions without regard to a centralized control center. Distributed routing is also more complex, but its advantages over the centralized approach have made it the preferred routing method in most communications networks.

2.1.5 MULTIPLEXING TECHNOLOGIES

Most of the emerged and emerging technologies use some form of multiplexing. Multiplexers accept low-speed voice or data signals from terminals, telephones, PCs, and user applications and combine them into one higher-speed stream for transmission efficiency. A receiving multiplexer demultiplexes and converts the combined stream into the original lower-speed signals. There are various multiplexing techniques:

Frequency Division Multiplexing (FDM)   This approach divides the transmission frequency range into channels. The channels are lower frequency bands, each of which is capable of carrying communication traffic, such as voice, data or video. FDM is widely used in telephone systems, radio systems, and cable television applications. It is also used in microwave and satellite carrier systems. FDM decreases the total bandwidth available to each user, but even the narrower bandwidth is usually sufficient to the users' applications. Isolating the bands from each other costs some bandwidth, but the simultaneous use outweighs this disadvantage.

Time Division Multiplexing (TDM)   This approach provides the full bandwidth to the user or application, but divides the channel into time slots. Each user or application is given a slot and the slots are rotated among the attached devices. The TDM multiplexer cyclically scans the input signals from the entry points. TDMs are working digitally. The slots are preassigned to users and applications. In case of no traffic at the entry points, the slots remain empty. This approach works well for constant bit rate applications, but leads to waste capacity for variable bit rate applications.

Statistical Time Division Multiplexing (STDM)   This approach allocates the time slots to each port on a statistical basis. Consequently, idle terminal time does not waste the capacity of the bandwidth. It is not unusual for two to five times as much traffic to be accomodated on lines using STDMs in comparison to a TDM solution. This approach can accomodate bursty traffic very well, but does not perform too well with continuous, nonbursty traffic.

Wave Division Multiplexing (WDM)   WDM is the optical equivalent of FDM. Lasers operating at different frequencies are used in the same fiber, thereby deriving multiple communications channels from one physical path.

2.1.6 ADDRESSING AND IDENTIFICATION SCHEMES

In order for user traffic to be sent to the proper destination, it must be associated with an identifier of the destination. Usually, there are two techniques in use:

An explicit address has a location associated with it. It may not refer to a specific geographical location but rather a name of a network or a device attached to a network. For example, the Internet Protocol (IP) address has a structure that permits the identification of the network, a subnetwork attached to the network, and a host device attached to the subnetwork. The ITU-T X.121 address has a structure which identifies the country, a network within that country, and a device within the network. Other entries are used with these addresses to identify protocols and applications running on the networks. Explicit addresses are used by switches, routers, and bridges as an entry into routing tables. These routing tables contain information about how to route the traffic to the destination nodes.

Another identifying scheme is known by the term of label, although other terms may be more widely used. Those terms are logical channel number (LCN) or virtual circuit identifier (VCI). A label contains no information about network identifiers or physical locations. It is simply a value that is assigned to a user's traffic that identifies each data unit of that user's traffic.

Almost all connectionless systems use explicit addresses, and the destination and source addresses must be provided with every PDU in order for it to be routed to the proper destination.

2.1.7 CONTROL AND CONGESTION MANAGEMENT

It is very important in communication networks to control the traffic at the ingress and egress points of the network. The operation by which user traffic is controlled by the network is called flow control. Flow control should assure that the traffic does not saturate the network or exceed the network's capacity. Thus flow control is used to manage congestion.

There are three flow control alternatives with emerged and emerging technologies:

- **Explicit flow control** This technique limits how much user traffic can enter the network. If the network issues an explicit flow control message to the user, the user has no choice but to stop sending traffic or to reduce traffic. Traffic can be sent again after the network has notified the user about the release of the limitations.
- **Implicit flow control** This technique does not restrict the flow absolutely. Rather, it recommends that the user reduce or stop traffic it is sending to the network if network capacity situations require a limitation. Typically, the implicit flow control message is a warning to the user that:
  — the user is violating its service level agreement with the internal or external supplier
  — the network is congested

In any case, if the user continues to send traffic, it risks having traffic discarded by the network.

- **No flow control** Flow control also may be established by not controlling the flow at all. Generally, an absense of flow control means that the network can discard any traffic that is creating problems. While this approach certainly provides superior congestion management from the standpoint of the network, it may not meet the performance expectations of the users.

2.2   EMERGED TECHNOLOGIES

This segment gives an overview on telecommunications technologies that have been in operation for the past two decades. These communication technologies cover T1/E1 carrier systems, voice networks, signaling system number 7, X.25-based packet switching networks and ISDN.

2.2.1   INTRODUCTION OF TECHNOLOGIES

T1/E1 Carrier Systems The T1/E1 carrier systems are high-capacity networks designed for the digital transmission of voice, data, and video. The original implementations digitalized voice signals in order to take advantage of the benefits of digital technology. The term T1 was devised by the telephone companies to describe a specific type of carrier equipment. Today it is used to define a general carrier system, a data rate, and various multiplexing and framing conventions. A more concise term is DS1, which describes a multiplexed digital signal which is carried by the T carrier. Typical rates are:

DS1	T1	1.544 Mbit/s
DS2	T2	6.312 Mbit/s
DS3	T3	44.736 Mbit/s
DS4	T4	274.176 Mbit/s

Europe and Japan use different throughput rates, but this does not change the basic characteristics of this technology.

Today, the majority of offerings digitize the voice signal through pulse code modulation (PCM), or adaptive differential pulse code modulation (ADPCM). Whatever the encoding technique, once the analog images are translated to digital bit streams, then many T1 systems are able to time division multiplex voice and data together in 24 user slots (30 in Europe) within each frame.

A typical topology is shown in Figure 2.3. Different communications forms can use one digital pipe for transmission. Data transmissions are terminated through a statistical time division multiplexer (STDM), which then uses the TDM to concentrate the traffic across the transmission line through a channel service unit (CSU); or other equipment, such as data service unit (DSU), or the DSU and CSU may be combined, as well.

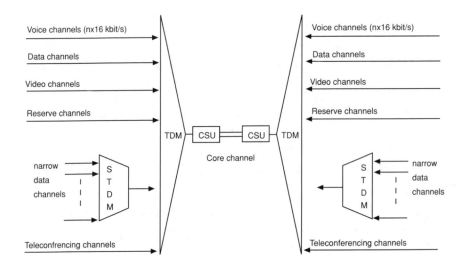

TDM = Time Division Multiplexers

STDM = Statistical Time Division Multiplexers

FIGURE 2.3   Configuration of T/E implementations.

The bandwidth of the channel or pipe can be divided into various T1 subrates. For example, a video system could utilize a 768 Kbit/s bandwidth, the STDM in turn could multiplex various data rates up to a 56 Kbit/s rate, and perhaps a CAD/CAM operation could utilize 128 Kbit/s of the bandwidth.

T/E-based carrier systems have been serving the industry well. But, they are quite limited in their management capabilities, and they provide very little support for customers for the provisioning of services. In the early days, the use of bandwidth control headers for network management was not encouraged due to the limited transmission capacity of the facilities to accomodate this overhead traffic. Today, the prevailing idea is to exploit the high capacity of optical fibers and the processors, and allocate a greater amount of bandwidth to support control, maintenance and management.

There are many opportunities to introduce additional services on the basis of clear channels. Examples are:

- Telex
- Teletex
- Telefax
- Electronic Mail
- Voice Mail
- Teleconferencing
- Videotex
- Electronic Data Interchange

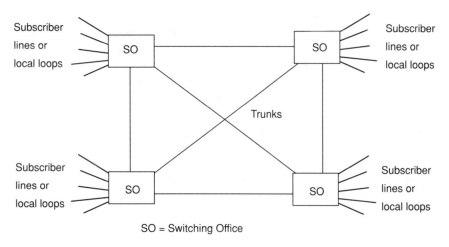

SO = Switching Office

FIGURE 2.4   Structure of typical voice networks.

Some of these services are losing their importance (e.g., telex, teletex, and videotex), others are gaining importance (e.g., mails, teleconferencing and EDI).

Voice Networks   A typical voice network is shown in Figure 2.4. The customer connects through the telephone system into the central office (CO), local exchange or end office. In generic terms, we can talk about switching offices. There are thousands of these offices in each country. Connection is provided to the switching office through a pair of wires or four wires called the local loop or subscriber loop, respectively.

The connections between the switching centers are hierarchically arranged. In the United States and Canada, there are toll centers, primary centers, sectional centers and regional centers. The use of direct or indirect connections between offices with tandem trunks or other tandem switching systems depends on several factors: distances between offices, the traffic volume between offices, and the potential for sharing facilities among the customers within the geographical area. In the case of intermediate traffic volumes or longer distances, the telephone system generally establishes a combination of direct and tandem lengths.

The connections are built around high-usage or high-volume trunks, which carry the bulk of the traffic. High-usage trunks are established when the volume of calls requires and justifies the installation of high-capacity channels between two offices. Consequently, trunk configurations vary depending on traffic volume between centers. The system attempts to switch the call down into the hierarchy, across the hierarchy, or up into the hierarchy. Routing the call ususally entails more intermediate switching, thereby increasing the connection delay.

Voice-oriented applications use the telephone network. In addition, many data communication systems use the telephone network for data transmission. The process is workable, but does not offer the highest quality. The majority of telephone local loops have been designed and sized to carry human speech signals and not digital data. Speech signals show the well-known analog waveform characteristics. But, servers and client devices are digital devices, and they require complex and expensive conversion systems. These differences will become insignificant when digitized and compressed voice transmission is going to be used.

The expansion of services in voice networks is continuous. There are multiple opportunities to increase the intelligence towards subscribers.

Virtual Private Networks (VPN) offer an alternative to corporate networks. Corporate networks operate private lines. VPN offers a private numbering plan to be determined by the customer. The telecommunications provider may offer multiple VPNs on the basis of a powerful physical network. The customer does not see the other VPN customers or details from the physical backbone. As part of customer network management, the provider may open a window into the VPN for the customer.

Intelligent Networks (IN) are based on the concept that certain events are received and evaluated by the Service Switching Points (SSP) in telephone networks. The results are forwarded to the Service Control Points (SCP) that are intelligent processors capable of making immediate decisions. There is an international capability set to be accepted and followed by providers. This set is the basis for generic and specified services. Some of the services are:

- Abbreviated Dialing
- Account Card Calling
- Automatic Alternative Billing
- Call Distribution
- Call Forwarding
- Conference Calling
- Credit Card Calling
- Mass Calling
- Televoting
- User Defined Routing

The prerequisite is that the SCPs are able to process a very large number of events within a very short period of time.

The TINA (Telecommunications Information Networking Architecture) Consortium is an international initiative formed by telecommunications operators and equipment suppliers, and computer vendors. Its main objective is to provide an architecture based on distributing computing technologies to enable telecommunications networks to support the rapid and flexible introduction of new services and the ability to manage both the services and the network in an integrated fashion. One of the main motivations for the TINA initiative was the modernization of IN. The IN concepts, architectures and specifications provide a means for designing, deploying, managing and operating telecommunications services. However, the services in question are mostly enhanced telephony services, based on the traditional call model where a user initiates a call to another user via a signaling mechanism. IN allows the control plane signals to be interpreted for calls other than just plain telephony. After intercepting the control plane signals, centralized service logic can translate the called number, or perform other operations depending on the service features required, before the control plane continues with the call set-up procedures.

IN operations are based entirely on control plane functions with protocol based interactions between the embedded control functions in the local switches and the centralized service logic and data. The IN techniques have been successful for implementing enhanced telephony services, but it is more difficult to introduce modern, advanced services such as multimedia, multiparty communications mechanisms to support applications such as workflow, joint document editing, intreractive distance learning, and others. Services such as these require advanced session management and control. More complex session control could be provided through signaling meachanisms, protocol based interactions, and centralized service logic functions, but traditional telecommunications engineering solutions such as IN are not as flexible as software engineering approaches based on object orientation and distributed systems.

TINA specifies a platform for the deployment and operation of advanced services. TINA adopts the Open Distributed Processing (ODP) framework for specifying a software platform for service logic, covering both service operation and service delivery. The idea is that instead of being limited by the IN architecture of the telephony call model and signaling protocols, new advanced services may be deployed directly on a Distributed Processing Environment (DPE) and may be designed and implemented according to object-oriented principles and distributed processing techniques. In this way, more flexible service design and implementation can be achieved by reusing software components while application interoperability is achieved through the services of the DPE. TINA has chosen OMG CORBA specification as the basis for its DPE. This initiative is characterized by its departure from protocol-based engineering principles toward software engineering techniques such as application programming interfaces and component interface specifications which are more closely related to programming languages used to implement the service logic.

TINA tries to overcome the polarization between services and their management. Since services need to be managed, management aspects should be an integral part of the service design. The service execution and management aspects should be unified in a common ODP-based environment, achieving reusability, uniformity, and economies of scale. The management architecture of TINA covers the principles and concepts for managing TINA systems. The architecture draws heavily on the TMN architecture and divides the overall management into five overall management functional groups such as configuration, fault, performance, security, and accounting management.

The hottest Internet-related subject is Internet telephony. Initiating telephone calls via the Internet remains a niche market. The quality of voice transmissions and the need for the called party to be at their computer and ready to receive a call has restricted the use of this technology to the already highly computer literate. The push to transmit voice over frame relay networks continues. Several vendors recently announced the availability of products that compress voice to as little as 4.8 Kbit/s with acceptable quality. Voice compression represents a prime market, as frame relay has become a highly popular, reliable, and efficient means of data transmission.

Speech recognition is another voice technology that has been approaching commercial acceptance for some time. The technology continues to improve, with numerous systems available that can recognize fluid speech, including accented speech. Although the public has generally accepted the use of the telephone touchpad for obtaining and conveying information, the approach has its limitations.

Signaling System Number 7   SS7 is the prevalent signaling system for telephone networks for setting up and clearing calls and furnishing different services, such as the use of toll free numbers. SS7 defines the procedures for the set-up, ongoing management, and clearing of a call between telephone users. It performs these functions by exchanging telephone control messages between the SS7 components that support the customers' connections.

The SS7 signaling data link is a full duplex, digital transmission channel operating at 64 Kbit/s. Occasionally, also an analog link can be used. The SS7 link operates on both terrestrial and satellite links. The actual digital signals on the link are derived from pulse code modulation multiplexing equipment or from equipment that employs a frame structure. The link must be dedicated to SS7. In accordance with clear channel signaling, no other transmission can be transferred with these signaling messages.

Figure 2.5 shows a typical SS7 topology. The subscriber lines are connected to the SS7 network through the Service Switching Points (SSP). The purpose of the SSPs is to receive the signals from the customer premises equipment (CPE) and perform call processing on behalf of the user. SSPs are implemented at switching centers. They serve as the source and destination for SS7 messages. SSPs initiate SS7 messages either to another SSP or to a signaling transfer point (STP). STPs are expected to translate the SS7 messages and to route those messages between network nodes and databases. The STPs are switches that relay messages between SSPs, STPs, and service control points (SCP). Their principal functions are very similar to the tasks of Layer 3 in the OSI model.

The SCPs contain software and databases for the management of the call. The SS7 components mentioned above can be implemented as discrete entities or in an integrated fashion. SS7 has multiple layers that show high conformance to the OSI layers. In particular, Layer 3 plays an important management role. The functional modules execute the following functions:

- Message routing
- Message distribution
- Signaling traffic management
- Signaling link management
- Signaling route management

SS7 has been a significant success in the telecommunications industry. It is implemented in public telephone networks by practically all carriers throughout the world. In addition, features of SS7 have found their way into other systems such as GSM and satellite signaling. SS7 is operating with emerging technologies as well.

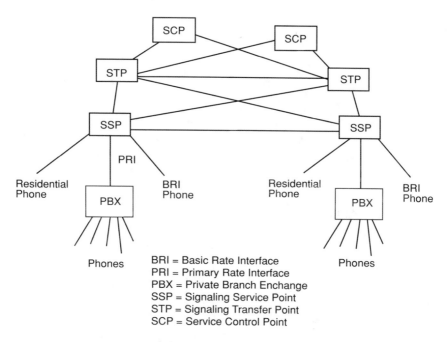

FIGURE 2.5   Typical structure of SS7.

X.25-Based Packet Switching Networks   The idea of the X.25 interface was to define rules about how a public packet data network would handle users' traffic and accomodate to various quality of service (QOS) features — called X.25 facilities — that were requested by the user. X.25 was designed to provide strict flow control on users' traffic and to provide substantial management services, such as sequencing and acknowledgment of traffic. Since its introduction in 1974, the standard has been expanded to include many options, services, and facilities, and several of the newer OSI protocols and service definitions operate with X.25. Unlike T1/E1, X.25 uses STDM techniques and is designed as a transport system for data.

The role and placement of X.25 is not always well understood. X.25 is not a packet switching specification. It is a packet network interface specification. X.25 does not specify operations within the network. X.25 is not aware whether the network is using adaptive or fixed-directory routing, or whether the internal operations of the network are connection-oriented or connectionless. Figure 2.6 shows a typical topology. It can be seen in this figure that X.25 defines the procedures for the exchange of data between a user device (DTE) and the network (DCE).

X.25 provides two options to establish and maintain communications between the user devices and the network:

1. Permanent virtual circuits (PVC) and
2. Switched virtual call (SVC).

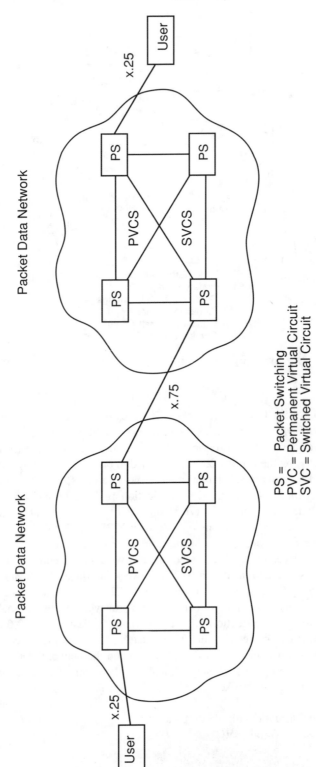

PS = Packet Switching
PVC = Permanent Virtual Circuit
SVC = Switched Virtual Circuit

FIGURE 2.6 Standards of packet data switches.

A sending PVC user is assured of obtaining a connection to the receiving user and of obtaining the required services of the network to support the user-to-user session. X.25 requires that a PVC be established before a session can begin. Consequently, an agreement must be reached by the two users and the network administration before the PVC is allocated. Among other things, agreement must be made about the reservation of logical channel numbers for the PVC session and the establishment of facilities. An SVC requires that the originating user device must transmit a call request packet to the network to start the connection operation. In turn, the network node relays this packet to the remote network node, which sends an incoming call packet to the called user device. If this receiving DTE chooses to acknowledge and accept the call, it transmits to the network a call accepted packet. The network then transports this packet to the requesting DTE in the form of a call connected packet. To terminate a session, a clear request packet is sent by either DTE. It is received as a clear indication packet and confirmed by the clear confirm packet.

X.25 also provides quality of service options to the user. These options are called facilities. These facilities are preprovisioned for PVCs or are provided during the call establishment phase of the SVC hook up. The user is allowed to obtain features such as call redirections, security features for closed user groups, throughput, networking delays, reverse charge, reverse charge prevention, and a wide variety of other useful application support operations.

X.25-based public and private packet switching networks are widely used. The technology is old. It was designed to support user traffic on error-prone low quality networks, with the assumption that most user devices were unintelligent. Moreover, X.25 was designed to operate on physical interfaces that are also old and slow, such as RS-232 and V.28.

But, despite these facts, X.25-usage continues to grow throughout the world for the following reasons:

- The technology is well understood
- It is available in off-the-shelf products
- There are extensive performance conformance tests available
- It is cost-effective for bursty, slow-speed applications.

Integrated Services Digital Networks (ISDN)   The initial purpose of ISDN was to provide a digital interface between a user and a network node for the transport of digitized voice and data images. It is now designed to support a wide range of services. Basically, all communication forms may be supported by ISDN. It has been implemented as an evolutionary technology on the basis of the telephone-based integrated digital network. Many digital techniques seen with T1 and E1 are used in ISDN. It includes signaling rates, transmission codes, and also physical plugs.

For ISDN, there are two important terms: functional groupings and reference points. Functional groupings are a set of capabilities needed in an ISDN user-access interface. Specific functions within a functional grouping may be performed by multiple pieces of equipment or software. Reference points are the interfaces dividing the functional groupings. Usually, a reference point corresponds to a physical interface between pieces of equipment. Figure 2.7 shows a typical structure of ISDN including all the reference points.

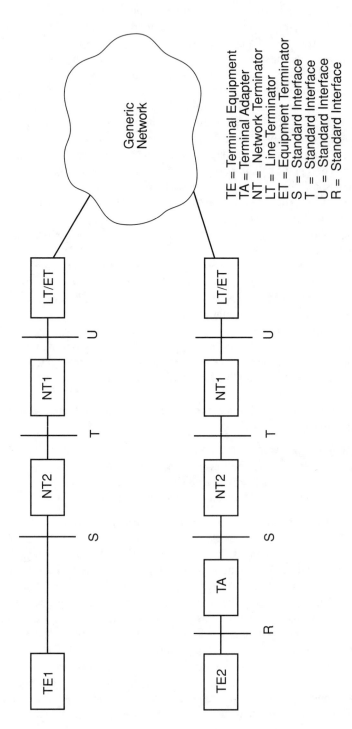

TE = Terminal Equipment
TA = Terminal Adapter
NT = Network Terminator
LT = Line Terminator
ET = Equipment Terminator
S = Standard Interface
T = Standard Interface
U = Standard Interface
R = Standard Interface

FIGURE 2.7 Typical topology of ISDN networks.

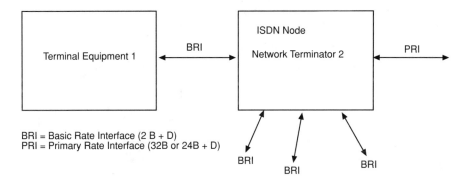

FIGURE 2.8   ISDN basic and primary rates.

The reference points labeled R, S, T, and U are logical interfaces between the functional groupings, which can be either a terminal type 1 (TE1), a terminal type 2 (TE2), or a network termination (NT1, NT2) grouping. The reference points help to define the responsibilities in communications networking based on ISDN.

Figure 2.8 gives more information about the throughput capabilities. Between a terminal equipment and an NT2 type ISDN node, the Basic Rate is used. The basic rate includes two B channels, each with 64 Kbit/s, and one D channel with 16 Kbit/s bandwidth. The Primary Rate is different in the United States and Europe. In the United States 23 B channels and 1D channel build the Primary Rate; Europe uses 31 B channels and 1 D channel for the Primary Rate.

The corresponding throughput rates are exactly the rates as with T1 and E1.

ISDN cannot be considered successful based on its performance since inception in the early 1980s. In North America, the progress has been much slower than in Europe because of the lack of cohesive implementation politics. This situation has changed in the past few years. RBOCs and Bellcore are agressively implementing ISDN, and preparing themselves for Broadband-ISDN (B-ISDN). Most European countries have implemented ISDN; there is even cooperation between countries to promote Euro-ISDN.

ISDN can be judged successful in another way. It has helped to implement the LAPD and the Q.931 messaging protocol. These two protocols can be found throughout the communication industry.

The technology is also useful for remote access to a data network. Telephone companies that market it well can win at ISDN, as well as those companies that can produce reasonably priced, easy-to-use customer access equipment. Its potential, however, is in competition with high-speed-modems and with much higher-capacity options such as digital subscriber loop and transmission of data via the broadband cable television network.

2.2.2   CONFORMANCE TO OSI LAYERS

All of the emerged technologies support a layered architecture. Evaluating the layers, we conclude that the depth of support is very different. Figure 2.9 shows the results for the four emerged technologies compared with OSI.

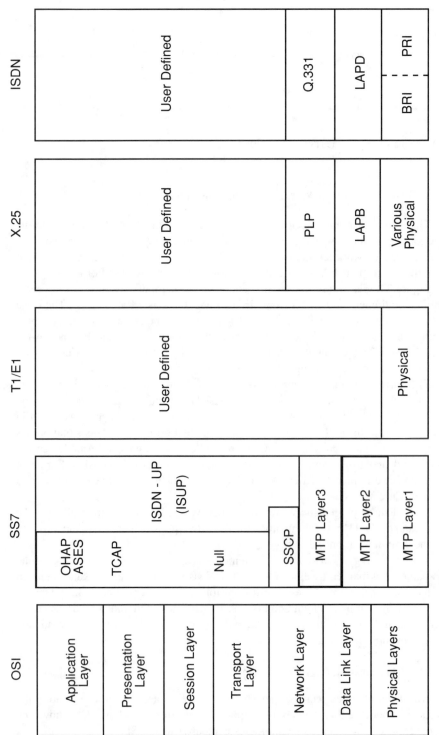

FIGURE 2.9 Layers of emerged technologies.

In case of T1/E1, just the physical layer is supported. The other six layers are expected to be defined by users. X.25 and ISDN show some similarities by fully defining the first 3 bottom layers. Layer 4 and higher are expected to be defined and maintained by the user. SS7 shows mixed results. Layers one to three are fully defined; in particular, layer 3 is rich on functionality. In addition, there are entries for layer 7 and ISDN-based alternatives for the remaining 3 layers between layer 4 and 6.

2.2.3 MANAGEMENT SOLUTIONS

The present status for emerged technologies can be summarized as follows:

- Proprietary solutions dominate: it means that the management protocols selected are controlled by the supplier of the equipment or facilities vendors.
- Reengineered by SNMP: many of the equipment vendors include SNMP-agents into their devices to meet the requirements of customers. The SNMP-agents provide information for performance management and reporting, but they usually do not change the real-time processing of status data within the devices.
- The management structures are very heterogeneous: in most cases, these structures are hierarchical including a manager of managers. This manager is using a proprietary architecture. Most of the interfaces to element managers and managed devices are proprietary.
- TMN is the result of penetration: suppliers recognizing the need for a generic standard, but they are not willing to invest heavily into supporting it. Some of the providers go as far as supporting the Q3-interface.
- Operating Support Systems are heavy: the legacy type OSSs support the emerged technologies on behalf of the suppliers, but they are not flexible enough to address future needs. They lack in separating operations functionality from operations data, in using flexible software, and in separating network management from service management.

2.3 EMERGING TECHNOLOGIES

This segment gives an overview on telecommunication technologies that are either in use, but still considered new technology or are considered for near-future implementation. These technologies include frame relay, FDDI, SMDS, ATM, SONET/SDH, and mobile and wireless communications. The capabilities of these technologies are presented using the same evaluation criteria. A few more innovative emerging technologies are discussed in addition to the previous ones, but not in depth. These include Digital Subscriber Line, CTI, and data and voice over cable. These technologies are emerging, but not yet mature.

2.3.1 FRAME RELAY

1. Name and description of the technology: Frame Relay

    The purpose of a frame relay network is to provide an end-user with a high-speed virtual private network (VPN) capable of supporting applications with large bit-rate transmission requirements. It gives a user T1/E1 access rates at a lesser cost than that which can be obtained by leasing comparable T1/E1 lines. It is actually a mesh network.

    The design of frame relay networks is based on the fact that data transmission systems today are experiencing far fewer errors and problems that they did decades ago. During that period, protocols were developed and implemented to cope with error-prone transmission circuits. However, with the increased use of optical fibers, protocols that expend resources dealing with errors become less important. Frame relay takes advantage of this improved reliability by eliminating many of the now unnecessary error checking and correction, editing, and retransmission features that have been part of many data networks for almost two decades.

2. How new is this technology?

    It has been working for many years. It represents a scaled-down version of LAPD. The flexibility of assigning bandwidth on-demand is somewhat new. Frame relay is one of the alternatives of fast packet switching. A typical topology is shown in Figure 2.10.

3. What are the application areas for this technology?

    Applications with bursty data with occasional high bandwidth requirements. Typical applications include:

    - LAN interconnections
    - Transmitting graphics and X-rays
    - Large database transfers
    - CAD/CAM transfers
    - Medium quality, still-image video transmissions
    - Connectionless data transfer
    - Connection-oriented data transfer
    - Voice transmission is in first testing phase

4. Topology and media dependence

    Actually no dependence from the topology. But, the most present implementations are point-to-point. Frame relay can operate over wire and over optical fiber.

5. Support of WANs, MANs, and LANs

    WAN-based. It is designed as a high-speed WAN to interconnect LANs and occasionally MANs. It is a typical interconnecting technology in the wide area. First of all data transmission is supported, but recently also voice is transmitted.

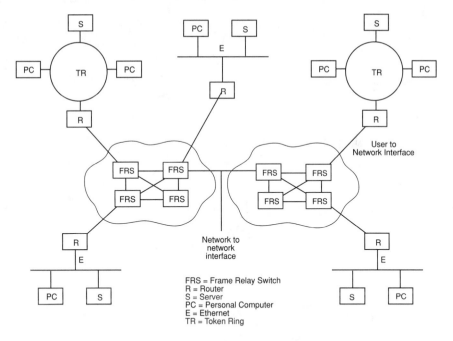

FIGURE 2.10   Structure of typical frame relay networks.

6. What other technologies this technology complements
    LAN internetworking over the wide area.

7. What other technologies this technology competes with
    Private leased lines, X.25-based packet switching networks, SMDS and ATM.

8. Basis of this technology
    It is frames-based. Its use is beneficial when traffic is bursty with unpredictable peaks.

9. What support connection management gets
    PVCs are fully supported. SVCs can be supported on an individual basis.

10. What flow control mechanisms are supported
    Implicit flow control is supported by FECN and BECN bits.

11. Bandwidth on demand
    It is supported by CIR, committed and excess burst rate operations. Usually, service level agreements are signed between the provider and the customer on CIR-basis.

12. Addressing schemes used by this technology
    It uses labels for identifying PVCs, which are called DLCIs.

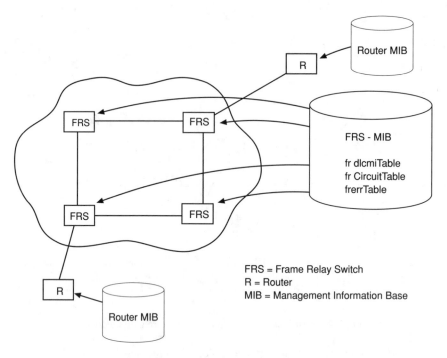

FIGURE 2.11   The Frame Relay MIB.

13. What are traffic discard options?
    It is supported with the DE bit.

14. MIB availability
    The frame relay objects are organized into three object groups:

    i. Data link connection management interface group
    ii. Circuit group
    iii. Error group

    These groups are stored in tables in the MIB and can be accessed by the SNMP manager. Figure 2.11 illustrates where SNMP operates, and lists the names of these three groups.
    The frDlcmi Table contains 10 objects. Their purpose is to identify each physical port at the UNI, its IP address, the size of the DLCI header that is used on this interface, timers for invoking status and status inquiry messages, the maximum number of DLCIs supported at the interface, whether or not the interface uses multicasting, and some other miscellaneous operations.
    The frCircuit Table contains 14 objects. Their purpose is to identify each PVC, its associated DLCI, if the DLCI is active, the number of BECNs and FECNs received since the PVC was created, statistics on the

number of frames and octets sent and received since the DLCI was created, the DLCI's Bc and Be, and other miscellaneous operations.

The third table is the frErrTable containing four objects. Their purpose is to store information on the types of errors that have occurred at the DLCI (unknown or illegal), and the time the error was detected. One object contains the header of the frame that created the error.

15. What are management opportunities?

SNMP-based and proprietary solutions compete for management. Basically, each physical and logical component can be managed by periodically polling the PDUs in the MIB. Any powerful management platform can accomodate frame-relay management. But, the polling overhead over the wide area should be controlled carefully.

2.3.2   FIBER DISTRIBUTED DATA INTERFACE

1. Name and description of the technology: *Fiber Distributed Data Interface (FDDI)*

FDDI was developed to support high-capacity LANs. To obtain this goal, the original FDDI specifications stipulated the use of optical fiber as the transport media, although it now is available on twisted pair cable (CDDI). FDDI has been deployed in many corporations to serve as a high-speed backbone network for other LANs, such as Ethernet and Token Ring.

Basically, the standard operates with 100 Mbps rate. Dual rings are provided for the LAN, so the full speed is actually 200 Mbps, although the second ring is used typically as a back-up to the primary ring. In practice, most installations have not been able to utilize the full bandwidth of FDDI. The standard defines multimode optical fiber, although single mode optical fiber can be used as well.

FDDI was designed to make transmission faster on an optical fiber transport. Due to the high capacity 100 Mbps technology, FDDI has a tenfold increase over the widely used Ethernet, and a substantial increase over Token Ring. FDDI was also to extend the distance of LAN interconnectivity. It permits the network topology to extend up to 20 km (14 miles).

FDDI II, which is able to incorporate voice, is not getting enough industry interest yet.

2. How new is this technology?

FDDI is actually not a new technology. But, internetworked FDDIs offer new alternatives in metropolitan areas competing with other technologies, such as frame relay and SMDS.

A typical topology is shown in Figure 2.12.

3. What are the application areas for this technology?

Any applicatiuon on a LAN. FDDI for data; FDDI II for voice and data.

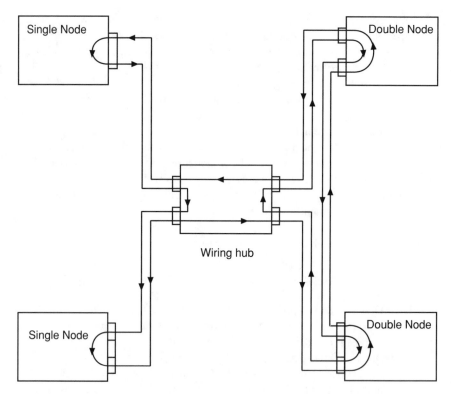

FIGURE 2.12   Wiring plan of FDDI.

4. Topology and media dependence

   Both dual and dual ring of trees are supported as topology. There is a definite media dependence at the PMD sublayer. But, FDDI can operate over different types of fiber as well as copper. No dependency at PHY and the MAC layers.

5. Support of WANs, MANs, and LANs

   This technology is LAN-based. It is designed to provide a high-speed LAN and to serve as a backbone to interconnect LANs.

6. What other technologies this technology complements

   Due to offering interconnectivity to LANs, all LAN technologies, such as Ethernet and Token Ring complement FDDI.

7. What other technologies this technology competes with

   ATM, and increasingly, the lower speed LANs, such as Ethernet and Token Ring that are upgraded by higher throughput by full-duplex operations or by LAN-switching. High speed Ethernet is in direct competition with FDDI.

8. Basis of this technology

    It is frame-based that are passed to the stations. Frame passing is controlled by a token rotation timer.

9. What support connection management gets

    No for FDDI I, which is connectionless. Yes for isochronous traffic on FDDI II.

10. What flow control mechanisms are supported

    The support is implicit with the token rotation timers and SA/ASA conventions, but no congestion notification operations.

11. Bandwidth on demand

    Somewhat supported because each station can be configured differently, allocation is dynamic as bandwidth is available.

12. Addressing schemes used by this technology

    It uses explicit addressing with the 48 bit MAC address.

13. What are traffic discard options?

    No support.

14. MIB availability

    The FDDI MIB has the following six groups:

    i. SMT Group: It contains a sequence of entries, one for each SMT implementation. The entries describe the station i.d., the operation i.d., the highest and lowest version i.d., the number of MACs in this station or concentrator and other configuration and status information.

    ii. MAC Group: It is a list of MAC tables, one for each MAC implementation across all SMTs. Each table describes the SMT index and the MIB-II IfIndex associated with this MAC, status and configuration information for the given MAC.

    iii. Enhanced MAC Counters Group: The MAC Counters table contains a sequence of MAC counters entries. Each entry stores information about the number of tokens received, number of times TVX expired, number of received frames, that has not been copied to the receive buffer, number of TRT expirations, number of times the ring entered operational status and threshold information.

    iv. PATH Group: It contains a sequence of PATH entries. Each entry starts with an index variable uniquely identifying the primary, secondary and local PATH object instances. The entries also store information on different min. and max. time value for MACs in the given PATH.

    v. PATH Configuration Table: It is a table of path configuration entries. This table contains a configuration entry for all the resources that may be in this Path.

vi. PORT Group: It contains a list of PORT entries. Each entry describes the SMT index associated with this port, a unique index for each Port within the given SMT, the PORT type, then neighbor type, connection policies, the list of permitted PATHs, the MAC number associated with the PORT (if any), the PMD entity class associated with this port and other capabilities and configuration information.

15. What are management opportunities?

FDDI has its own management capabilities, defined in SMT. But, it has never really taken off. Instead, suppliers are concentrating on SNMP capabilities. Using the MIB of FDDI agents, any SNMP manager can be used to manage FDDI.

2.3.3   SWITCHED MULTI-MEGABIT DATA SERVICE

1. Name and description of the technology: Switched Multi-Megabit Data Service (SMDS)

SMDS is a high-speed connectionless packet switching service which extends LAN-like performance beyond a subscriber's location. Its purpose is to ease the geographic limitations that exist with low-speed wide area networks. SMDS is designed to connect LANs, MANs, and also WANs.

The major goals of SMDS are to provide high-speed interfaces into customer systems, and at the same time, allow customers to take advantage of their current equipment and hardware. Therefore, the SMDS operations are not performed by the end user machine; they are performed by CPE, such as a router.

2. How new is this technology?

SMDS is positioned as a service. If SMDS is considered unto itself, it is not a new technology; it offers no new method for designing or building networks. This statement is emphasized because SMDS uses the technology of Dual Queue Dual Bus (DQDB), and then offers a variety of value-added services, such as bandwidth on demand, to the SMDS customer.

SMDS is targeted for large customers and sophisticated applications that need a lot of bandwidth, but not permanently. Generally, SMDS is targeted for data applications that transfer a lot of information in a bursty manner.

However, applications that use SMDS can be interactive. For example, two applications can exchange information interacting through SMDS, such as an X-ray, a document, etc. The restriction of SMDS is based on the fact that SMDS is not designed for real-time, full-motion video applications. Notwithstanding, it does support an interactive dialog between users, and allows them to exchange large amounts of information in a very short time. For example, it takes only one to two seconds for a high quality color graphic image to be sent over an SMDS network. For many applications, this speed is certainly adequate.

A typical topology is shown in Figure 2.13.

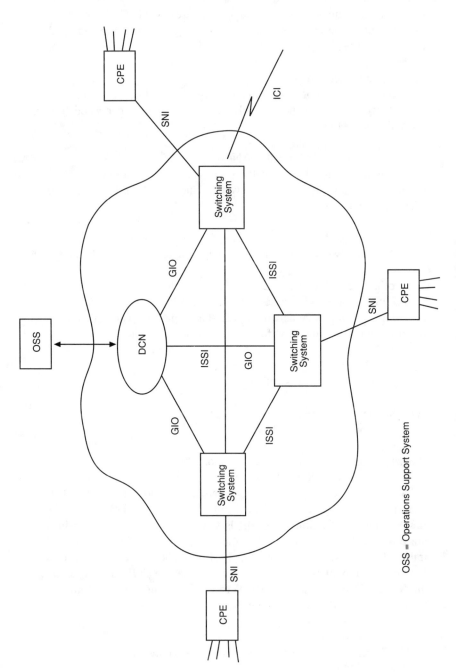

OSS = Operations Support System

FIGURE 2.13   Typical topology of SMDS.

3. What are the application areas for this technology?

    SMDS targets high-speed data transfers and internetworking LANs. Included are graphics, X-rays, large databases, CAD/CAM, medium-quality, still-image video. Practically everything but voice and high-quality video can be transferred.

4. Topology and media dependence

    If physical layer of IEEE 802.6 is used, then there is a topology dependence. Otherwise no dependence. There is no dependence from transmission media.

5. Support of WANs, MANs, and LANs

    It cannot be defined specifically. In most cases, it is used as interconnecting technology for LANs or even for individual workstations with heavy occasional traffic.

6. What other technologies this technology complements

    LAN internetworking technologies can complement this technology.

7. What other technologies this technology competes with

    SMDS competes with leased lines, frame relay at lower transmission rates, high-speed private networks, X.25-based packet switching at lower trasnmisison rates, and ATM for higher transmission rates.

8. Basis of this technology

    It is cell-based which is called a slot.

9. What support connection management gets

    No support; it is connectionless.

10. What flow control mechanisms are supported

    It is implicit with SIR. The SMDS traffic management operations are founded on the concept called sustained information rate (SIR). This concept is similar to the committed information rate (CIR) used in the frame realy specifications. The SIR is founded on access classes which are provided for DS3 circuits. Each access class identifies the different traffic characteristics for varying applications. The access class places a limit on the amount of sustained information that the CPE can send across the System Network Interface (SNI) to the SS. It also places a limit on the burstiness of the transfer from the CPE to the SS.

11. Bandwidth on demand

    This is one of the major advantages of this technology.

12. Addressing schemes used by this technology

    SMDS uses ISDN addresses.

13. What are traffic discard options?
    It is supported.

14. MIB availability
    Figure 2.14 shows the location of the MIB in relation to the SMDS network. The Standard Interface Protocol (SIP) layers are also known in this figure to aid in reading the following material. The MIB is organized around managed objects which are contained in major groups. The groups, in turn, are defined in tables. As shown in the figure at the bottom, the major entries in the MIB are the SMDS address, which is the conventional 60-bit address preceded by the 4-bits to signify individual group addresses. Thereafter, the groups are listed with their object identifier name. These names will be used in this segment to further describe the entries.

    The sipL3Table contains the layer 3 (L3-PDU) parameters used to manage each SIP port. It contains entries such as port numbers, statistics on received traffic, information on errors such as unrecognized addresses, as well as various errors that have occurred at this interface.

    The sipL2Table, as its name implies, contains information about layer 2 (L2-PDU) parameters and the state variables associated with information on the amount of the number of level 2 PDUs processed, error information such as violation of length of protocol data unit (PDU), sequence number errors, MID errors, etc.

    The sipDS1PLCPTable contains information on DS1 parameters and state variables for each port. The entries in the table contain error information such as DS1 severity erred framing seconds (SEFS), alarms, unavailable seconds encountered by the PLPC, etc.

    The sipDS3PLPCTable contains information about DS3 interfaces and state variables for each SIP port. Like its counterpart in DS1, this table contains information on severity erred framing seconds, alarms, unavailable seconds, etc.

    The ipOverSMDSTable contains information relating to operations of IP running on top of SMDS. It contains information such as the IP address, the SMDS address relevant to the IP station, addresses for ARP resolution, etc.

    The smdsCarrierSelection group contains information on the interexchange carrier selection for the transport of traffic between LATAs.

    Finally, the sipL3PDUErrorTable contains information about errors encountered in processing the layer 3 PDU. Entries such as destination error, source error, invalid BAsize, invalid header extention, invalid PAD error, BEtag mismatch, etc. form the basis for this group.

    As mentioned earlier, SNMP is used to monitor the MIBs and report on alarm conditions that have occurred based on the definitions in the MIBs.

15. What are management opportunities?
    SMDS management is complex. Switching systems are intelligent devices with the need of real-time decision making. In such situations, SNMP is not necessarily the right choice. However, SNMP may be used to transmit MIB-entries to the manager for performance reporting.

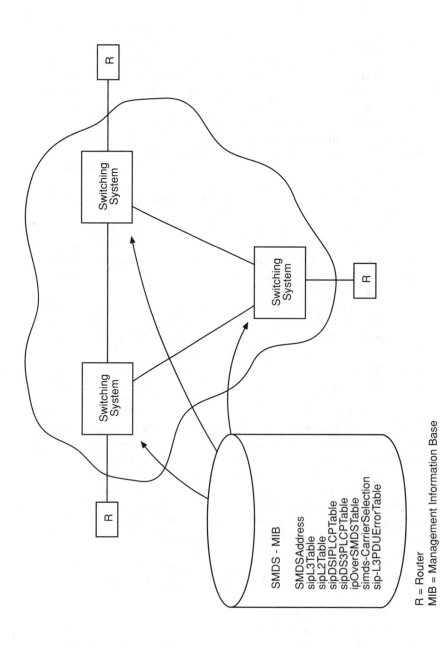

FIGURE 2.14 The SMDS-MIB.

SMDS - MIB

SMDSAddress
sipL3Table
sipL2Table
sipDSIPLCPTable
sipDS3PLCPTable
ipOverSMDSTable
simds-CarrierSelection
sip-L3PDUErrorTable

R = Router
MIB = Management Information Base

2.3.4   ASYNCHRONOUS TRANSFER MODE

1. Name and description of the technology: Asynchronous Transfer Mode
   (ATM)

   The purpose of ATM is to provide a high-speed, low-delay, multiplexing and switching network to support any type of user traffic, such as
   voice, data, or video applications. ATM is one of four fast relay services.
   ATM segments and multiplexes user traffic into small, fixed-length units
   called cells. The cell is 53-octets, with 5-octets reserved for the cell header.
   Each cell is identified with virtual circuit identifiers that are contained in
   the cell header. An ATM network uses these identifiers to relay the traffic
   through high-speed switches from the sending customer premises equipment (CPE) to the receiving CPE.

   ATM provides limited error detection operations. It provides no retrans-
   mission services, and few operations are performed on the small header.
   The intention of this approach — small cells and with minimal services
   performed is to implement a network that is fast enough to support multimegabit transfer rates.

2. How new is this technology?

   ATM is a new technology. ATM is supposed to be the foundation of
   providing the convergence, multiplexing, and switching operations. ATM
   resides on top of the physical layer. A likely ATM topology is shown in
   Figure 2.15.

3. What are application areas for this technology?

   Practically, all kinds of applications can be supported. It is a multimedia
   technology. All applications supported by frame relay and SMDS can be
   supported by ATM, too.

4. Topology and media dependence

   Strictly speaking, ATM is not topology dependent, but current implementations are point-to-point. ATM can operate over wire and optical fiber.

5. Support of WANs, MANs and LANs

   ATM may be used as a WAN-backbone, connecting MANs and LANs.
   It also can be used in the MAN- and LAN-area as a local backbone.
   Although they are compatible, there are two different ATMs: local area
   ATM and wide area ATM. ATM in the LAN is, and will continue to be,
   moving faster than ATM in the WAN in number of installations. This is
   because it requires only an internal decision to implement ATM in a
   campus or corporate network. However, all of the RBOCs are committed
   to ATM in their backbones and it will become available directly.

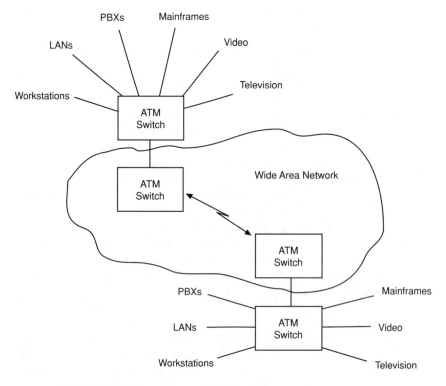

FIGURE 2.15   Typical use of ATM switches in WANs and LANs.

6. What other technologies this technology complements
   SONET/SDH can be used in combination with ATM as the basis of B-ISDN.

7. What other technologies this technology competes with
   Private leased lines, X.25-based packet switching networks, SMDS, frame relay, and high-speed LAN backbone networks, such FDDI and fast Ethernet.

8. Basis of this technology
   It is a cell-based technology.

9. What support connection management gets
   It is supported by PVCs, with SVCs stipulated in later releases by standards groups.

10. What flow control mechanisms are supported
    The support is implicit using the flow control field in the ATM header.

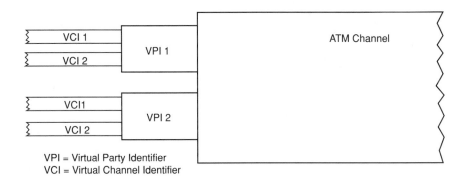

FIGURE 2.16   Connection identifiers for ATM

11. Bandwidth on demand

It is supported within the service level management agreement and network bandwidth availability.

12. Addressing schemes used by this technology

ATM is using labels for identifying connections, which are called virtual channel identifiers (VCIs) and vitrual path identifiers (VPIs). Figure 2.16 shows these two mechanisms.

13. What are traffic discard options?

It is supported by the Cell Loss Priority (CLP) bit. The CLP bit will be defined in service level agreements with customers.

14. MIB availability

The ATM Forum has published a management information base (MIB) as part of its Interim Local Management Interface Specification (ILMI) (Figure 2.17). The ATM MIB is registered under the enterprise node of the standard SMI in accordance with the Internet. MIB objects are therefore prefixed with 1.3.6.1.4.1.353 and MR interface specifications Af-nm-0058 and AF-nm-0027.

Each physical link (port) at the UNI has a MIB entry that is defined in the atmfPortTable. This table contains a unique value for each port, an address, the type of port (DS3, SONET, etc.) media type (coaxial cable, fiber, etc.), status of port, (in service, out of service, etc.) and other miscellaneous objects.

The atmfAtmlayer Table contains information about the UNIs physical interface. The table contains the port i.d., maximum number of VCCs, VPCs supported and configured on this UNI, active VCI/VPI bits on the UNI, and a description of public or private for the UNI.

The atmVpcTable and atmVccTable contain similar entries for the VPCs and VCCs respectively on the UNI. These tables contain the port

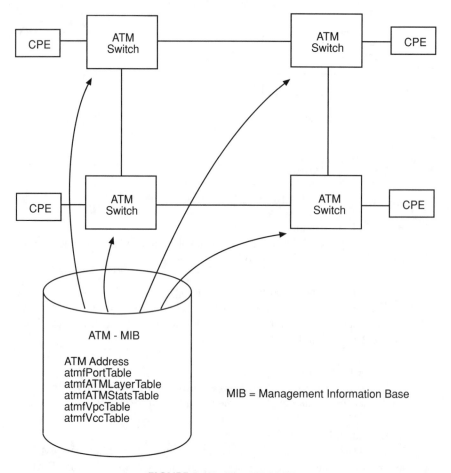

FIGURE 2.17   The ATM MIB.

i.d., VPI or VCI values for each connection, operational status (up, down, etc.), traffic shaping and policing descriptors (to describe the type of traffic management applicable to the traffic), and any applicable QoS that is applicable to the VPI or VCI.

15. What are the management opportunities?
    The ATM Forum has defined two aspects of user network interface (UNI) network management:

    • ATM layer management at the M plane, and
    • Interim Local Management Interface (ILMI) specification.

    M-Plane management:
    • Most of the functions for ATM M-plane management are performed with the SONET F1, F2, and F3 information flows. ATM is concerned with F3 and F4 information flows.

- **ILMI** Because the ITU-T and the ANSI have focused on C-plane and U-plane procedures, the ATM Forum has published an interim specification called ILMI. The major aspects of ILMI are the use of SNMP and a MIB. The ILMI stipulates the following procedures. First, each ATM device supports the ILMI, and one UNI ILMI MIB instance for each UNI. The ILMI communication protocol stack can be SNMP/UDP/IP/AAL over a well known VPI/VCI value. SNMP is employed to monitor ATM traffic and the UNI VCC/VPC connections beased on the ATM MIB with the SNMP get, Get-Next, Set, and Trap operations.

2.3.5 SONET AND SDH

1. Name and description of the technology: Synchronous Optical Network (SONET)/Synchronous Digital Hierarchy (SDH)

   SONET/SDH is an optical-based carrier (transport) network utilizing synchronous operations between the network components. The term SONET is used in North America, and SDH is used in Europe and Japan. Attributes of this technology are:

   - A transport technology that provides high availability with self-healing topologies
   - A multivendor which allows multivendor connections without conversions between the vendors' systems
   - A network that uses synchronous operations with powerful multiplexing and demultiplexing capabilities
   - A system that provides extensive operation, administration, maintenance, and provisioning (OAM&P) services to the network user and administrator

   SONET/SDH provides a number of attractive features when compared with current technology. *First*, it is an integrated network standard on which all types of traffic can be transported. *Second*, the SONET/SDH standard is based on the optical fiber technology which provides superior performance in comparison to microwave and cable systems. *Third*, because SONET/SDH is a worldwide standard, it is now possible for different vendors to interface their equipment without conversion.

   Fourth, SONET/SDH efficiently combines, consolidates, and segregates traffic from different locations through one facility. This concept, known as grooming, eliminates back hauling and other inefficient techniques currently being used in carrier networks. Back hauling is a technique in which user payload is carried past a switch that has a line to the user and sent to another endpoint. Then, the traffic to the another user is dropped, and the first user's payload is sent back to the switch and relayed back to the first user. In present configurations, grooming eliminates, but it requires expensive configurations, such as back-to-back multiplexers that are connected with cables, panels, or electronic cross-connect equipment.

   Fifth, SONET/SDH eliminates back-to-back multiplexing overhead by using new techniques in the grooming process. These techniques are implemented in an add-drop multiplexer (ADM).

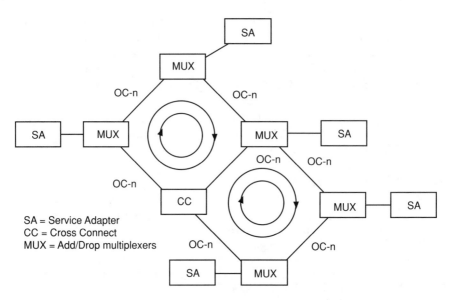

FIGURE 2.18   Typical topology of SDH/SONET rings.

Sixth, the synchronous aspect of SONET/SDH makes for more stable network operations. These types of networks experience fewer errors than the older asynchronous networks, and provide much better techniques for multiplexing and grooming payloads.

Seventh, SONET/SDH has notably improved OAM&P features relative to current technology. Approximately 5% of the bandwidth is devoted to management and maintenance.

Eighth, SONET/SDH employs digital transmission schemes. Thus, the traffic is relatively immune to noise and other impairments on the communications channel, and the system can use efficient time division multiplexing (TDM) operations.

2. How new is this technology?

    SONET has been around a couple of years. The technology is not completely new, but its implementation is new. The typical topology is shown in Figure 2.18.

    Extensive operation, administration, and maintenance (OAM) in carrier networks is new, as well.

3. What are application areas for this technology?

    SONET/SDH is application independent. It is simply a physical carrier technology — a transport service for ATM, SMDS, frame relay, T1, E1, and others.

4. Topology and media dependence

    No topology dependence, but dual rings provide the reliability and performance. Optical fiber is needed for the higher bit rates.

5. Support of WANs, MANs and LANs
   It is a WAN-based technology.

6. What other technologies this technology complements
   All carrier transport systems.

7. What other technologies this technology competes with
   All existent carrier transport systems, in particular private leased lines networks that are T1 or E1 based.

8. Basis of this technology
   The basis is an envelope that resembles a frame.

9. What support connection management gets
   No support.

10. What flow control mechanisms are supported
    No support.

11. Bandwidth on demand
    No support.

12. Addressing schemes used by this technology
    No support.

13. What are traffic discard options?
    No support.

14. MIB availability
    The SONET/SDH MIB consists of eight groups. Each of the following groups have two tables except the Medium group: one is the Current Table, the other one is the Interval Table.
    The SONET/SDH XXX Current Table contains various statistics that are being collected for the current 15-minute interval. The SONET/SDH XXX Interval Table contains various statistics that are being collected by each system over a maximum of the previous 24 hours of operation. The past 24 hours may be broken into 96 completed 15-minute intervals. A system is required to store at least four completed 15-minute intervals. The default value is 32 intervals.

    • *The SONET/SDH Medium Group*   SONET/SDH interfaces for some applications may be electrical interfaces and not optical interfaces. This group handles the configuration information for both optical SONET/SDH interfaces and electrical SONET/SDH interfaces, such as signal type, line coding, line type, and the like.

- *The SONET/SDH Section Group* This group consists of two tables: The SONET/SDH Section Current Table and the SONET/SDH Section Interval Table.

    These tables contain information on interface status, counters on errored seconds, severely errored seconds, severely errored framing seconds and coding violations.

- *The SONET/SDH Line Group* This group consists of two tables: The SONET/SDH Line Current Table and the SONET/SDH Line Interval Table.

    These tables contain information on line status, counters on errored seconds, severely errored seconds, severely errored framing seconds and unavailable seconds.

- *The SONET/SDH Far End Line Group* This group may only be implemented by SONET/SDH Life Terminating Equipment (LTE) systems that provide for a far end block error (FEBE) information at the SONET/SDH Line Layer. This group consists of two tables: The SONET/SDH Far End Line Table and the SONET/SDH Far End Line Interval Table.

- *The SONET/SDH Path Group* This group consists of two tables: The SONET/SDH Path Current Table and the SONET/SDH Path Interval Table.

    These tables contain information on interface status, counters on errored seconds, severely errored seconds, severely errored framing seconds and coding violations.

- *The SONET/SDH Far End Path Group* This group consists of two tables: The SONET/SDH Far End Path Current Table and the SONET/SDH Far End Path Interval Table.

- *The SONET/SDH Virtual Tributary (VT) Group* This group consists of two tables: The SONET/SDH VT Current Table and the SONET/SDH VT Interval Table.

For SDH signals, virtual tributaries are called VCs instead of VTs.

VT1.5	VC11
VT2	VC12
VT3	none
VT6	VC3

These tables contain information on virtual tributaries width and status, counters on errored seconds, severely errored seconds, severely errored framing seconds and unavailable seconds.

- *The SONET/SDH Far End VT Group* This group consists of two tables: The SONET/SDH Far End VT Current Table and the SONET/SDH Far End VT Interval Table.

15. What are management opportunities?

    The OAM functions are associated with the hierarchical, layered design of SONET/SDH. Figure 2.19 shows the five levels of the corresponding OAM operations, which are labeled F1, F2, F3, F4, and F5. F1, F2, and F3 functions reside at the physical layer; F4 and F5 functions reside at the ATM layer.

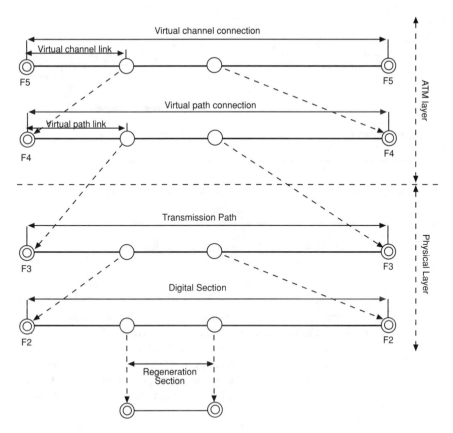

FIGURE 2.19 Relationships in ATM layers.

The Fn tags depict where the OAM information flows between two points as shown in Figure 2.20.

The five OAM flows are:

F5 OAM information flows between network elements performing VC functions. From the perspective of a B-ISDN configuration, F5 OAM operations are conducted between B-NT2/B-NT1 endpoints. F5 deals with degraded VC performance, such as late arriving cells, lost cells, cell insertion problems, etc.

F4 OAM information flows between network elements performing VP functions. From the perspective of a B-ISDN configuration, F4 OAM flows between B-NT2 and ET. F4 OAM reports on a unavailable path or a virtual path that cannot be guaranteed.

F3 OAM information flows between elements that perform the assembling and disassembling of payload, header error control (HEC) operations, and cell delineation. From the perspective of a B-ISDN configuration, F3 OAM flows between B-NT2 and virtual path cross connect and exchange termaination (ET).

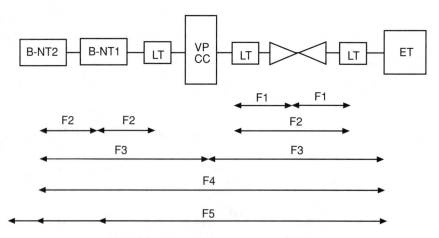

FIGURE 2.20 Information flows in B-ISDN.

F2 OAM information flows between elements that terminate section endpoints. It detects and reports on loss of frame synchronization and degraded error performance. From the perspective of a B-ISDN configuration, F2 OAM flows between B-NT2, B-NT1, and logical terminal (LT), as well as from LT to LT.

F1 OAM information flows between regenerator sections. It detects and reports on loss of frame and degraded error performance. From the perspective of a B-ISDN, F1 OAM flows between LT and regenerators.

2.3.6 MOBILE AND WIRELESS COMMUNICATION

1. Name and description of the technology: Mobile and wireless communications

   The purpose of a mobile communications system is the provision for telecommunications services between mobile stations and fixed land locations, or between two mobile stations. There are two forms of mobile communications:

   - Cellular
   - Cordless

   The best approach is to examine their major attributes and compare them with each other. First, a cellular system usually has a completely defined network which includes protocols for setting up and clearing calls and tracking mobile units through a wide geographical area. So, in a sense, it defines a user network interface (UNI) and an network node interface (NNI). With cordless personal communications, the focus is on access methods in comparison to a closely located transceiver — usually within a building. That is to say, it defines a geographically limited UNI.

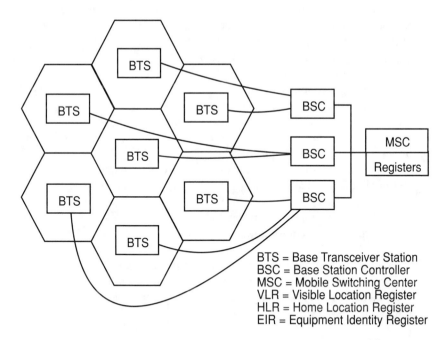

<figure>FIGURE 2.21   Structure of wireless networks.</figure>

Cellular systems operate with higher power than do the cordless personal communications systems. Therefore, cellular systems can communicate within large cells with a radius in the kilometer range. In contrast, cordless cellular communication cells are quite small, usually in the order of 100 meters.

Cellular will continue to be a preferred medium for the consumer and business market. Personal Communiations Service (PCS) will emerge as a driver for wireless. Telecommunications providers will use PCS to help reduce cellular churn, joining the customers closer to the vendor by providing end-to-end service. Wireless data connectivity, driven by lighter and smaller equipment capable of being carried by humans as well as in vehicles, will drive wireless connectivity requirements deeper and further into the network infrastructure.

2. How new is this technology?

The technology is not new, but implementation took some time. Cordless systems are undergoing rapid technological changes. Different protocols and standards are used.

A typical cellular system topology is shown in Figure 2.21. This figure concentrates on GSM (Groupe Speciale Mobile).

3. What are the application areas for this technology?

Mobile telephone (voice), with increasing use for data. Basically, this technology is suitable for any low-speed data application, voice, and low-to-medium quality video.

Cellular Digital Packet Data (CDPD) uses the existing cellular network for the sole purpose of moving bursty data, or data that is transmitted back and forth only occasionally and in quick, short spurts. CDPD is being rapidly deployed by cellular providers.

4. Topology and media dependence
   The technology is dependent on cell topology. It relies on wireless, radio media.

5. Support of WANs, MANs, and LANs
   The technology is best suited for MAN by using large cells over a metropolitan area, and cordless by using small cells in buildings.

6. What other technologies this technology complements
   Any application requiring transient nodes.

7. What other technologies this technology competes with
   It competes with hard-wired systems, including local loops and wired LANs and MANs. If capacity increases, it will become competitive to frame relay, SMDS, and ATM.

8. Basis of this technology
   If digital, it is frame based. Most frequently, Time Division Multiplexing (TDM) slots are used. In the majority of cases, it is still analog.

9. What support connection management gets
   The solutions are similar to phone calls.

10. What flow control mechanisms are supported
    No direct support.

11. Bandwidth on demand
    No support.

12. Addressing schemes used by this technology
    It is using telephone numbers, user IDs and OSI-based Network Service Access Points (NSAPs).

13. What are traffic discard options?
    No support.

14. MIB availability
    The premise of the mobile MIB is the following:

    • Network managers require more control over remote network workstations.

- To manage these remote machines, the network manager requires data on remote network access.
- Some of this data can be supplied via SNMP from the remote workstation.
- Other important information is available only on the local network at the point where the connection was created.
- This requires, therefore, two important SNMP agents with associated mobile MIB components: one at the remote workstation and the other at the local network connection point.

The following groups of information have been created:

- *The Mobile Client Group*   It contains information to be relayed from the remote workstation to a network management system on the attached network. This group stores information on mobile client name, description, location, phone number, power management configuration, connection hardware and software, client type (CPU, RAM, disk, video), system software, date, time, network adapter used, and configuration and statistics.
- *The Mobile Server Group*   It contains information to be relayed from the local network server to a network management system on the same network. This group stores information about the network server where remote connections can be originated, such as server's name, remote network hardware, slot and port number, server uptime, connection speed, service type, and traffic statistics.
- *The Trap Group*   This group desctibes SNMP trap types for remote workstations, such as mobile computer docked, undocked, suspended, resumed, Personal Computer Memory Card International Association (PCMCIA) inserted and a trap table. This is a table of alerts, which can be sent to the specified Internet Protocol (IP) address using the specified protocol by setting the value of the mobileTrapTrigger object to the index of an entry in this table.
- *Adapter Group*   It contains information about network adapters, including hardware information and type of connection.
- *Link MIB*   It provides data about mobile network links, such as link status and link performance.

15. What are the management opportunities?

     Management is more difficult than with wirelines due to the fact more information must be processed in real-time or near real-time. In most cases, proprietary solutions dominate. The implementation of TMN and other standards is extremely slow.

2.3.7   OTHER EMERGING TECHNOLOGIES

The Internet, Intranets, and Extranets are penetrating the enterprise. Both parties, the suppliers and the users, must be aware of the opportunities and limitations of this technology. The basis is the IP-infrastructure that may be built as an overlay on existing network structures. IP can and should utilize all emerged and emerging technologies that are offered for network connectivity and switching, and for the physical transmissions. IPv2, v4 and v6 is the basic infrastructure for transport services, such as Transmission Control Protocol (TCP) and User Datagram Protocol (UDP), for application support, such as Domain Name and Security Services, and finally for applications, such as remote access, e-mail, WWW, and electronic commerce. The proliferation of these applications drives the need for broader bandwidth in the backbone and in access networks.

Digital subscriber line (xDSL) The enabling technology is digital subscriber line (xDSL), a scheme that allows mixing data, voice, and video over phone lines. There are, however, different types of DSL to choose from, each suited for different applications. All DSL technologies run on existing copper phone lines and use special and sophisticated modulation to increase transmission rates (ABER97).

Asymmetric digital subscriber line (ADSL) is the most publicized of the DSL schemes and is commonly used as an ideal transport for linking branch offices and telecommuters in need of high-speed intranet and Internet access. The word *asymmetric* refers to the fact that it allows more bandwidth downstream (to the consumer), than upstream (from the consumer). Downstream, ADSL supports speeds of 1.5 to 8 Mbit/s, depending on the line quality, distance, and wire gauge. Upstream rates range between 16 and 640 Kbit/s, again depending on line quality, distance, and wire gauge. For up to 18,000 feet, ADSL can move data at T1 using standard 24-gauge wire. At distances of 12,000 feet or less, the maximum speed is 8 Mbit/s.

ADSL delivers a couple of other principal benefits. First, ADSL equipment being installed at carrier's central offices offloads overburdened voice switches by moving data traffic off the public switched telephone network and onto data networks — a critical problem resulting from Internet use. Second, the power for ADSL is sent by the carrier over the copper wire with the result that the line works even when the local power fails. This is an advantage over ISDN, which requires a local power supply and thus a separate phone line for comparable service guarantees. The third benefit over ISDN is that ADSL furnishes three information channels — two for data and one for voice. Thus, data performance is not impacted by voice calls. Rollout plans are very aggressive with this service. Its widespread availability is expected for the end of this decade. Figure 2.22 shows a generic structure of ADSL in use.

Rate-adaptive digital subscriber line (RADSL) has the same transmission limits as ADSL. But as its name suggests, it adjusts transmission speed according to the length and quality of the local line. Connection speed is established when the line synchs up or is set by a signal from the central office. RADSL devices poll the line before transmitting: standards bodies are deciding if products will constantly maintain the line speed. RADSL applications are the same as with ADSL and include Internet, intranets, video-on-demand, database access, remote LAN access, and lifeline phone services.

High-bit-rate digital subscriber line (HDSL) technology is symmetric, meaning that it furnishes the same amount of bandwidth both upstream and downstream. The most mature of the xDSL approaches, HDSL already has been implemented in the

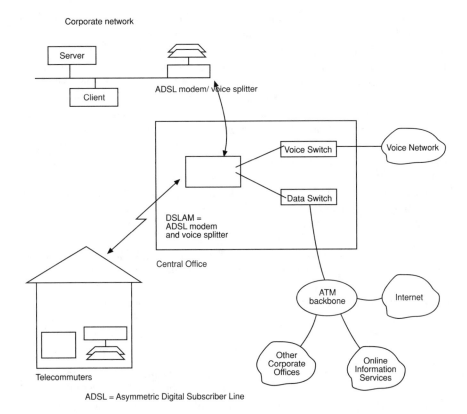

FIGURE 2.22   ADSL in use for multimedia transmissions.

telco feeder plant — the lines that extend from central office to remote nodes —
and also in campus environments. Because of its speed — T1 over two twisted pairs
of wiring, and E1 over three — telcos commonly deploy HDSL as an alternative to
T1/E1 with repeaters. At 15,000 feet, HDSL operating distance is shorter than
ADSLs, but carriers can install signal repeaters to extend its useful range (typically
by 3,000 to 4,000 feet). Its reliance on two or three wire-pairs makes it ideal for
connecting PBXs, interecxhange carrier point of presence (POPs), Internet servers,
and campus networks. In addition, carriers are starting to offer HDSL to carry digital
traffic in the local loop, between two telco central offices and customer premises.
HDSL's symmetry makes this an attractive option for high-bandwidth services like
multimedia, but availability is still very limited.

Single-line digital subscriber line (SDSL) is essentially the same as HDSL with two
exceptions: It uses a single wire pair and has a maximum operating range of 10,000
feet. Since it is symmetric and needs only one twisted pair, SDSL is suitable for
applications like video teleconferencing or collaborative computing with identical
downstream and upstream speeds. SDSL standards still are under development.

Very high-bit rate digital subscriber line (VDSL) is the fastest DSL technology. It
delivers downstream rates of 13 to 52 Mbit/s and upstream rates of 1.5 to 2.3 Mbit/s

over a single wire pair. But, the maximum operating distance is only 1,000 to 4,000 feet. In addition to supporting the same applications as ADSL, VDSL, with its additional bandwidth, could potentially enable carriers to deliver high-definition television (HDTV). VDSL still is in the definition stage.

A number of critical issues must be resolved before DSL technologies achieve widespread commercial deployment. Standards still are under development. Modulation techniques, such as carrierless amplitude phase (CAP) and discrete multitone (DMT) have been separated by standards bodies. Some other problems include interoperability, security, eliminating interference with ham radio signals and lowering power systems requirements from the present 8 to 12 watts down to 2 to 3 watts. A nontechnical, but important factor will be how well carriers can translate the successes they realized in their xDSL technology trials to market trials and then to commercial deployments.

Although cable systems offer an attractive, high-bandwidth channel to carry data as well as video signals, the implementation speed is slow. The targets can be both businesses and residential areas. At this time, entertainment distribution is unidirectional. In order to support communication in both directions, bandwidth should be partitioned also in the opposite direction from the perspective of the sender. The technology is available to build cable modems. In North America, there is an agreement between suppliers. It covers how cable television (CATV) systems will carry digital video and data in standard 6 MHZ cable channels. Reliability of cables, power supply and management seem to be the critical issues with this emerging technology.

Computer-telephone integration (CTI) technology takes many forms, from the simplest dial-the-phone from the PC to sophisticated links between computers and telephone systems in call center environments. Essentially, this technology describes any link between a computer and phone, enabling such functions as unified messaging, or having a single PC-based interface to all of voice, e-mail, and fax messages. On the high end, a call center could receive incoming calls, identify callers either through caller identification or by asking them to touch-tone an account number. The interest for this technology is great, but it would need more corporate investment.

High-speed LANs   *Fast Ethernet* is an umbrella for many technologies that increase the bandwidth of present implementations. The following technological alternatives are in use:

Full duplex LANs   The throughput can be increased when transmission is supported in both directions simultaneously. In most cases, cabling does not need to be changed at all. If two cable pairs are available between stations and the hubs, full-duplex operation can be supported with the result of doubling the potential throughput. It is a tactical solution for network managers, but very helpful in heavily loaded segments. In particular, server-hub-connections can benefit from this solution. This migration needs adapter cards, hub boards, and bridge-parts supporting full duplex operations. But, the implementation may include full duplex fast Ethernet, full duplex FDDI, and also full duplex Token Ring. Using fiber, two separate cables are necessary, each supporting unidirectional transmission. The higher speed does not cause emission problems, because the individual cables are operated at the "old" speed. They are better utilized.

100Base-T   This supports 100 Mbps of bandwidth using existing Ethernet Media Access Control (MAC) sublayers operating at 100 mbps instead of 10 Mbps. Cabling supported in the first release of this standard includes four-pair Category-3 unshielded twisted pair, two-pair Category-5 unshielded twisted pair, two-pair shielded twisted pair, and multimode fiber.

Standard bodies are busy to modify CSMA/CD to keep and increase its efficiency at higher speeds. Cabling is star supported by centrally located repeater hubs. But, the number of hubs is limited to 2 between any end-user-devices. Due to the length of the collision window and as a result due to the round-trip-delay, the distance of 100Base-T is limited to 210 meters and to 100 meters between hubs and end-user-devices. The majority of market leaders are supporting this technology. This technology, however, does not offer a breakthrough; it is a combination of shared media and a not too efficient access method.

100VG AnyLAN   This supports Ethernet and Token Ring frame formats at 100 Mbps. It operates over four-pair Category-3 unshielded twisted pair, two-pair shielded twisted pair, and multimode fiber. The physical topology is star; stations are connected to the central hub of the segments. The hub hosts the polling authority allowing the implementation of demand priority access schemes. The hub polls each station whether they want to send or not. Usually, polling is implemented in a sequential order. But, stations and applications may receive higher priorities by changing the polling table. Hubs function as repeaters; as an additional feature, they also can support security schemes by filtering incoming and outgoing packets by source and destination addresses.

The AnyLAN indicates that the hub — intelligence is assumed — supports both Ethernet and Token Ring. The limitation of distance is again approximately 100 meters between hubs and end-user-devices. Due to using demand priorities, the utilization levels are expected to be higher than in the case of 100Base-T with CSMA/CD. The support on behalf of manufacturers is still very limited.

Also LAN switching may be implemented to provide the full LAN bandwidth to the communicating partners. The next step is a gigabit Ethernet that uses the same protocols while boosting transmission speeds to a billion bits per second. It will serve first as backbone, interconnecting multiple switches. This technology will provide an upgrade path for applications with over-average bandwidth demand.

Internet applications   The Internet has received a lot of attention as a technology for unification, simplification, and cost savings for practically all communication applications. The value of the Internet to businesses and consumers continues to grow. Few major businesses can be found that do not have some level or type of presence on the Internet. Also intranets are very popular as companies use Internet-technology to communicate across both the local and wide area. Users also can use their Web browser to navigate the networks of the company. Departments or individual users can set up their own home pages on the intranet, enabling all departments to find information they need from each other, or to send information to each other via e-mail. Telecommunications and intranet providers are expected to size the backbone and access networks properly.

Tag switching A new tag-based switching technique aims to make it easier for service providers to keep up with the Internet's and intranets's explosive growth.

Tag switching combines the performance and traffic management capabilities of data link layer switching with the proven scalability of network layer routing. The technology provides three key benefits (DOWN96).

First, by combining ATM and network layer routing, tag switching eliminates the scalability problem of a pure data layer network ringed by a router overlay. Intranets can take advantage of the high perfomance of ATM switches, as well as the ability to provide Internet and ATM/frame relay services on the same platform. Second, tag switching simplifies traffic management in router-based intranets by integrating Layer 2 circuit capabilities. The ability to control the flow of packets across a Layer 2 infrastructure to support load balancing has been one of the benefits of using ATM or frame relay switches in Internet or intranet cores. Tag switching supports this capability, but, more importantly, extends these capabilities to intranets that are built completely from Layer 3 routers. Third, tag switching enables higher performance platforms by simplifying packet forwarding and switching decisions.

Tag switching is based on the concept of label swapping: units of data — a packet or a cell — carry a short, fixed-length label that provides processing information. Tag switches assign tags to multiprotocol frames for transport across packet or cell-switched networks. A tag switching internetwork consists of tag edge routers, tag switches, and the tag distribution protocol.

2.3.8 CONFORMANCE TO OSI LAYERS

All of the emerging technologies support a layered architecture. Evaluating the layers we conclude that the depth of support is very different. Figure 2.23 shows the results for the six emerged technologies compared with OSI.

In case of frame relay, just the physical and half of the data link layer are supported. The other five and a half layers are expected to be defined by users. FDDI supports the first two bottom layers leaving the rest to be defined and maintained by the user. SMDS fully defines the first three bottom layers. Layer 4 and higher are expected to be defined and maintained by the user. ATM defines the first two bottom layers, only. The remaining five layers must be defined and maintained by the users. SDH/SONET and wireless technologies support the physical layers only. In case of GSM, SS7 signaling may be utilized. In this case, more functionality is supported.

2.3.9 MANAGEMENT SOLUTIONS

The present status for emerging technologies can be summarized as follows:

- *Proprietary solutions dominate* This means that the management protocols selected are controlled by the supplier of the equipment or facilities vendors. Working groups do not exist with the exception of the ATM Forum.
- *Reenginered by SNMP* Many of the equipment vendors include SNMP-agents into their devices to meet the requirements of customers. The

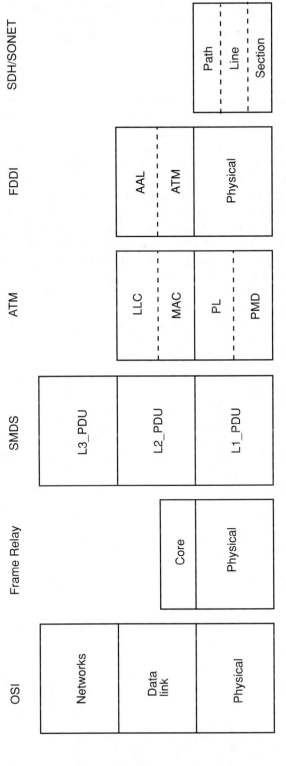

FIGURE 2.23 Overview of what communication layers are supported by emerged technologies.

SNMP-agents provide information for performance management and reporting, but they usually do not change the real-time processing of status data within the devices. Most equipment is extremely intelligent. SNMP is simply too slow for certain decisions.

- *The management structures are heterogeneous*   In most cases, domains are managed. Domain boundaries may be defined by the geography or by the family of managed objects. Most of the interfaces to element managers and managed devices are still proprietary. But, at least, the interfaces are going to be defined clearly by TMN.

- *TMN has a very low penetration*   Suppliers have recognized the need for a generic standard, but they are not willing to invest heavily into supporting it. Some of the providers go as far as supporting the Q3-interface. There are no implementation examples for the higher layers of TMN. The network element and network management layers are well understood.

- *Definition of MIBs*   Each technology requires the definition of public and private MIBs. It is merely a question of time until each of the emerging technologies will be using MIBs as the basis of management.

- *Operating Support Systems are in the transitioning process*   The legacy type OSSs will give place to the new OSSs that are based on separating operations data from operating, on the implementation of very flexible software, and on separating network management from service management.

2.4   SUMMARY

This chapter has introduced emerged and emerging technologies. In both cases, management solutions have been addressed. Management structures are heterogeneous consisting of proprietary and SNMP solutions. The TMN-concept is accepted, but little progress is seen to support service and business management applications. Web-based management is still too early for these technologies, but it may be considered for the new OSSs after the reengineering phase. Powerful gateways to network elements and element management systems are needed to interpret, consolidate and forward events, messages, and alarms to the centralized or distributed umbrella managers.

In the following chapters, emerging technologies and their management will be addressed in some detail.

3 Understanding TMN

CONTENTS

FIGURES

Telecommunications Management Network (TMN) standards are receiving increasing industry attention. Unfortunately, a number of myths and misunderstandings about these standards have emerged as telecommunication companies attempt to determine how to apply them to business needs. This chapter clarifies the basic principles behind TMN standards and addresses myths about their implementation.

3.1   WHAT IS TMN?

TMN is an acronym for *Telecommunications Management Network*. TMN standards efforts generally refer to the functional architecture model and interface specifications created and supported by the International Telecommunictions Union — Telecommunications Divsion (ITU-T) and other national and international standards organizations. The term "TMN" refers to a collection of systems that support the management needs of network and service providers for planning, provisioning, installing, maintaining,

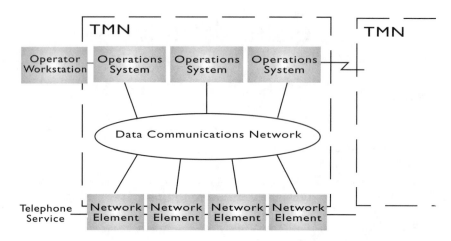

FIGURE 3.1   Relationship between TMN and telecommunications networks.

operating, and administering telecommunications networks and services. In short, a TMN provides overall management functions for telecommunications networks and services. The systems that make up a TMN can communicate with each other, with network equipment, or with other systems in other TMNs.

The general relationship between a TMN and a telecommunications network is shown in Figure 3.1. The TMN, shown enclosed in a dashed line, consists of related operations functions necessary to manage Network Elements (NEs) and services. Operations Support Systems (OSSs) in a TMN can communicate with each other, with NEs, or with counterpart systems in other TMNs. Different TMNs could be owned or operated by one or more telecommunications providers.

The basic concept behind a TMN is to provide an organized architecture to enable various types of OSSs and/or telecommunications equipment to work together for the exchange of management information. In defining this concept, TMN standards bodies recognize that many network and service providers have large infrastructures of existing OSSs, networks, and equipment already in place which must be accommodated within the architecture. The TMN functional architecture has been designed intentionally to accommodate a broad range of operations solutions.

The ITU-T (Telecommunications Standardization Sector of the International Telecommunications Union) is responsible for studying technical, operating, and tariff questions and issuing recommendations on them with a view to standardizing telecommunications, worldwide.

3.2   FUNCTIONAL AREAS AND LAYERS

The TMN principles include a categorization of management functions into five general areas. These areas identify classes of common management functions:

1. *Configuration Management* provides functions which exercise control over, identify, collect data from, and provide data to NEs.
2. *Fault Management* enables the detection, isolation, and correction of abnormal operation of the telecommunications network and its environment.

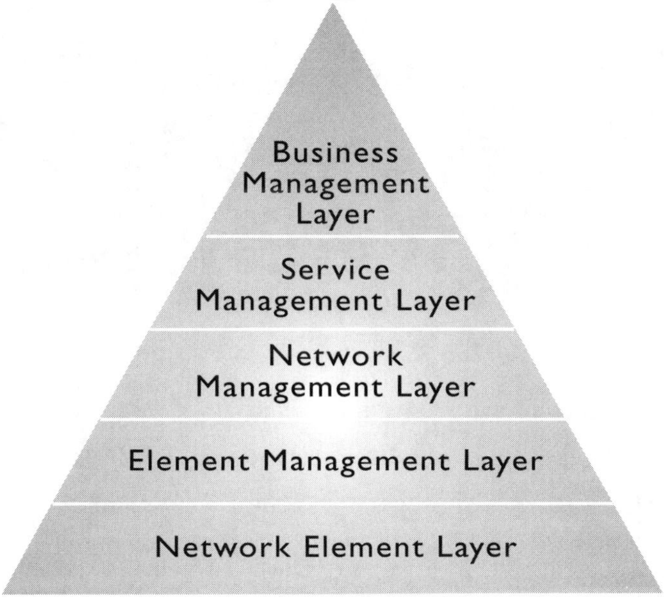

FIGURE 3.2 TMN functional layers.

3. *Performance Management* evaluates and reports on the behavior of telecommunications equipment and the effectiveness of the network or NEs.
4. *Accounting Management* measures network service usage and determines the costs for such usage.
5. *Security Management* facilitates the prevention and detection of improper use of network resources and services and the containment of and recovery from security breaches.

TMN efforts also have created the notion of distinct management layering. TMN layers categorize different levels of management functionality into a hierarchy, ranging from low-level functions for specific NEs to high-level functions for the business enterprise. The layers, which have grown to be widely recognized, usually are depicted in a pyramid, as shown in Figure 3.2, and consist of the following:

- *Business Management Layer* manages the total enterprise. It usually encompasses goal setting, budgeting, planning, and similar functions.
- *Service Management Layer* manages customer service functions, including identification, usage reporting, and quality of service maintenance/reporting.
- *Network Management Layer* manages all NEs, both individually and as a set. It is not concerned with how a particular element provides services internally.
- *Element Management Layer* manages NEs on an individual basis and supports an abstraction of the functions provided by the Network Element Layer.
- *Network Element Layer* contains the NEs themselves.

FUNCTIONAL LAYERS	FUNCTIONAL AREAS				
	Configuration	Fault	Performance	Accounting	Security
Business					
Service					
Network					
Element					
NE					

FIGURE 3.3   Functional layers and areas.

Functional Areas and Layers: Myths vs. Reality

Myth   The TMN layers are the defining characteristic of TMN standards efforts.

Reality   TMN standards primarily are concerned with information models and interfaces and apply to a variety of layering approaches. Originally these layers were introduced only to illustrate an example of functional architecture.

Myth   The TMN layers define a physical architecture model for OSSs.

Reality   The TMN layers, with the functional areas described above, define a functional model intended to help categorize management functions. These layers can apply to any operations architecture. Any OSS — new or legacy — can fit within this layered model.

TMN standards bodies have intended that the management layers and areas would be used as tools in categorizing functional operations architectures, not for dictating a physical architecture.

Combined, the functional layers and areas create a complete matrix for defining and describing the management capabilities offered by any OSS in any TMN. This matrix is a valuable tool for identifying management functions needed in new OSSs, and for categorizing existing operations functionality in legacy OSSs. The matrix is shown in Figure 3.3.

Myth   To be TMN compliant, an OSS must provide functionality that fits within a single box of this matrix. In other words, it must perform functions in one area for one layer (for example, fault management at the network management layer).

Reality   TMN standards clearly recognize that management systems may perform functions that cross more than one functional area or layer. Since this matrix is intended to define a functional mapping, not a physical architecture, it does not matter that physical systems overlap the functional boundaries.

3.3   TMN ARCHITECTURES

The TMN architectures define a set of components as basic building blocks for the management of networks as well as the means for their interaction. TMN standards identify these in terms of functional, physical, and informational architectures.

The functional architecture identifies the basic functional building blocks which enable a TMN to perform management functions. These include OSS functions, mediation functions, workstation functions, NE functions, and Q-adaptor functions.

The functional building blocks are directly reflected in the TMN physical architecture, which includes a variety of configurations for implementing the functional building blocks.

The physical TMN architecture building blocks provide these functions:

- An OSS performs management functions for the TMN.
- The mediation device processes information passed between OSSs and NEs. It may store, adapt, filter, threshold, and condense information.
- The workstation provides the means to interface with, visualize, and interpret TMN information by the management user.
- NEs provide telecommunications support functions, and communicate with the TMN for monitoring and/or control.
- Q-adaptors connect with and translate between TMN and non-TMN systems.

As shown in Figure 3.4, these functional blocks may be implemented as discrete systems and connected together directly, or through a data communications network. It is important to remember that the TMN architecture is intended to be flexible and to support many possible implementations. This figure represents a simplified implementation of these functions and interfaces.

Note: The interface points shown in Figure 3.4 (F, Qx, Q3, X, and M) are described in the next section, "TMN Interfaces."

The third aspect of TMN architecture is the information architecture. This defines an object-oriented approach for communicating between a managing system and the system or NE being managed (Figure 3.5). In general, the managing system is referred to as the manager, while the managed system is referred to as the agent. An example of a managing system is an OSS. An example of a managed system is an NE or another OSS.

Object orientation and manager/agent communications are the two primary concepts of the TMN information architecture. Information models used to define the object-oriented view of a managed system through TMN manager/agent interfaces are described in the following section.

TMN standards tend to get complicated in the area of management information models. These highly flexible, generic models are intended to represent the structure of management information that is passed between TMN systems in an object-oriented manner. A variety of these standardized information models represent common (technology independent) management information. There are also several models to represent specific network technologies such as ATM and SONET.

TMN Architectures: Myth vs. Reality

Myth Standard information models clearly define how a system NE is to be managed.

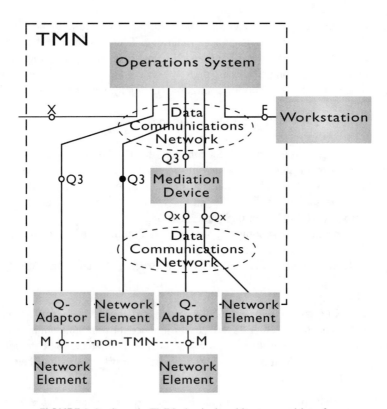

FIGURE 3.4   Sample TMN physical architecture and interfaces.

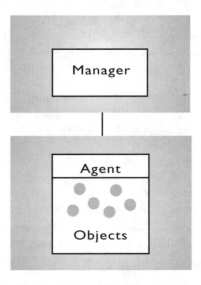

FIGURE 3.5   Manager/agent relationship with object orientation.

Reality   Information models clearly define the structure of information passed between systems in a TMN. They specify the microscopic behavior of objects in the modeled system. However, they don't define how the system is managed as a whole — the macroscopic behavior of the NE or OSS. This is left to the management application.

3.4   TMN INTERFACES

The TMN architecture identifies specific functions and their interfaces. These functions are what allow a TMN to perform its management activities. The TMN architecture provides flexibility in building a management system by allowing certain functions to be combined within a physical entity. The following function blocks, along with their typical methods of physical realization, are defined within the TMN specifications:

Operations Systems Functions (OSF)   This monitors, coordinates, and controls the TMN entities. It is a TMN-compliant management system or set of management applications. The system has to make it possible to perform general activities, such as management of performance, faults, configuration, accounting, and security. In addition, specific capabilities, for planning of operations, administration, maintenance, and provisioning of communications networks and systems should be available. These capabilities are realized in an operations system. The operations systems can be implemented in many different ways. One possibility is a descending abstraction (e.g., business, service, and network) wherein the overall business needs of the enterprise are met by coordinating the underlying services. In turn, the individual services are realized through coordinating the network resources.

Work Station Function (WSF)   The WSF provides the TMN information to the user. This typically consists of things such as access control, topological map display, and graphical interfaces. These functions are realized in a workstation.

Mediation Function (MF)   This function acts on information passing between an OSF and a network element function (NEF) or Q adapter function (QAF) to ensure the data the MF emits complies with the needs and capabilities of the receiver. MFs can store, adapt, filter, threshold, and condense information. The MFs provide the abstract view necessary to treat dissimilar elements in a similar manner. MFs may also provide local management to their associated NEFs (in other words, the MF may include an element manager). The MF function is realized in a mediation device. Mediation can be implemented as a hierarchy of cascaded devices, using standard interfaces. The cascading of mediation devices and the various interconnections to network elements provide a TMN with a great deal of flexibility. This also allows for future design of new equipment to support a greater level of processing within the network element without the need to redesign an existing TMN.

Q Adapter Function (QAF)   This connects non-TMN-compliant NEFs to the TMN environment and is realized in a Q adapter. A Q adapter allows legacy devices

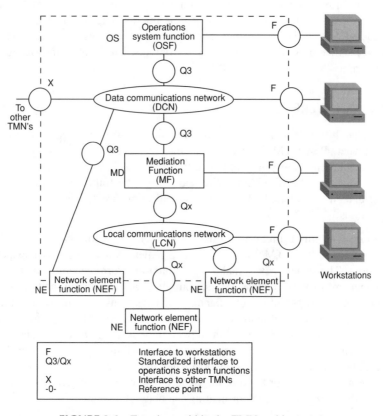

FIGURE 3.6   Functions within the TMN architecture.

(i.e., those that do not support the TMN management protocols, including SNMP devices) to be accommodated within a TMN. A Q adapter typically performs interface conversion functions (i.e., it acts as a proxy).

Network Element Function (NEF)   This function is realized in the network elements themselves. They can present a TMN-compliant or noncompliant interface. This would include such things as physical elements (switches), logical elements (virtual circuit connections), and services (operations systems software applications). Figure 3.6 illustrates the functions within a TMN environment. The portions that are outside the TMN environment are not subject to standardization. For example, the human-interface portion of the workstation function is not specified in the TMN standard.

In order for OSSs in a TMN to communicate with NEs and other TMN systems, standards bodies have defined a set of interfaces to ensure the compatibility of interconnected items. Communication protocols, data representation models for messages, and generic message definitions must all be compatible.

Interface points in the TMN architecture are shown in Figure 3.4 and include the following types:

Q3 is the TMN interface between OSSs, NEs, mediation devices, Q-adaptors, and other OSSs.

Qx interfaces between a mediation device and an NE, Q-adaptor, or another mediation device.

X is the standard interface between OSSs in two different TMNs. (The sharing of trouble information over this interface between two telecommunications providers is often called electronic bonding.)

F interfaces between a workstation and an OSS or mediation device. (This interface has not yet been standardized.)

M is a nonstandard interface to an NE.

TMN standard interfaces consist of recommended communications protocols combined with a management information model (also referred to as an object model) for the specific implementation of that interface. The management information model presents an abstraction of the management aspects of network resources and the related support management activities. The model determines the scope of the information that can be exchanged across the standardized interface.

Management protocols specified for TMN include the Common Management Information Protocol (CMIP), and the File Transfer, Access, and Management protocol. These protocols are organized into protocol suites or profiles for specific TMN interfaces. CMIP is recommended for Q3 and X interfaces and may be supported over Open System Interconnection (OSI) or TCP/IP (via RFC 1006) communications stacks. CMIP is a robust, object-oriented communications protocol well suited for managing complex systems.

Supporting a standard TMN interface involves meeting a suite of functionality recommendations, information models, communications protocols, and other components. A complete TMN interface consists of the following elements:

- An architectural definition of the communicating TMN entities, encompassing their respective functional roles and their interrelationships, for example, the OSS and NE in their roles as managing system and managed system and their physical interconnection via a Q3 interface.
- The Operations, Administration, Maintenance, and Provisioning (OAM&P) functionality to be supported by the communication.
- Management application messages and information models to support the OAM&P functional requirements.
- Resource information models that provide an abstraction of the telecommunication network resources to be managed in the form of generic or technology/service-specific managed objects.
- Communication protocols to transport the messages between TMN communicating entities.
- Conformance requirements that identify and characterize, as required or optional, each aspect of an interface specification that could be implemented in a system.

- Profiles that assemble implementable and marketable packages of functionality from among the various mandatory and optional aspects of the standards.

There are many possible TMN interface configurations with much flexibility in their implementation. For example, consider two equivalent NEs that are supplied by different vendors, both supporting Q3 interfaces and the same standard information models. While implementing what should be the same standard interface, these vendors can support different subsets of the same information models. They may apply the models to their equipment differently. Some vendors use nonstandard extensions or modify the standard models to simplify their implementation. This can result in two different and incompatible implementations of the same standards.

TMN Interfaces: Myths vs. Reality

Myth   Q3 defines a single, unique interface between all OSSs and all NEs.

Reality   No single, unique TMN interface has been defined for use between all OSSs and all NEs. Q3 includes a group of communications protocols that may be applied differently. In addition to these options, many information models may be used to represent the information passed between systems for management purposes. Different NEs and OSSs will support different information models based on specific functionality.

Myth   TMN interface standards allow full plug-and-play operations.

Reality   Although this is possible, it often is not the case. The broadness of the standards allows for flexibility in their implementation. Therefore, the same set of interface standards may be implemented differently. While interface standards can allow low-level plug-and-play communication, they don't guarantee that an OSS meeting a standard interface will actually be able to manage an NE just because it supports the same standards.

3.5   COMMON TMN QUESTIONS

3.5.1   WHY IS TMN IMPORTANT?

TMN efforts are important for several reasons. The most basic of these is the process of industry-wide collaboration that is defining and categorizing telecommunications network management. This collaboration has already provided a common language useful in identifying, specifying, developing, and purchasing management functionality and systems across the industry.

Another important purpose for TMN efforts is to provide a common framework for OSSs to communicate with NEs and with each other. This requires a broad set of common functional definitions, interface protocols, and information models. Common use of standards will increase the interoperability between systems and reduce the number of inflexible, proprietary, and ad hoc operations solutions.

The long-range promise of this work is, ultimately, to reduce the cost of and time to implement management functionality while increasing the flexibility, extensibility, and interoperability of operations solutions.

3.5.2 What is a TMN-Compliant System?

TMN recommendations have been so broadly defined that they encompass many aspects of network and service management. Also, some elements of networking have not yet been completely standardized. This makes it difficult to know just what makes a network "TMN compliant."

Functional and interface aspects are the simplest gauges of TMN compliance. The initial building blocks for TMN compliance can be found in the general TMN principles identified in recommendations such as "M.3010, Principles for a Telecommunications Management Network" and "M.3400, TMN Management Functions" (see *TMN Standards Bibliography*). However, simply following these guidelines will not guarantee TMN compliance.

3.5.3 Why does TMN Seem so Complex?

The primary source of complexity in TMN standards can be found in the standard information models used to define management interfaces. Complexity itself is not bad: a complex model is often needed to represent a complex system or behavior. However, it is often argued that many of the fundamental information models are far more complex than they need to be.

3.5.4 Common TMN Myths vs. Reality

Myth A layered management architecture, consisting of business, service, network, and element management layers, is inherently TMN compliant. Any OSS in that network is a TMN-compliant system.

Reality These layers (shown in Figure 3.2) provide a useful categorization of management functionality, but do not determine TMN compliance.

TMN-compliant systems also must support standard interfaces. Although TMN efforts have grown to encompass a broad range of management aspects and operations functionality, the essence of TMN standardization lies in its detailed interface specifications. As discussed above, however, because these standards are so broad, there are many ways to meet them. For example, a unique implementation of one set of standards may be claimed to be "TMN compliant."

Myth TMN information models are too complex to implement practically.

Reality Standard information models can be very complex. However, as the telecommunications industry makes strides in developing these standards, more powerful tools are gradually being made available to simplify their implementation.

As industry experience with TMN standards increases, their relative complexity is being handled in several ways. For example, OSI's NetExpert framework helps reduce this complexity through its Graphical User Interface (GUI)-based, rapid-development toolset. NetExpert enables the implementation of interfaces and applications without traditional low-level software development efforts. In addition, it allows complex information models used for Q3 interfaces to be readily translated into simpler object models for internal management applications. This also enables a common management application to support a variety of interfaces, standard and nonstandard.

3.6   IMPLEMENTING TMN STANDARDS

Because TMN standards are so broad, there are many ways to meet them. A network may be TMN-compliant in one area and not in another. The result of the variability and flexibility in the way standards can be implemented has led to a concern within the industry about the complexity of the TMN models. However, as the telecommunications industry makes strides in developing these standards, more powerful tools that simplify the implementation of the standards are gradually becoming available. Ultimately, when widely implemented, TMN standards are intended to reduce both the cost of and time to implement management functions and to increase the flexibility, extensibility, and interoperability of operations solutions. In the long run, standardization of the telecommunications industry will make network management more cost effective and practical.

3.7   SUMMARY

The full implementation of TMN is a very long process. In fact, full implementation may never happen. But telecommunications suppliers are expected to position their services and products in accordance with TMN models. The same is true for management framework providers.

4   Industry Standards for Network Management

CONTENTS

FIGURES

TABLES

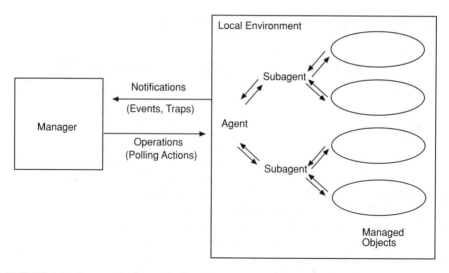

FIGURE 4.1   Communications paths between manager, agents, subagents and managed objects.

Industry standards help suppliers, vendors and customers to communicate with each other more efficiently. But to find the common denominator of all interests is not easy. Service creation, provisioning, assurance, delivery and service management cannot wait for fully completed standards. This chapter summarizes the most important management standards for multimedia communications. Besides TMN, CMIP, SNMP, RMON, OMA, and DMTF, also Web-based standards are outlined that may help to unify and simplify management frameworks and applications. The Network Management Forum may help to improve the interoperability between multiple suppliers by providing practical guides, such as specifications, solution sets, and conformance documents.

4.1   MANAGER-AGENT RELATIONSHIP

Protocols always represent an agreement between the communicating parties. In the area of network management, management/agent relationships have been frequently implemented. Figure 4.1 shows this relationship. The agent side may be hierarchical by implementing subagents into specific devices. Depending on the nature of the management protocol, either the manager or the agent starts the dialog. There are always exceptions for high-priority networking events.

In the case of CMIP, eventing techniques are used. Assuming that the agents are intelligent enough to capture, interpret, filter and process events, they will notify the manager about alarming conditions. Of course, the manager is allowed to interrupt and send inquiry-types of messages to the agents.

In the case of SNMP, the manager polls the agents periodically. The agents respond by sending information on device status and performance. Usually, the agents wait for the poll unless unusual events occur in the network. For such events, special traps can be defined and implemented.

4.2 COMMON MANAGEMENT INFORMATION PROTOCOL (CMIP)

TMN protocols include OSI protocols such as CMIP and FTMP, ISDN and Signaling System Number 7 protocols. They are organized into protocol suites or profiles for specific TMN interfaces. Functions and protocols support TMN services which include:

- Traffic management
- Customer management
- Switching
- Management of transport networks
- Management of intelligent networks
- Tariffing and changing

The primary protocol is Common Management Information Protocol (CMIP). The estimated overhead scares both vendors and users away with the exception of the telecommunications industry, where separate channels can be used for management. CMIP is event driven assuming processing capabilities at the agent level. Once fault and performance thresholds are exceeded, the manager is alarmed by the agent. This is similar to SNMP traps.

Open Systems Interconnected network management follows an object-oriented model; physical and logical real resources are managed through abstractions known as managed objects (MO). Most are handled by applications in agent roles and are accessed by applications in manager roles in order to implement management policies. The global collection of management information is consolidated in MIBs; each agent handles a part of it in its Management Information Tree. Information in the manager-agent interactions is conveyed through the management service/protocols CMIS and CMIP. In the context, agent, managed system and managed node are synonymous; manager, managing application, and management stations are synonymous as well. CMIP is part of the OSI management framework. Its elements are:

- CMIS/CMIP defines the services provided to management applications and the protocol providing the information exchange capability
- Systems management functions specifying all the functions to be supported by management
- Management information model that defines the MIB
- Layer management that defines management information, service, and functions to specific layers

Based on the seven-layer model, network-management applications are implemented in layer seven. Layers one through six contribute to network management by offering the standard services to carry network management related information. Figure 4.2 shows the structure for the application layer. In particular, four System Management Application Entities (SMAE) are very useful:

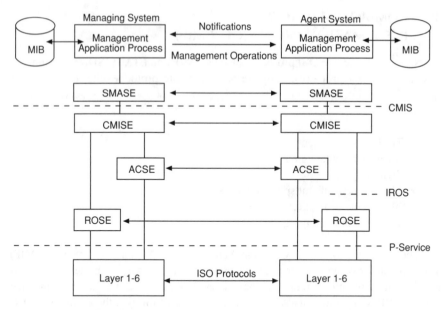

FIGURE 4.2   Application layer protocols and services.

For generic use   Association Control Service Element (ACSE) and Remote Operations Service Element (ROSE).

For specific use   Common Management Information Service Element (CMISE) and System Management Application Service Element (SMASE).

Communication services in the OSI management model are provided by the Common Management Information Services (CMIS). The service is realized through CMIP over a full or lightweight OSI stack. In the OSI world, there are two mappings defined:

- A service over a full OSI stack (CMIP)
- A connectionless one over Logical Link Control 1 (LLC1)(CMOL)

In the Internet world, there is a third mapping to provide the service over TCP/UDP using a lightweight presentation protocol (CMOT). CMOT and CMIP applications are portable on each other's stack with the same API, but will not work over CMOL.

Figure 4.3 shows the complete structure of CMISEs and SMFs. The overall goal is to support typical network management functions, such as fault, configuration, performance, security, and accounting management. SMFs, or a group of SMFs, support specific management functions. For the communication between entities, CMISE is used. Figure 4.4 summarizes these service elements between SMASE and CMISE. Each of the service elements are defined in Table 4.1.

Further details can be found in (TERP96) — *Effective Management of LANs*, McGraw-Hill.

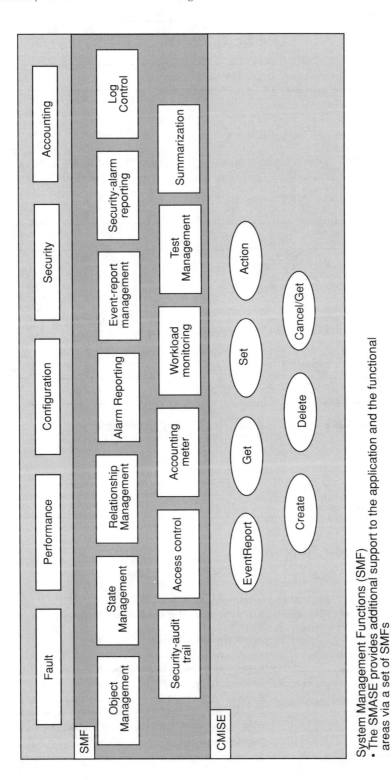

System Management Functions (SMF)
• The SMASE provides additional support to the application and the functional
  areas via a set of SMFs

FIGURE 4.3   Overview of OSI management structure.

CMISE Services

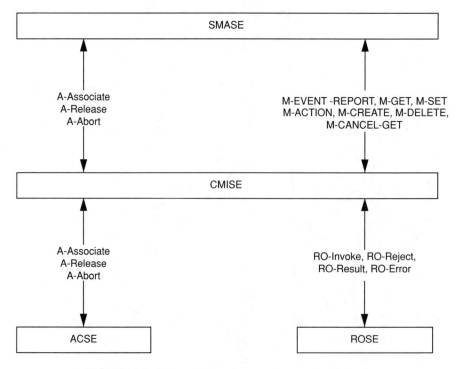

FIGURE 4.4   Information exchange using service elements.

TABLE 4.1
Definition of Service Elements

Service	Type	Definition
Management Notification Service		
M-even-report	Confirmed/unconfirmed	Reports on an event to a peer CMISE service user
Management Operation Service		
M-get	Confirmed	Requests retrieval of information from peer CMISE user
M-set	Confirmed/unconfirmed	Requests modification of information from peer CMISE user
M-action	Confirmed/unconfirmed	Requests that peer CMISE user perform some action
M-create	Confirmed	Requests that peer CMISE user create an instance of an MO
M-delete	Confirmed	Requests that peer CMISE user delete an instance of an MO
M-cancel-get	Confirmed	Requests that peer CMISE user cancel outstanding invocation of M-get service

The strengths of CMIP include:

- General and exensible object-oriented approach
- Support from the telecommunication industry
- Support for manager-to-manager communications
- Support for a framework for automation

Weaknesses of CMIP are:

- It is complex and multilayered
- High overhead is the price of many confirmations
- Few CMIP-based management systems are shipping
- Few CMIP-based agents are in use

CMIP prefers the use of the OSI-stack for exchanging CMIP protocol data units, but CMIP also can run on other stacks. In layer seven, there are also other applications which may be combined with CMIP.

4.3 SNMPv1 AND SNMPv2 (SIMPLE NETWORK MANAGEMENT PROTOCOL)

In the SNMP environment, the manager can obtain information (Figure 4.5) from the agent by polling managed objects periodically. Agents can transmit unsolicited event messages, called "traps," to the manager. The management data exchanged between managers and agents is called the management information base (MIB). The data definitions outlined in SMI must be understood by both managers and agents.

The manager is a software program housed within the management station. The manager has the ability to query agents using various SNMP commands. The management station is also in charge to interpret MIB data, construct views of the systems and networks, compress data, and maintain data in relational or object-oriented databases.

In terms of MIBs, there are a lot of changes. In addition to standard MIBs, such as MIB I and II (Table 4.2), the IETF has defined a number of adjunct MIBs covering hosts, bridges, hubs, repeaters, FDDI networks, AppleTalk networks, and frame relay networks.

In terms of SNMP, the following trends are expected. SNMP agent-level support will be provided by even a greater number of vendors. SNMP manager-level support will be provided by only a few leading vendors in the form of several widely-accepted platforms. Management platforms provide basic services, leaving customization to vendors and users.

Wider use of intelligent agents is also expected. Intelligent agents are capable of responding to a manager's request for information and performing certain manager-like functions, including testing for thresholds, filtering, and processing management data. Intelligent agents enable localized polling and filtering on servers, workstations, and hubs, for example. Thus, these agents reduce polling overhead and management data traffic, forwarding only the most critical alerts and processed data to the SNMP manager.

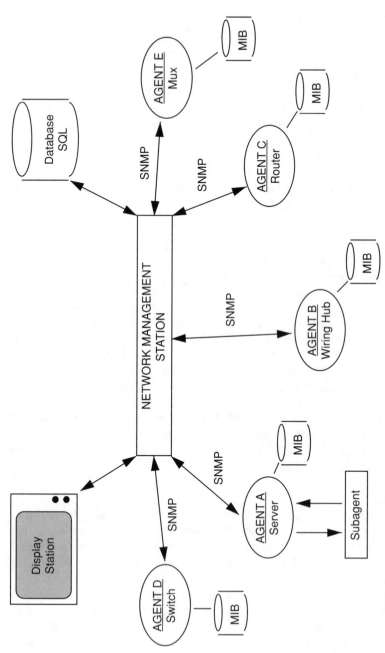

MIB=Management Information Base
SNMP=Simple Network Management Protocol

FIGURE 4.5   Structure of SNMP-based management services.

TABLE 4.2
MIB II Structure

11 Categories of Management (2) Subtree	Information in the Category
System (1)	Network device operating system
Interfaces (2)	Network interface specific
Address translation (3)	Address mappings
IP (4)	Internet protocol specific
ICMP (5)	Internet control message protocol specific
TCP (6)	Transmission control protocol specific
UDP (7)	User datagram protocol specific
EGP (8)	Exterior gateway protocol specific
CMOT (9)	Common management information services on TCP specific
Transmission (10)	Transmission protocol specific
SNMP (11)	SNMP specific

RMON MIB will help to bridge the gap between the limited services provided by management platforms and the rich sets of data and statistics provided by traffic monitors and analyzers.

The strengths of SNMP include:

- Agents are widely implemented
- Simple to implement
- Agent-level overhead is minimal
- Polling approach is good for LAN-based managed objects
- Robust and extensible
- Offers the best direct manager-agent interface

SNMP met a critical need. It was available and implementable at the right time. The weaknesses of SNMP include:

- Too simple, does not scale well
- No object-oriented data view
- Unique semantics make integration with other approaches difficult
- High communication overhead due to polling
- Many implementation-specific (private MIB) extentions
- No standard control definition
- Small agent (one agent per device) may be inappropriate for systems management.

SNMP is being continuously improved and extended. SNMPv2 addresses many of the shortcomings of version 1. SNMPv2 can support either a highly centralized management strategy or a distributed one. In the latter case, some systems operate both in the role of manager and agent. In its agent role, such a system will accept commands from a superior manager; these commands may deal with access of information stored locally at the intermediate manager or may require the intermediate manager to provide summary information about subagents. The principal enhancements to SNMPv1 provided by version 2 fall into the following categories: (STAL96)

- Structure of management information is being expanded in several ways. The macro used to define object types has been expanded to include several new data types and to enhance the documentation associated with an object. Noticeable is the change that a new convention has been provided for creating and deleting conceptual rows in a table. The origin of this capability is from RMON.
- Transport mappings help to use different protocol stacks to transport the SNMP information including user datagram protocol, OSI connectionless-mode protocol, Novell internetwork (IPX) protocol, and Appletalk.
- Protocol operations with the most noticeable change include two new PDUs. The GetBulkRequest PDU enables the manager to retrieve efficiently large blocks of data. In particular, it is powerful in retrieving multiple rows in a table. The InformRequest PDU enables one manager to send trap type information to another.
- MIB extentions contain basic traffic information about the operation of the SNMPv2 protocol; this is identical to SNMP MIB II. The SNMPv2 MIB also contains other information related to the configuration of SNMPv2 manager to agent.
- Manager-to-manager capability is specified in a special MIB, called M2M. It provides functionality similar to the RMON MIB. In this case, the M2M MIB may be used to allow an intermediate manager to function as a remote monitor of network media traffic. Also reporting is supported. Two major groups, Alarm and Event, are supported.
- SNMPv2 security includes a wrapper containing authentication and privacy information as a header to PDUs.

The SNMPv2 framework is derived from the SNMP framework. It is intended that the evolution from SNMP to SNMP2 be seamless. The easiest way to accomplish this is to upgrade the manager to support SNMPv2 in a way that allows the coexistence of SNMPv2 managers, SNMPv2 agents, and SNMP agents. In order to map commands mutually into the target protocol, proxy agents are used (STAL96). The actual implementation of the proxy agent depends on the vendor; it could be implemented into the agent or into the manager.

4.4   RMON1 AND RMON2 (REMOTE MONITORING)

The Remote MONitoring (RMON) MIB will help to bridge the gap between the limited services provided by management platforms and the rich sets of data and statistics provided by traffic monitors and analyzers. RMON defines the next generation of network monitoring with more comprehensive network fault diagnosis, planning, and performance tuning features than any current monitoring solution. The design goals for RMON are: (STAL96)

- **Offline operation**   In order to reduce overhead over communication links, it may be necessary to limit or halt polling of a monitor by the manager. In general, the monitor should collect fault, performance, and configuration information continuously, even if it is not being polled by a manager. The monitor simply continues to accumulate statistics that may be retrieved by the manager at a later time. The monitor may also attempt to notify the manager if an exceptional event occurs.
- **Preemptive monitoring**   If the monitor has sufficient resources, and the process is not disruptive, the monitor can continuously run diagnostics and log performance. In the event of a failure somewhere in the network, the monitor may be able to notify the manager and provide useful information for diagnosing the failure.
- **Problem detection and reporting**   Preemptive monitoring involves an active probing of the network and the consumption of network resources to check for error and exception conditions. Alternatively, the monitor — without polling — passively can recognize certain error conditions and other conditions, such as congestions and collisions, on the basis of the traffic that it observes. The monitor can be configured to continuously check for such conditions. When one of these conditions occurs, the monitor can log the condition and notify the manager.
- **Value-added data**   The network monitor can perform analyses specific to the data collected on its subnetworks, thus relieving the manager of this responsibility. The monitor, for instance, can observe which station generates the most traffic or errors in network segments. This type of information is not otherwise accessible to the manager that is not directly attached to the segment.
- **Multiple managers**   An internet working configuration may have more than one manager in order to achieve reliability, perform different functions, and provide management capability to different units within an organization. The monitor can be configured to deal with more than one manager concurrently.

Table 4.3 summarizes the RMON MIB groups for Ethernet segments. Table 4.4 defines the RMON MIB groups for Token Ring segments. At the present time, there are just a few monitors that can measure both types of segments using the same probe.

RMON is very rich in features and there is the very real risk of overloading the monitor, the communication links, and the manager when all the details are recorded, processed, and reported. The preferred solution is to do as much of the analysis as possible locally, at the monitor, and to send just the aggregated data to the manager. This assumes powerful monitors. In other applications, monitors may be reprogrammed during operations by the managers. This is very useful when diagnosing problems. Even if the manager can define specific RMON requests, it is still necessary to be aware of the trade-offs involved. A complex filter will allow the monitor

TABLE 4.3
RMON MIB Groups for Ethernet

Statistics group	Features a table that tracks about 20 different characteristics of traffic on Ethernet LAN segment, including total octets and packets, oversized packets, and errors.
History group	Allows a manager to establish the frequency and duration of traffic-observation intervals, called *buckets*. The agent then can record the characteristic of traffic according to those bucket intervals.
Alarm group	Permits the user to establish the criteria and thresholds that will prompt the agent to issue alarms.
Host group	Organizes traffic statistics by each LAN node, based on time intervals set by the manager.
HostTopN group	Allows the user to set up ordered lists and reports based on the highest statistics generated by the host group.
Matrix group	Maintains two tables of traffic statistics based on pairs of communicating nodes: one is organized by sending node addresses, the other by receiving node addresses.
Filter group	Allows a manager to define, by channel, particular characteristics of packets. A filter might instruct the agent, for example, to record packets with a value that indicates they contain DECnet messages.
Packet capture group	This group works with the Filter group and lets the manager specifiy the memory resources to be used for recording packets that meet filter criteria.
Event group	Allows the manager to specifiy a set of parameters or conditions to be observed by the agent. Whenever these parameters or conditions occur, the agent will record an event into the log.

to capture and report a limited amount of data, thus avoiding overhead on the network. However, complex filters consume processing power at the monitor; if too many filters are implemented, the monitor will become overloaded. This is particularly true if the network segments are busy, which is probably the time when measurements are most valuable.

Figure 4.6 shows the RMON probes in the segments. The probes can be implemented as stand-alone hardware probes, or in routers and hubs embedded hardware probes, or as software probes implemented in PCs.

The existing and widely used RMON-version is basically a MAC-standard. It does not give LAN managers visibility into conversations across the network or connectivity between various network segments. The extended standard is targeting the network-layer and higher. It will give visibility across the enterprise. With remote access and distributed workgroups, there is a substantial intersegment traffic. The following functionalities are included:

- Protocol distribution
- Address mapping
- Network layer host table
- Network layer matrix table
- Application layer host table

TABLE 4.4.
RMON MIB Groups for Token Ring

Statistics group	This group includes packets, octets, broadcasts, dropped packets, soft errors, and packet distribution statistics. Statistics are at two levels: MAC for the protocol level and LLC to measure traffic flow.
History group	Long-term historical data for segment trend analysis. Histories include both MAC and LLC statistics.
Host group	Collects information on each host discovered on the segment.
HostTopN group	Provides sorted statistics that allow reduction of network overhead by looking only at the most active hosts on each segment.
Matrix group	Reports on traffic errors between any host pair for correlating conversations on the most active nodes.
Ring station group	Collects general ring information and specific information for each station. General information includes: ring state (normal, beacon, claim token, purge); active monitor; number of action stations. Ring station information includes a variety of error counters, station status, insertion time, and last enter/exit time.
Ring station order	Maps station MAC addresses to their order in the ring.
Source routing statistics	In source-routing bridges, information is provided on the number of frames and octets transmitted to and from the local ring. Other data includes broadcasts per route and frame counter per hop.
Alarm group	Reports changes in network characteristics based on thresholds. Events may be used to inititiate functions such as data capture or instance counts to isolate specific segments on the network.
Event group	Logging of events on the basis of thresholds. Events may be used to initiate functions such as data capture or instance counts to isolate specific segments of the network.
Filter group	Definitions of packet matches for selective information capture. These include logical operators (and, or, not) so network events can be specified for data capture, alarms, and statistics.
Packet capture group	Stores packets that match filtering specifications.

- Application layer matrix table
- User history collection table
- Protocol configuration
- Protocol distribution and protocol directory table

The issue was how to provide a mechanism that would support the large number of protocols running on any one network. Current implementations of RMON employ a protocol filter which analyzes only the essential protocols. RMON2, however, will employ a protocol directory system which will allow an RMON2 application to define which protocols an agent will employ. The Protocol Directory Table will specify the various protocols an RMON2 probe can interpret.

Address Mapping   This feature matches each network address with a specific port to which the hosts are attached. It also identifies traffic-generating nodes/hosts by MAC, Token Ring or Ethernet address. It helps identify specific patterns of network

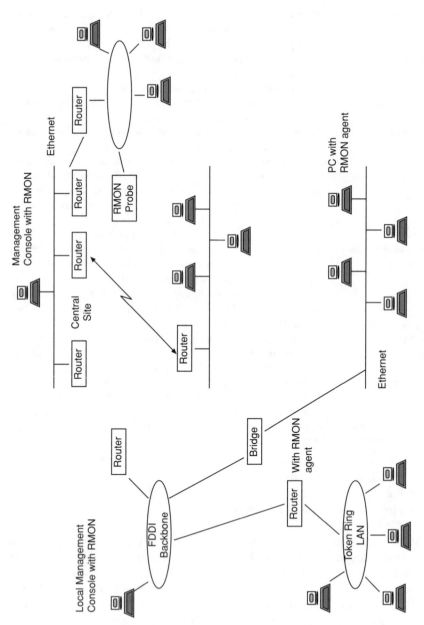

FIGURE 4.6 RMON probes in LAN segments.

traffic useful in node discovery and network topology configurations. In addition, the address translation feature adds duplicate IP address detection resolving a common troublespot with network routers and virtual LANs.

Network-Layer Host Table   Tracks packets, errors, and bytes for each host according to a network-layer protocol. It permits decoding of packets based on their network layer address, in essence permitting network managers to look beyond the router at each of the hosts configured on the network.

Network-Layer Matrix Table   Tracks the number of packets sent between a pair of hosts by network layer protocol. The network manager can identify network problems quicker using this matrix table which shows the protocol-specific traffic between communicating pairs of systems.

Application-Layer Host Table   Tracks packets, errors and bytes by host on an application-specific basis (e.g., Lotus Notes, E-mail, Web, etc.). Both the application-layer host table and matrix table trace packet activity of a particular application. This feature can be used by network managers to charge users on the basis of how much network bandwidth was used by their applications.

Application-Layer Matrix Table   Tracks packet activity between pairs of hosts by application (e.g., pairs of hosts exchanging internet information).

Probe Configuration   Currently, vendors offer a variety of proprietary means for configuring and controlling their respective probes. This complicates interoperability. The Probe Configuration Specification, based on the Aspen MIB, defines standard parameters for remotely configuring probes — parameters such as network address, SNMP error trap destinations, modem communications with probes, serial line information, and downloading of data to probes. It provides enhanced interoperability between probes by specifying standard parameters for operations, permitting one vendor's RMON application the ability to remotely configure another vendor's RMON probe.

User History Collection Group   The RMON2 history group polls, filters, and stores statistics based on user-defined variables creating a log of the data for use as a historical tracking tool. This is in contrast to RMON1 where historical data is gathered on a predefined set of statistics.

After implementation, more and more complete information will be available for performance analysis and capacity planning.

4.5   INTEROPERABILITY GUIDELINES AND PROJECTS FROM NM/FORUM

Of most importance to the strategist, interoperability guidelines provide a practical way to implement standards across management sysems from multiple suppliers. Interoperability guidelines attempt to embrace both existing and emerging management standards and technologies. Consequently, the guidelines go beyond TMN by including concepts from DCE, DME, TINA (Telecommunications Intelligent Network Architecture), and CORBA (Common Object Request Broker Architecture) for implementation and software interoperability purposes. As these technologies mature further, they will be more fully utilized by future guidelines.

Inoperability guidelines can be characterized by the following attributes:

- Bring together the many standards and specifications from all parts of the telecommunications and computing industries, turning what would be numerous divergent outputs into a coordinated set of tools that will work together to achieve a specific business goal. Besides traditional management work based on ISO and IU, Inoperability guidelines take into consideration the need for service providers to interface with their customers' management systems. It requires to work with the Internet Engineering Task Force to include SNMP. Similarly, because inoperability guidelines are focused on implementation, the work of X/Open and the Object Management Group (see next segment 4.6) to advance computing platform technology is considered critical.
- Deliver an implementable set of specifications, and it provides the means for multiple standards and specifications to work together in systems by providing internetworking strategies and specifications.
- Through a series of controlled releases, inoperability guidelines provide "freeze" points at which developers can build to a specific version of a standard with confidence that their investment will not be worthless in a couple of months. Inoperability guidelines releases are expected to be updated in 24- to 30-month cycles; migration strategies make the upgrading job to the next release relatively easy.
- Inoperability guidelines are aimed at relating the many standards and technologies to specific management needs. Thus, additional specifications such as ensembles are applied to the profiled standards in order to deliver the capability to develop value-added applications, and standards and specifications are grouped in ways that reflect actual product characteristics. Procurement and development support, including educational guides and tool kits, make it possible to use the results of the program effectively.

Two kinds of documentation are valuable to users. These are *functional specifications* and *ensembles*. The documentation includes the following functional specifications:

- Testing management allows a managing sysem to invoke, control, monitor, and obtain results of tests on a remote system through an interoperable interface.
- Scheduling management allows the scheduled triggering of a function within a managed resource via an interoperable interface.
- Path tracing identifies all of the managed resources involved in an end-to-end communications path.
- Trouble management provides for trouble reporting and tracking between systems cooperating toward the resolution of a problem via an interoper-

able interface. Inoperability guidelines provide cross-jurisdictional trouble management.
* Security management provides for the exchange of security information across an interoperable interface, such as authentication, access control, security alarm, and security audit trail.

The end product of the new service concept is the ensemble, a package of specifications intended to solve one specific management problem by addressing it from a functional point of view. This is the documentation form to provide in one place all information needed to specify a new NM/Forum-compliant service. It includes requirements, examples, scenarios, information model definitions, references to managed object classes in libraries, and references to all applicable standards. To ensure interoperability, it also references conformance documents. The interoperability guidelines include two ensembles relating to Reconfigurable Circuit Service.

In order to work with recommendations, software portability is extremely important. interoperability guidelines specify the following APIs to isolate an application from particular implementations of underlying services:

* XMP (X/Open Management Protocol) API specifies the protocol interface between an application and a CMIP or SNMP manager, making the application portable from one manager implementation to another. It is used to access the service elements of CMIP and SNMP agents in the managed system.
* XOM (X/Open OSI-abstract-data Manipulation) API specifies the data interface between an application and a CMIP or SNMP manager, making the application portable from one manager implementation to another. It is used to generate the data arguments and parameters in conjunction with XMP to access attribute data in managed objects.

CORBA (Common Object Request Broker Architecture) specifies the interface beween an application object and an Object Request Broker (ORB), making the objects portable from one ORB implementation to another. An ORB allows one object to use another without regard to their remote physical locations in a distributed object environment.

A working team within the Network Management Forum called SPIRIT (Service Providers Integrated Requirements for Information Technology) is attempting to define general purpose software platform requirements to establish a broader basis for network management application portability, interoperability, and modularity.

In principle, for general computing requirements, SPIRIT will follow the work of X/Open guidelines. For application as either a Management Platform or as a Managed Agent, the interoperability guidelines specifications will be followed.

4.6 OBJECT MANAGEMENT ARCHITECTURE (OMA) FROM OMG

OMA permits the cooperation of distributed objects independently from their locations. OMA has not been designed and structured for network management only. It is a general purpose standard. In particular, it helps to design and implement distributed applications. This architecture has gained interest because the OMA technology is going to be increasingly used for end-user-devices.

The organizational model is using a peer-to-peer approach with the result that communicating objects are equivalent in their importance. Figure 4.7 shows this architecture.

The communication model is based on CORBA (Common Object Request Broker Architecture). The Object Request Broker is the coordinator between distributed objects. The broker receives messages, inquiries, and results from objects and routes them to their destinations. If the objects are in a heterogeneous environment, multiple brokers are required. They will talk to each other in the future by a new protocol based on TCP/IP. There is no information model available; no operations are predefined for objects. But, an object exists containing all the necessary interfaces to the object request broker. For the description, Interface Definition Language (IDL) is being used. There are no detailed MIBs for objects because OMA is not management specific.

The functional model consists of the Object Services Architecture. It delivers the framework for defining objects, services, and functions. Examples for services are naming, storing objects' attributes, and the distribution/receipt of events and notification. CORBA services and facilities represent more generic services; they can occur in multiple applications or can be used in specific applications. The driving force beyond designing common facilities for systems management is the X/Open Systems Management Working Group. The Managed Set Service, defined by this Group, encourages grouping of objects in accordance with their management needs resulting in easier administration. In the future, more services are expected to be defined; the next is an Event Management Service that expands the present Object Event Service by a flexible mechanism of event filtering.

4.7 TELECOMMUNICATIONS INFORMATION NETWORKING ARCHITECTURE (TINA)

Telecommunications Information Networking Architecture (TINA) is based on the concept that call processing in networks, and its control and management are separated from each other. TINA is actually a concept-integrator from IN, TMN, ODP (Open Distributed Processing) from ISO and CORBA from OMG. The core is OSI-based network management, expanded by the layered structure of TMN. The emphasis with TINA is not on the management of network elements, however, but on the network-and services-layers. TINA is going to be standardized by a consortium consisting of telecommunications suppliers, computer and software vendors.

4.8 DMTF (DESKTOP MANAGEMENT TASK FORCE)

Basically, SNMP may be utilized to manage systems assuming system components accommodate SNMP agents. But there are no MIBs yet that describe principal indicators for management purposes. An important emerging standard for desktop

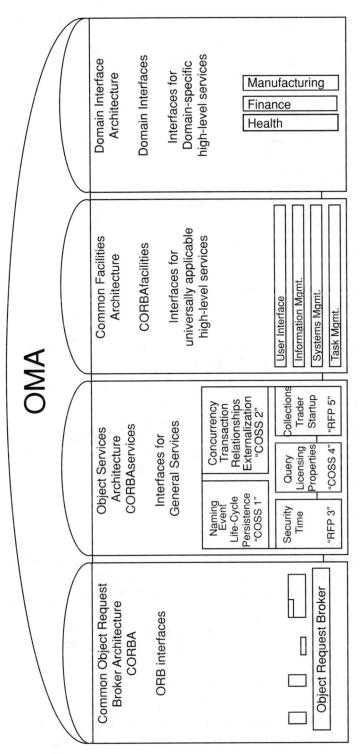

FIGURE 4.7 Open management architecture from OMG.

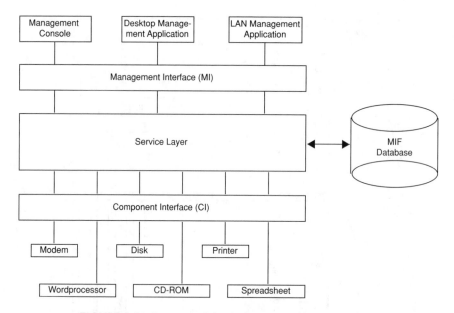

FIGURE 4.8   Structure of the desktop management interface.

management is the Desktop Management Interface (DMI). The Desktop Management Task Force (DMTF) had defined the DMI to accomplish the following goals:

- Enable and facilitate desktop, local, and network management
- Solve software overlap and storage problems
- Create a standard method for management of hardware and software components using MIFs (Management Information Format)
- Provide a common interface for managing desktop computers and their components
- Provide a simple method to describe, access, and manage desktop components.

The scope of management under DMTF includes CPUs, I/Os, motherboards, video cards, network interface cards, faxes, modems, mass storage devices, printers, and applications. Figure 4.8 shows the structure of this standard. There is a clear separation between the managed components (CI) and the services offered for management (MI). The commands are similar to SNMP, but not identical.

4.9   WEBIFICATION BASICS FOR SYSTEMS AND NETWORKS MANAGEMENT

The goals with webification are always twofold:

- Managing Web-servers and Web-browsers
- Use of the Internet and/or intranet technologies to manage systems and networks

4.9.1   Web Technology

The World Wide Web is an Internet technology that is layered on top of basic TCP/IP services. The Web is now the most popular Internet application next to electronic mail. Like most successful Internet technologies, the underlying central functionality of the Web is rather simple:

- A file naming mechanism — the Universal Resource Locator (URL)
- A typed, stateless retrieval protocol — the Hypertext Transfer Protocol (HTTP)
- A minimal formatting language with hypertext links — the Hypertext Markup Language (HTML)

The Universal Resource Locator (URL) is part of a larger family of file naming mechanisms called Universal Resource Names (URN) that are used to designate objects within the WWW. URLs name the physical location of an object; URNs name the identity without regard to location. Uniform Resource Citations (URCs) describe properties of an object. At this time, only URLs are in widespread use.

URL is the homepage-address, such as: http://www.snmp.com/.

Like TCP/IP, SNMP, and other popular protocols, URLs were originally considered to be temporary solutions until more powerful mechanisms could be developed. The simplicity and intuitive nature of URLs no doubt contributed to their rapid acceptance.

Hypertext Transfer Protocol (HTTP)   HTTP is the protocol used for access and retrieval of Web pages. As such, it is widely viewed as the core Web protocol. It is an application-level protocol used almost exclusively with TCP. The client, typically a Web browser, asks the Web server for some information via a "Get request." The information exchanged by HTTP can be any data type and is not limited to HTML.

HTTP usage has already surpassed that of older Internet access and retrieval mechanisms such as file transfer protocol (ftp), telnet, and gopher. However, these older services often co-exist with and are supported by HTTP-based Web browsers.

HTTP is a simple protocol; its clients and servers are said to be stateless because they do not have to remember anything beyond the transfer of a single document. However, HTTP's simplicity results in inefficiency; for a typical HTML page, the client first retrieves the page itself, then discovers potentially dozens of images contained within the page, and issues a separate HTTP request for each. Each HTTP request requires a separate TCP connection. HTTP pages are not real time. To retrieve new network status, the user must call up the Web page again. HTTP/Web is only good for monitoring one device at a time. This is the reason that Java is considered necessary for continuous monitoring.

To overcome this multi-step process, typical Web browsers may open several TCP connections at once. However, this practice may overload slower-speed communication links. HTTP is a textual protocol — all headers are transferred as mostly ASCII text — simplifying the writing of simple browsers.

HyperText Markup Language (HTML)   Web browsers have become widely popular because they all share understanding of a simple media type — HTML formatting language. HTML is easy to understand, and can be written by hand or generated from other text formats by translators. HTML is actually a simple document type of the Standardized Generalized Markup Language (SGML).

HTML is simpler than nroff and other document languages in that it is not programmable. As a result, the descriptive capabilities of HTML are limited to low-level constructs, such as emphasis or indented lists. However, because HTML parsers are rather forgiving of HTML coding violations, many Web pages contain coding mistakes used purposely to achieve particular layout effects on popular browsers.

HTML is optimized for display rather than printing or storage. HTML has no notion of pages, making formatted printing difficult.

Web browsers function as clients, asking Web servers for information by using the HTTP protocol. Each request is handled by its own TCP connection and is independent of each previous request. As noted earlier, just the retrieval of one HTML page may require establishing several TCP connections. Consequently, network managers need to be aware of the resource limitations of their Internet or intranet infrastucture when rolling out Web applications, since Web usage is significantly resource consumptive.

Web browsers increasingly are being used to support internally-developed corporate Web applications ranging from company job postings and notices about benefit policy updates to supporting Lotus Notes-based groupware activities.

Web server functions are storing, maintaining, and distributing information to clients using the HTTP protocol. Web servers contain the Web pages that may be individually designed and maintained by the owners of the Web servers. There are numerous hardware and software platforms to support Web servers.

4.9.2   MANAGING NETWORK AND SYSTEM COMPONENTS

The rapid deployment of the TCP/IP protocols has meant that most networking vendors feel compelled to support the TCP/IP open standards as well as proprietary protocols such as SNA, DECnet, IPX/SPX, NetBeui, and AppleTalk. As a result, the TCP/IP protocols today are the most widely used protocols on any computing platform with any operating system. Personal computers, minicomputers, mainframes, supercomputers, personal digital assistants, as well as dedicated network devices such as routers, hubs, switches, gateways, and printers all support the TCP/IP protocol suite.

One of the challenges facing a network vendor is management of the networked device. Besides providing the device functions, such as routing or printing, a vendor has to provide a method of managing the device. To date, there have been a number of approaches. The first approach used mostly in the early days of networked devices is to provide support for a "dumb" terminal such as a DEC VT100 or IBM 3270 to attach to the device through a dedicated serial port. This approach minimizes development expense, in that standard terminals or PC-based terminal emulators are used for the management application. In general, however, these text oriented displays have not been as well received as graphical user interfaces.

This led to the second approach, which is to develop a proprietary management application. This approach generally leads to the most efficient management of the device with the strongest support for the user, because the vendor has complete control of the management application. One of the problems with this approach is the cost of development and quality assurance, especially if multiple platforms are supported. An additional problem is that of ongoing maintenance. As client computing platforms change, the networking vendor must not only develop and maintain their product capabilities, but they also must make sure their management applications continue to work.

The third approach to device management has been the use of SNMP. This protocol is part of the TCP/IP family and has been widely adopted by device manufacturers. While SNMP is heavily used in the world's largest corporations, it has not been received as warmly by medium-sized to smaller companies. One of the factors limiting appeal is the cost of dedicated SNMP management consoles. Another problem with the SNMP approach is that the presentation of the management information is often not as user-friendly as a vendor would like.

A fourth approach to device management is now being implemented by some leading edge manfacturers. Recognizing the growth of the intranet with its widespread deployment of Web browsers, these vendors have seen that the reasons for intranet growth also provide an attractive solution to network management. By incorporating Web servers into their products, these vendors have given their customers an intuitive, user-friendly way to manage their devices using open protocols. This approach has all the benefits of the older dumb terminal approach for minimizing development costs, but provides a consistent cross-platform graphical user interface for an enhanced user experience. In addition, the Web protocols provide the ability for a vendor's customer to link back to the vendor's Web site for new product information, product documentation, or customer support. These vendors and others all believe that Web protocols represent a good approach to today's network management since they provide enhanced device management and reduced vendor costs.

There are additional Web-based services being offered. Using Web protocols, it is possible to enhance the customer support process by building pages that automatically gather device configuration information when a customer fills out a trouble report. Using Java applets delivered from a built-in Web server, we can display dynamic graphs of device performance or sound audio alarms.

4.9.3 STANDARDS FOR WEBIFICATION

In July 1996, five major vendors announced an initiative to define da-facto standards for Web-based Enterprise Management (WBEM). This effort, spearheaded by Microsoft, Compaq, Cisco, BMC, and Intel, was publicly endorsed by more than

50 other vendors, as well. The initial announcement called for defining the following specifications:

- *HyperMedia Management Schema (HMMS)*   An extensible data description for representing the managed environment that was to be further defined by the Desktop Management Task Force (DMTF).
- *HyperMedia Object Manager (HMOM)*   A data model consolidating management data from different sources; a C++ reference implementation and specification, defined by Microsoft and Compaq, to be placed in the public domain.
- *HyperMedia Management Protocol (HMMP)*   A communication protocol embodying HMMS, running over HTTP and with interfaces to SNMP and DMI.

SunSoft also has announced a programming environment for developing Web-based network and systems management software. This environment, called Solstice Workshop, consists of a Java Management API, (JaMAPI), a small footprint database, and a Java programming environment. Solstice Workshop's big drawing card is its extensibility and the popularity of Java's "write once, run anywhere" appeal. JaMAPI requires Java, whereas HMMP/HMMS/HMOM specifies HTML/HTTP, altrough Java is not specifically excluded.

Prior to these announcements, three developers from Hewlett Packard produced an Internet draft proposal proposing use of port 280 for exchanging HTTP management data. This same Internet draft describes a very lightweight HTTP Manageable MIB as well as a tunnelling facility for SNMP over HTTP.

Among these three efforts, the WBEM is certainly the broadest in scope, addressing not only protocol issues, but also data modeling and extensible data description as well. While JaMAPI includes object class definitions, it does not go as far as data description.

The complete solution is envisioned in Figure 4.9 including both major directions of webification.

The initial euphoria over WBEM and JaMAPI is starting to wear off and it is time for doing the hard work of pounding out specifications and, more importantly, building products. Customer demand will push Web-based management to its limits, but disillusionment is sure to set in if a lack of progress becomes obvious on the standards front. However, there are several emerging products, developed with an eye for supporting current and future standards, that bring to market a practical approach to take advantage of Web-based management.

4.10   SUMMARY

There are multiple standards for network management. All of them have advantages and disadvantages, and different application and implementation areas. Telecommunications suppliers and customers will have to live with multiple standards. The

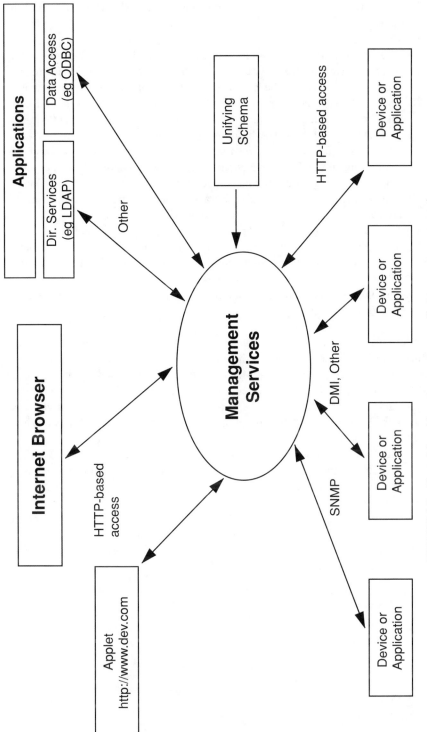

FIGURE 4.9 Use of the object manager to coordinate management services.

question is how these standards can interoperate seamlessly. There are basically three alternatives:

- *Management gateway*   The interoperability is realized by a special system responsible for translating management information and management protocols. Looking at the practical realization of such a gateway, it is important to target the use of OMA for both OSI- and Internet-based management. Many existing object specifications for management could be taken over by the OMA-based management.
- *Platforms with multiple architecture*   The interoperabiltity is realized by a multilingual platform, understanding multiple protocols. Protocol conversion is not necessary. Management information can be interpreted and transformed by the platform or by applications. Different architectures are supported simultaneously, but without deep integration.
- *Agent with multiple architectures*   The interoperability is realized at the agent level. In this case, the management agent understands multiple protocols and languages. It requires some intelligence for the agent. If selected, agent software must be implemented in many, practically in all, networking components. This number is considerably higher than in the case of management platforms.

There is a new group — Joint X/Open NMF Inter-Domain Management Group — that addresses in particular the interoperability between OSI-Management, Internet-Management, and OMG-OMA. This type of work takes a lot of time. In the meantime, practical solutions are absolutely necessary. In most cases, gateways deliver the quickest solutions. Such a possible solution with management gateways is shown in Figure 4.10. CORBA plays an important role in both gateways, related to OSI-CMIP and Internet-SNMP.

Standardization is absolutely necessary to ensure interoperability of various components of communication systems. This chapter has laid down the basics. Management frameworks and platforms may support some of the standards, but there is no product which is supporting all of them.

The following chapters will answer the question of how NetExpert supports these standards.

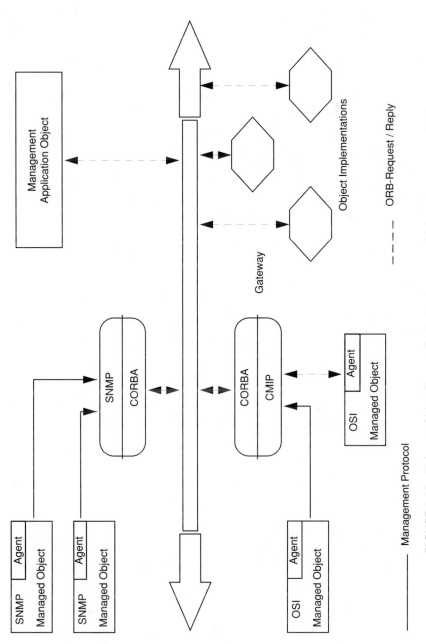

FIGURE 4.10 Using an Object Request Broker to connect multiple management protocols.

5 NetExpert Framework Technical Overview

CONTENTS

FIGURES

The information-communications service industry is undergoing exponential change. New technologies are evolving at an unprecedented rate, deregulation and privatization around the world are creating new markets, and consumer demand for the latest services seems insatiable. As a result, traditional service providers, such as long distance and regional telephone companies, and emerging carriers, including cable-television operators, wireless providers, Internet Service Providers, and utility companies, share the need to rapidly deploy and easily alter management systems to meet evolving business and network requirements.

These requirements include the ability to integrate existing network equipment and systems with new elements and applications, to customize systems and applications, to provide operations support systems and network management functions tailored to business models, and to scale management solutions to accommodate swiftly growing networks and exploding traffic volumes. Service providers are seeking a total solution that has the ability to manage complexity by integrating heterogeneous platforms and tools.

In response to these requirements, Objective Systems Integrators (OSI) has created a software framework that allows service providers to deploy reliable, scalable solutions that reduce risks and provide layered application creation environments. The NetExpert framework is a set of high-level building blocks that are integrated to create a complete support structure for development and delivery of total network management and operations support solutions.

About the NetExpert Framework This framework consists of a set of integrated software modules and graphical user interface (GUI) development tools to simplify and speed the creation and deployment of complex management solutions. The object-oriented architecture of the NetExpert framework provides the building blocks to implement operations support and management systems expeditiously using high-level tools. It frees network analysts and engineers from working with cumbersome, low-level program language to work on the added value of the application, focusing on meeting the needs of the business — improving, expanding, and differentiating service.

The NetExpert framework is founded on open systems and object-oriented methodology, which enables it to adapt well to changing standards, transmission protocols, and equipment data models. NetExpert is based on the Telecommunications Management Network (TMN) architecture created by the Telecommunications Standardization Sector of the International Telecommunications Union. It supports the development and deployment of applications for the main TMN management areas — fault, configuration, accounting, performance, and security — and the implementation of layered management architectures. In addition, the NetExpert framework employs expert rules that replace complex programming languages and enable network analysts to model desired system behaviors by using simple GUI-based rule editors.

5.1 THE FRAMEWORK ADVANTAGE

The primary advantage of the NetExpert framework is that all of the building blocks needed for a complete management solution are already in place — integrated, deployed, and tested in over 250 OSI customer installations. This gives service providers a strategic advantage by shifting the distribution of development time and effort from application programming interfaces (APIs) and system drivers to the functional application layer. Providers can devote development resources to core business areas that are crucial to remaining competitive.

Cross-Platform Compatibility   The NetExpert framework has been designed to simplify the integration of new and existing applications. It eliminates dependence on hard-coded support systems and facilitates the creation of a truly integrated management solution. The power of the NetExpert framework allows dynamic change of not only the network, but also the organizations that must manage it.

NetExpert's flexible object modeling environment provides support for both standards-compliant and legacy network components. All elements of a network, existing as objects in NetExpert, can be added or deleted with virtually no disruption of normal network and management environments.

Flexibility   The NetExpert framework consists of several components that can be configured in many ways to support a variety of management solutions. For example, NetExpert can be used to manage a cellular telephone network or a bank's automatic teller network — two very different configurations — with equal success.

A NetExpert-based OSS can employ a few or all of the component parts, as required. Initial investment can be minimized by purchasing only required components. As needs change, new components can be added.

Distribution and Scalability   NetExpert solutions can be centralized or distributed. In a small network, all required software can be located on a single machine, reducing hardware costs. As the network grows, NetExpert's software modules can be distributed to support high message volumes and large numbers of devices or applications, as is the case in the world's largest SONET networks.

Operator Productivity   Automatic procedures can be defined to perform routine fault, configuration, and performance management tasks, thus freeing operators to deal with more complex functions. Many of these procedures can be initiated directly from the graphic view of the network element affected. Because procedures are automated, functional differences in network elements are transparent to the operator.

Operator workstation features such as the Online Advisor and predefined dialogs enable operators to respond to network problems in a prescribed expert manner. The use of a common GUI interface for management applications can simplify operator training and reduce training costs.

Implementation Speed/Development Productivity　NetExpert's framework components and rapid-development tools allow developers and integrators to focus directly on implementing the business solution by reducing the need to develop and integrate the supporting software infrastructure.

The NetExpert framework includes object-oriented, GUI-based, fourth-generation language (4GL) tools that enable rapid development of management applications. These tools, in conjunction with the integrated framework components, greatly increase development and integration productivity.

Extensibility　The NetExpert framework is readily extensible through a set of distributed, object-oriented APIs and a variety of application/device-oriented packaged rulesets. These allow users and third-party developers to add functionality and integrate new framework building blocks.

Reusability　Components that are part of the NetExpert framework can be reused for many applications. Rapid-development tools also allow reusable NetExpert applications to be packaged and custom tailored — not custom built.

Open Architecture　The NetExpert framework is built with an open architecture, allowing applications to be easily customized and new functionality to be incorporated by customers and third parties.

Vendor Independence　The NetExpert framework supports products from a variety of major computer hardware and database suppliers, allowing management solutions to be independent of those vendors.

5.2　THE NetExpert RUNTIME SYSTEM

The integrated software processes that form the core of the NetExpert framework provide a runtime system that covers all the basic functions any distributed network management/operations support solution needs. This includes gateways linking network elements and legacy applications to the management system; intelligence to decode messages received from the network elements, or from other OSSs, such as order entry, and to act on them; and workstation tools that enable an operator to keep the network running, manage traffic flows, perform customer provisioning, and execute other commands supported by connected network elements.

The NetExpert framework has been designed to manage external application systems that include network elements, network element management systems, legacy Operations Support Systems (OSSs), CMIP agents/managers, SNMP agents/managers, COBRA-compliant systems, user applications, databases, and even people. In NetExpert all of these elements are defined as objects. Throughout this document the term "managed object" refers to any element of the system or network that is managed by NetExpert.The Runtime System is Shown in Figure 5.1.

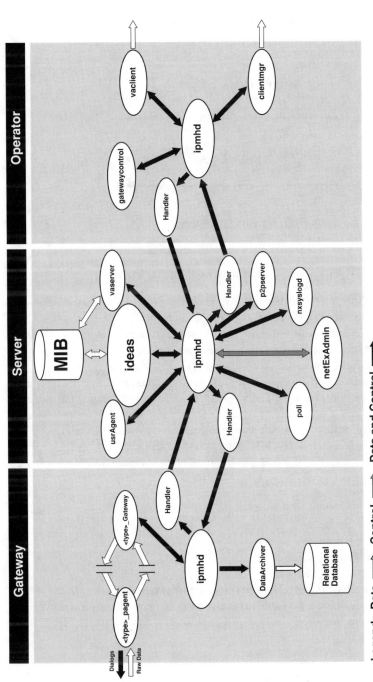

Legend: Data ⟹ **Control** ⟹ **Data and Control** ➡

Note: This diagram shows a NetExpert runtime system running each module on different host machines.

FIGURE 5.1 NetExpert runtime system.

The processes that make up the runtime NetExpert system can be divided into three functional areas or modules: gateways, servers, and operator workstations. Processes in these functional areas all can run on the same UNIX platform or can be distributed among different

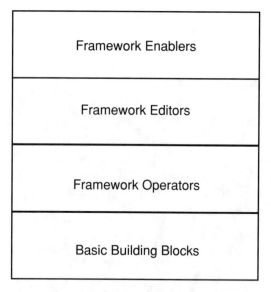

FIGURE 5.2   Layers of framework building blocks.

platforms throughout a TCP/IP network to provide the robust foundation required to handle continuous and large volumes of events and traffic.

In 1998, NetExpert's platform base was expanded to include Windows 95 and Windows NT. Processes ported to these platforms include gateways for the NT platform and the WebOperator for both 95 and NT.

TMN standards define functional and physical architectures that include the components that perform required TMN management functions. These components can be configured in a variety of ways and can include several discrete systems connected directly or through a data communications network. TMN components also can be distributed to reflect TMN management layering.

The framework consists of a series of building blocks (Figure 5.2). They can be characterized by four layers:

1. Basic building blocks for support features
2. Framework operators
3. Framework editors
4. Framework enablers

These building blocks have been developed and matured to provide the most comprehensive suite of tools available to manage evolving communication networks. The physical architecture of the NetExpert Framework is represented in Figure 5.3.

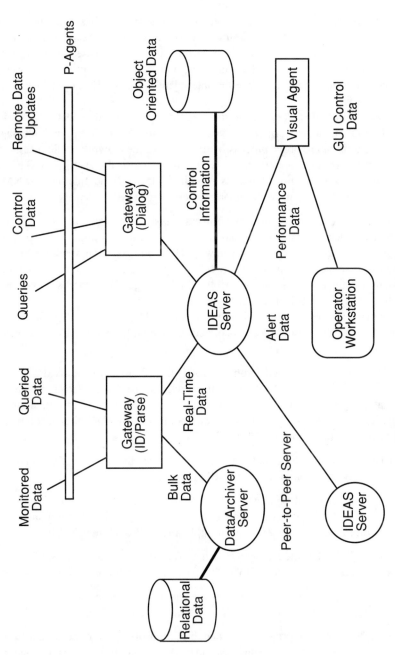

FIGURE 5.3 Basic NetExpert physical architecture.

5.3   COMPONENTS OF THE ARCHITECTURE

5.3.1   GATEWAYS

As NetExpert has evolved, more and more intelligence has been built into gateway processes. Currently, processing functions available at the gateway include event thresholding, dialog initiation, and polling. In future releases, gateway processes will become complete analysis engines, with complex event processing and object manipulation capabilities, like the Intelligent Dynamic Event Analysis Subsystem (IDEAS).

Gateways and their associated protocol agents can be located centrally or distributed throughout a network. Distribution of protocol translation and message filtering functions can improve processing time and reduce network traffic overhead.

NetExpert currently supports three types of gateways: the Intelligent Generic Gateway for legacy and non-standards-compliant devices, the Intelligent SNMP Gateway, and the Intelligent Multiplexing Gateway for CMIP devices.

Protocol Agents   The protocol agent process translates between an external system's (device or legacy application) native protocol and the common data representation used by the associated generic gateway. Thus, one common, reusable NetExpert gateway is able to communicate bidirectionally with a number of different types of external systems using various protocols. This leverages NetExpert's easy-to-use Rule Editor over a number of managed network environments.

Protocol agents are usually used with NetExpert's Intelligent Generic Gateways. Following are the most common protocol agents used with NetExpert generic gateway processes.

FIFO   Data can be sent directly from an ASCII text file to any gateway designated as using the FIFO protocol. The data goes directly to the *FIFO.MOUT* file and is not translated by a protocol agent. This method is an easy way to test rules, simulate protocol agents, or create objects automatically. Simply use the UNIX `cat` command to send or receive data from the FIFO files.

Serial   The Serial protocol agent translates for devices that are directly connected to NetExpert asynchronously via a tty (RS232) port.

Shell   The Shell protocol agent uses csh, TELNET services, or any port designated in */etc/services*. It can be used to execute dialogs consisting of UNIX commands and to execute commands to other systems or applications through the TCP/IP network.

TCP/IP   The TCP/IP protocol agent translates for asynchronous devices connected to NetExpert by a dedicated socket-to-socket connection or through a terminal server.

X.25   The X.25 protocol agent translates for native X.25 devices or asynchronous devices connected to NetExpert via a X.25 Packet Assembler/Dissembler.

Database   The Database protocol agent enables multiple synchronous connections to Informix, Oracle, and Sybase databases and provides a facility for users to connect, disconnect, and send SQL commands to selected databases. This protocol agent also

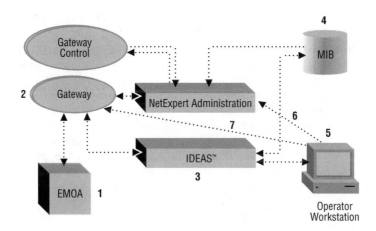

FIGURE 5.4   NetExpert system — event data flow overview.

can receive asynchronous updates from Oracle and Sybase database servers via triggers embedded in the database. Each database protocol agent can be configured for multiple synchronous connections or a single asynchronous connection.

SONET   The SONET protocol agent allows direct connections to SONET network elements. This protocol agent uses the seven-layer OSI stack with Target Access Resolution Protocol support.

Other Protocol Agents   Additional protocol agents are available, including Wellfleet BCN, Datakit, and SVAAG. Custom protocol agents can be developed by OSI's Professional Services group or by users via the Protocol Agent Toolkit.

Intelligent Gateways   A gateway is a NetExpert process that attaches to a physical or logical port to communicate with an External Managed Object Agent (EMOA) that owns all managed objects that it controls. Note that an EMOA may be an external application subsystem as well; that is, a DBMS or a legacy OSS. The gateway process is active on the host where it is started. If a gateway is started on the server, the process is active on the server. If a gateway is started on a client machine, the process is active on the client. A separate process is started for each gateway started.

Anything directly connected to a NetExpert gateway is an EMOA . In Figure 5.4, network element data flows from an EMOA to a "Gateway **2**," where it is analyzed. Data that needs additional processing is passed on to "IDEAS **3**" and the "NetExpert Management Information Base (MIB) **4**." IDEAS processes the data it receives and sends alarm information to the "Operator Workstation **5**."

Management commands flow back to the EMOA from the gateway. These commands can originate at the gateway, in IDEAS, or from the Operator Workstation. In addition, operators can send queries and commands back to EMOAs via "NetExpert's Command and Response System (CARS) **6**" or "VisualAgent DialogForms **7**."

NetExAdmin acts as the administrator for the system and interacts with Gateway Control and the MIB to control gateways in the system and manage CARS interactions.

Gateways also control the execution of dialogs and polls. Dialogs are automated scripts used to communicate directly with a managed object. They are used to query the status of a device within the network, to automatically resolve device or network problems, and to issue device commands. Dialogs also can be used to replace routine and repetitive interactions such as user logins, device restarts, and paging of operators or to replace a manual command/response interaction between an operator and a device.

Dialogs can be invoked manually by an operator or can be executed by IDEAS as the result of rule analysis; on a scheduled basis, or at scheduled intervals by polling; by the gateway on startup or shutdown; or through gateway analysis rules.

Polls are dialogs used to communicate with managed objects at predetermined times. Gateways can initiate polls on demand, in response to specific network events or defined gateway analysis rules, or via a time-based or recurring schedule. Polls can be used for provisioning, performance monitoring, querying managed objects, or initiating routine dialogs and events.

Intelligent Generic Gateway   As previously discussed, protocol agents are positioned between the external system (legacy network elements, nonstandard elements, management systems, databases, etc.) and the Intelligent Generic Gateway to provide a bidirectional interface. The Intelligent Generic Gateway applies the identification and parse rules defined in the Rule Editor to the incoming data and uses those rules to filter incoming messages, identify important events, and parse out data to be stored in the database or used in other processes. The gateway also executes specific object-oriented, server method analysis rules. Multiple object and event identifications can occur from the same or overlapping segments of incoming data.

Gateway rules are defined in the Rule Editor and loaded from the NetExpert database into memory when a gateway is started. Gateways can be individually reinitialized to load new or modified rules without shutting down NetExpert. In addition, gateways can be configured to share memory with other gateways that are connected to the same class of device, thus reducing the load on system resources.

Identifying Events   Event identification rules describe the characteristics of the incoming data and provide recognition of relevant messages from the data stream. These rules are used to package incoming device messages into events for further processing. Multiple events may be identified from one inbound device message.

Rules can search for beginning and ending string identifiers, positional statements, and unique strings. ID rules can be defined as inclusive (x and y and z), exclusive (x or y or z), or some combination (x and [y or z]). In addition, ID rules can be specified as "default" to capture data that doesn't match existing ID rules.

Event hierarchies are built to speed message identification. Rules defined for a parent event are passed on to its child events. In addition, information obtained (parsed) by a parent event is passed on to all identified child events.

Parsing Data   Once an ID rule has been satisfied, important pieces of data may need to be parsed out and passed on to the server module for processing. Parse rules identify message data, such as device names, counts, and states, to be stored in memory as NetExpert attribute values. Multiple attribute values may be parsed from one data stream and there is no restriction on the type of data that can be identified. Normally each attribute contains one value. However, if an attribute is defined as a type "SetOf," more than one value may be parsed into it.

Analyzing Data   Gateway analysis rules provide a subset of IDEAS functionality. This built-in intelligence enables thresholding, data completion, and value transformations at the gateway. Rules also can be defined to generate and clear events, run dialogs, and write information to log files.

While NetExpert offers a CORBA access product which supports user-programmed "gateway" interfaces to a provided NetExpert CORBA IDL, in 1998 OSI will release an Intelligent CORBA Gateway that automatically generates NetExpert MIB components and interface definitions from any CORBA IDL. Thus, CORBA server and client behavior can be provided by NetExpert without programming languages.

Intelligent SNMP Gateway   The SNMP gateway process provides bidirectional communication between the NetExpert system and SNMP agents in the managed network. This bidirectional communication consists of Get/Set requests and Get/Set responses. One SNMP Gateway process is run for each SNMP community that is to be managed.

NetExpert uses a shared conceptual schema to map SNMP objects, variables, tables, lists, traps, and nodes into NetExpert's attributes, events, and managed object definitions. This translation of SNMP objects and traps into a common object model enables NetExpert to apply consistent management behaviors to both SNMP and non-SNMP objects. The Intelligent SNMP Gateway maps SNMP nodes to NetExpert managed object names. The SNMP Trap Daemon works in conjunction with the Intelligent SNMP Gateway to map traps received from the SNMP server to corresponding events within NetExpert. The SNMP MIB Loader loads the SNMP schema into the NetExpert MIB.

The SNMP server process functions like a protocol agent. It accepts unsolicited events (traps) from SNMP agents and translates requests from NetExpert's SNMP Client API format into SNMP Protocol Data Unit format and vice versa. The SNMP Client API supports SNMPv2.

The Intelligent SNMP Gateway and Trap Daemon do not use ID and parse rules. The SNMP Client API supports the ability to associate attributes, DataArchiver filter objects, and named gateway analysis scripts (events) with an SNMP request. A request that flows through the SNMP gateway can generate the associated event. Thus, any given request can be associated with a predetermined action. This feature facilitates the filtering of traps and enables the SNMP gateway to forward data received from Get requests to the DataArchiver process for storage.

The SNMP GUI process provides a user interface to the Intelligent SNMP Gateway. It includes a MIB browser that provides an overview of the MIB tree, displays the values of selected SNMP variables for the specified host and SNMP Community, and launches Get/Set requests. The SNMP GUI also includes applications that allow users to configure, initiate, and cancel polls; control polling intervals; select and monitor thresholds graphically; and display the results of real-time Get operations in tables, lists, and graphs. Graph plot types provided include line, surface, histogram, bar, stacked bar, horizontal histogram, and horizontal stacked bar.

Intelligent Multiplexing Gateway (CMIP/Q3)   The Intelligent Multiplexing Gateway communicates between the NetExpert system and the CMIP/Q3 protocol agent. It extracts information from incoming CMIP messages, mediates between the CMIP interface information model and the operations application information model, and executes bidirectional communication between operations applications and CMIP systems. The CMIP/Q3 Protocol Agent accepts CMIP requests from multiple NetExpert dialogs. Individual requests are multiplexed through the gateway, which controls the generation of CMIP invoke IDs. There can be one or more associations between the CMIP/Q3 Protocol Agent and each CMIP device.

The NetExpert CMIP/Q3 Protocol Agent supports request/indication and confirmation/response communications with agents complying with ISO standards. It can communicate with multiple CMIP agents to receive and optionally confirm CMIP event reports.

In addition, the CMIP/Q3 Protocol Agent can function as a CMIP agent. It can generate M-EVENT-REPORT messages for transmission to an external CMIP manager and can respond to CREATE, DELETE, GET, SET, and ACTION indications received from CMIP managers. The CMIP/Q3 Protocol Agent enables a rich support of CMIP, including full association handling; operations services; scoping, filtering, and synchronization; linked replies; and access control.

The CMIP/Q3 Protocol Agent can be coupled with application-specific programs. This logic is provided through handlers that extend the functions of the protocol agent to enable mapping between NetExpert managed object names and CMIP distinguished names, translation of FIFO information to and from the Gateway and between different character formats, and encryption and decryption (or other generation and decoding) of the CMIP accessControl parameter.

Interactions with the Intelligent Multiplexing Gateway are designed for synchronous (blocked) operations. However, it also is possible for analysis rules to be written so that multiple requests can be simultaneously outstanding.

Intelligent CORBA Gateway   CORBA is becoming very important for interfacing loosely coupled OSS components found above the TMN Element Management Layer. Most CORBA software tools are designed for professional software programmers. In 1998, NetExpert will offer an Intelligent CORBA Gateway™ wizard that will take any given CORBA IDL object-oriented interface specification and automatically generate the client-server NetExpert Gateway that may be accessed (client and/or server) by user-written external application subsystems at runtime. The OSI provided CORBA wizard also will generate NetExpert MIB definitions automatically for CORBA IDL defined objects.

5.3.2   Servers

The servers module of the NetExpert framework contains processes for system administration, interprocess communications, the core analysis engine, operator security management, graphics management, scheduling, and process error monitoring and auditing.

NetExpert error reporting uses the UNIX syslogd facility. Error messages and informational messages can be routed to various locations as specified in the UNIX syslog configuration file.

NetExpert also includes a new framework component, the Object Server, which removes object services from IDEAS, supporting more scalability by separating data from behavior. Object services such as naming, events, life cycle, persistence, and concurrency control are performed for managed objects that include classes of objects built in to NetExpert, such as gateways and alerts, as well as classes of custom-defined managed objects. All objects can be found, created, named, changed, and deleted by IDEAS, by any gateway, or by user-developed application components using a CORBA standard interface.

Another new component is the Optimizer. This process will provide automated optimization of routing between two or more endpoints based on current or planned availability of bandwidth and least cost. The Optimizer will handle point-to-point, mesh, and ring architectures. A GUI will be included to enable runtime updates of cost functions, math constraints, and weightings.

NetExAdmin is the runtime administration process for NetExpert. NetExAdmin acts as the controller for the system and interacts with Gateway Control and the MIB to control gateways in the system. NetExAdmin performs several other tasks, including: starting all defined subsystem processes, dispatching dialogs to gateways, establishing CARS dialog activities, answering requests for gateway information from IDEAS to do dialog or SNMP Get/Set operations, administering ALIVE_AND_WELL functions, and shutting down NetExpert.

IPMH NetExpert processes communicate with each other via messages sent through the Interprocess Message Handler Daemon (IPMHD). IPMHD is a mature, stable, remote procedure, call-based message handler that has been continuously improved since 1989 to increase performance and fault tolerance.

There are two types of messages transmitted by the IPMHD process: NetExpert Vectors (NXV) and NetExpert's encoded Common Management Information Protocol (NXCMIP). NXV is a platform-independent record format composed of a variable-length string containing an ordered set of fields of variable lengths. NXV messages are identified by class and type.

Messages are either directly addressed or published. Directly addressed messages are sent to IPMHD with the address of the destination process and forwarded by IPMHD to that address. Published messages are sent to IPMHD without a complete destination address. They are forwarded by IPMHD based upon message class and subscription lists.

IPMHD contains support for client recovery, daemon recovery, network outage recovery, dynamic-message buffer size, and message buffer overflow to disk. IPMH uses the NetExpert configuration file (*.netexpertrc*) and UNIX environment variables to specify configuration parameters.

IPMHD continues to evolve with emphasis on robustness and guaranteed message-delivery services to NetExpert processes. Object-oriented abstractions will simplify the task of switching to different network protocols such as tcp, pop3, etc. Connectivity to Windows, CORBA-enabled services, and HTTP also are supported .

The Handler Process enables multiple IPMH daemons to communicate across multiple hosts. The handler process, therefore, handles all host-to-host communication within a NetExpert system. A handler process is a child process of IPMHD.

The handler is unidirectional. It forwards message traffic from the local system outward to the remote system. Therefore, two handlers are required for complete

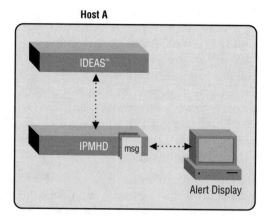

FIGURE 5.5 IPMH architecture — all NetExpert processes (IDEAS, Alert Display) residing on Host A.

bidirectional communication between any two IPMH daemons, one on each of the interacting hosts. However, depending upon configuration parameters, a handler on one host can communicate with multiple IPMH daemons on other hosts.

IPMH Architecture   Figure 5.5 identifies the basic IPMH architecture. In this example, IPMH messages are exchanged between two IPMHD client processes (IDEAS and Alert Display) through the use of the IPMHD server process.

This figure shows a simplified version of message queues used by IPMHD. Message queues preserve messages until they are delivered. If a client is connected via a slow link or if a client exits abnormally, the number of messages stored for the client increases.

Server to Remote Host   Figure 5.6 illustrates the IPMH architecture when IPMHD clients reside on multiple host computers. Arrows show message flow from NetExpert server to remote NetExpert clients.

On the NetExpert server computer shown in Figure 5.6, a separate handler process exists for each remote host computer. However, the handler process can be configured to communicate with more than one remote host — so the server could have just one handler communicating with both workstations. Each handler has a one-message queue that contains the message that the handler is currently sending to the remote IPMHD.

On each remote host is an IPMHD process that distributes messages to and from clients that are co-located with it. This IPMHD process has the same architecture and queues as the other IPMHD process. Each has its own separate configuration parameter values.

Remote Host to Server   Figure 5.7 illustrates the IPMH message flow when remote NetExpert clients communicate with the NetExpert server. This example resembles the previous example, except that the handlers responsible for message delivery reside on the remote host computers.

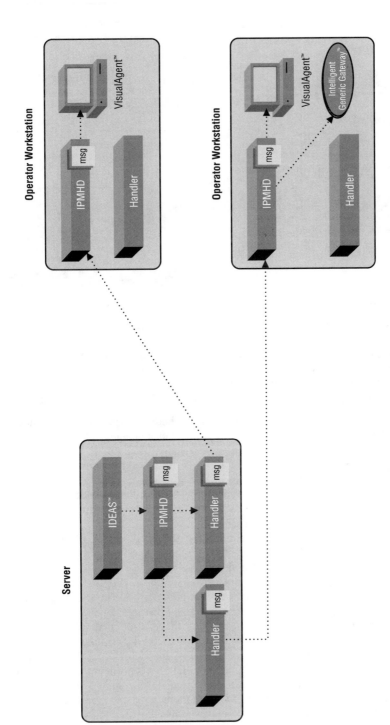

FIGURE 5.6 IPMH architecture — server to remote host communication.

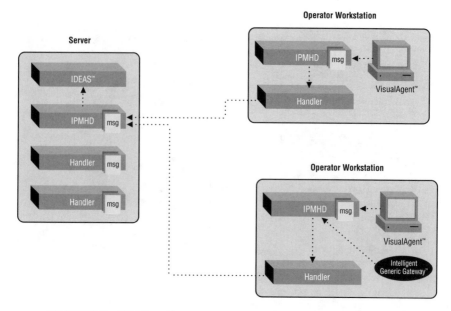

FIGURE 5.7   IPMH architecture — remote host to server communication.

The **ipmhTool** is an interactive utility that performs a variety of functions involving NetExpert interprocess messaging. *ipmhTool* can be used to display interprocess message information and configuration, status, and runtime statistics; to send and receive messages; and to set process aliases and local and remote connections.

IDEAS and IdeasLog   IDEAS provides an embedded expert system capable of diagnosing and resolving problems in real time (fault management); setting parameters, service thresholds, and managed object states (configuration management); analyzing load and usage patterns and resource utilization (performance management); and transforming billing data (accounting management).

This expert system provides a fast and efficient tool that includes the ability to handle arithmetic operations, thresholds, timers, branched or conditional logic, Boolean operations, and array (multivalued) attributes.

No programming experience is required for rule development. Instead, the Rule and Dialog Editors provide easy-to-use forms, menus, and free-text fields for developing rules and associating actions and responses.

IDEAS processes events coming from different gateways and different managed objects. It first identifies or automatically creates (discovers) a managed object (affected managed object) related to the event being processed using attribute values parsed by the gateway. IDEAS then executes the user-defined rules (an object-oriented method defined for the managed-object class) specified as the event's behavior. This rule/method execution may thread through many related managed-object rules/methods before reaching completion. Execution may be suspended several times, allowing other rule/method executions to be started and resumed.

Actions defined as an event's behavior can include correlating the data received with other data received from related objects; suppressing secondary events; sending root-cause alerts to the operator workstation module for display; creating, deleting, or updating objects in the database; changing object attribute values; setting watch-dogs and timers; and writing data to log files. IDEAS also can execute dialogs to initiate automatic fixes, request additional information from a device or application, and execute commands (for example, to run test equipment, page service represen-tatives, or use provisioning commands).

Alerts are displayed in the Alert Display window on the operator workstation. Alerts are color-coded by five industry-standard severity levels — critical, major, minor, warning, and indeterminate. Severity levels can be automatically upgraded as a result of dialog responses, thresholding, or additional analysis.

Other features of NetExpert's event-analysis engine include the ability to identify chronic problems or degrading service situations by specifying event thresholds. Thresholds can be based on event occurrences over a fixed-time interval, sliding-time interval, or specified-time duration, with the ability to perform analysis and commands when specific alerts are acknowledged or cleared.

ideasLog is a process that writes to log tables in the NetExpert database, such as eventlog and alertlog. When *ideasLog* is enabled, these table writes are off-loaded from IDEAS, thus improving performance. The *ideasLog* process can be located on a separate host from IDEAS to help balance resources.

Currently, IDEAS is sequentially multiple threaded. There is no limit to the number of threads that IDEAS can start and have suspended at any point in time. Robust parallel multiple threading is now supported across platforms on which NetExpert is ported. A future release of NetExpert will take advantage of this.

Object Server   In 1998, OSI was to release a new type of NetExpert process known as the Object Server that will manage all transient and persistent NetExpert Managed Objects (MIB) including Events and Alerts. A given NetExpert system may deploy two or more of these servers to scale and distribute the object-oriented data manip-ulated by one or more IDEAS-type processes as well as external systems via NetEx-pert Gateways and user-defined CORBA systems.

Event Analysis   IDEAS represents the perfect blend of easy-to-use fourth-generation technology, expert-system capabilities, and object-oriented architecture. Analysts use the spreadsheet-like Rule Editor to compose object-oriented methods (behaviors) for managed object classes. These behaviors are composed from a large set of high-level operators included with IDEAS. Also user-extensible IDEAS operators are supported.

In a 1998 release, OSI will offer extended object-oriented IDEAS operators, based on separately built, shared libraries of highly refined, carefully designed, reusable C++ code. These new Rule Operators will be designed for constructing specific TMN applications as well as supporting end-to-end TMN application inte-gration. All new Rule Operators will be made available, within NetExpert's Rule Editor, to the TMN application expert who no longer needs to be a software engineer. Extended Operators will reduce the complexity and size of the rulesets that TMN application experts must use to implement the end-to-end solutions that will differ-entiate the service provider in the marketplace. The design quality of the Extended Operators will give NetExpert its runtime performance advantage over similar sys-tems that must be coded using programmer improvement tools.

AuthAgent is the process that handles the interface between IDEAS and the operator workstation processes. AuthAgent keeps track of all users that are logged into the system, checks authorizations, and routes alerts to authorized users.

When an alert is generated by IDEAS, a message is sent to AuthAgent. AuthAgent tests alert data against the authorizations for the operator associated with each running *alertdisplay* process. The alert is then sent to each *alertdisplay* with appropriate authorizations.

Authorizations can be set up to control all aspects of an operator's capabilities, including which objects can be viewed and what actions can be performed. They also can be used to create partitioned views of complex networks. Authorizations are set up in the Authorizations Editor.

TroubleTicketAgent is the process that handles the interface between the trouble ticket process and IDEAS and the Alert Display. TroubleTicketAgent creates trouble tickets, correlates trouble tickets to alerts, and publishes the messages that populate trouble ticket status and number into the Alert Display window.

WebOperator Server is a Java-based, multiplatform Internet application that uses a browser to provide existing NetExpert operator functionality over Internet and the corporate intranet.

WebOperator Server is a stand-alone Java application that runs on the same host machine as the user's internal web server. It interfaces with NetExpert and the NetExpert database to provide NetExpert services and data to WebOperator clients.

More WebOperator   WebOperator 1.0, running on low-cost platforms, provides network managers the ability to remotely administer their network over a low-bandwidth, dial-up Intranet connection. This capability makes fault monitoring and identification practical on the network even when the network manager is at a remote site. Also, the ability to view NetExpert's user interface from a PC or workstation makes WebOperator a cost-effective solution at NetExpert sites that have a large number of users.

Web Operator also simplifies installing NetExpert upgrades in systems where administration of the entire network is distributed over multiple workstations. Web-Operator consists of both client and server software components. The server component is a Java applications, and thin-client Java applets are downloaded at runtime. WebOperator 1.0 operates within a Netscape 3.01 Java-enabled Internet browser on Windows 95, Windows NT, and UNIX-based systems.

Polling is the process that starts and manages the scheduled execution of dialogs and events. Polling is configured by selecting the Polling option from the NetExpert Administration Editor.

Paging Daemon is the process responsible for monitoring the current status of alerts, as well as reading and processing the alert paging tables to determine if and when it is appropriate to generate a page or other automatic notification.

Paging Daemon runs a shell script, *page.sh*, that can be configured to run a variety of UNIX communication commands such as `cu` and `mail`. Depending on the contents of *page.sh*, Paging Daemon can generate calls to digital pagers, send email notifications, or issue user notifications through NetExpert's *ipmhTool* utility.

When an alert is received, if paging is enabled for the alert's affected object, then the line in the contact list is notified. Additional occurrences of an alert do not trigger a page if the first page has not yet been acknowledged.

Additionally, an interval timer can be set to step through the list of pager contacts for the managed object. If the interval time expires and the alert has not been cleared or acknowledged, the next pager contact number (in top-down order) is notified.

The timer is reset and the sequence continues until the alert is cleared or acknowledged. If the end of the list is reached before the alert is cleared or acknowledged, the sequence continues from the top of the list.

Paging is configured by selecting the Paging Editor from the Client Manager menu in the operator workstation. Refer to the "Operator Workstation" section for more information.

The **System Log Daemon** is a process that reads and forwards system messages to a log file. It also logs operator-initiated actions. A new system log file is created each day. The location of the log file and the severity level of forwarded messages can be specified.

The **cleandb utility** is used to remove cleared alerts, events, and trouble tickets from the NetExpert database. Removed data can be deleted, stored in database tables, or stored in flat ASCII files.

High Availability is a hot-backup, fully redundant solution that provides monitoring, detection, and restart capabilities for NetExpert server processes. In cases of critical (IDEAS, *replica*, or *heartbeat*) or total process failure (such as a server crash), High Availability will activate the designated secondary system.

The secondary system maintains a synchronized backup of the primary system's database so that data loss is minimized.

The basic components of High Availability are:

- Shared memory segment (*Heartbeat*)
- Process monitor (*ProcessMonitor*)
- Database replication (*Replica*)
- Database synchronization (*DBsync*)

The Heartbeat and ProcessMonitor processes also are packaged separately under the name ProcessMonitor for users who wish to monitor process status on a single server system but do not need complete database replication. Replica and DBsync are not included in the ProcessMonitor package.

Shared Memory Segment (Heartbeat)   The shared memory segment (Figure 5.8) is established and controlled by the Heartbeat process. This memory segment is used to store specific information about the status of NetExpert application processes. A GUI window that displays state information for all monitored processes is included.

A resource file (*hrt.rc*) is used to establish the shared memory segment for Heartbeat. This file also is used to set the parameters to manage individual memory segments (or slots). The resource file is configured using the High Availability Editor (*HAConfig*).

ProcessMonitor   The High Availability ProcessMonitor uses IPMHD to query specified NetExpert processes by sending an ALIVE_&_WELL message. This process is similar to the UNIX `ping` command, which sends a request to another host on the network. If the other host responds to the request (or ping), it is said to be "alive."

A resource file (*nxp.rc*) is used to specify the NetExpert application processes that ProcessMonitor is to monitor. The results of the ALIVE_&_WELL queries are stored by ProcessMonitor in the local portion of the shared memory segment that is originally established by Heartbeat.

If ProcessMonitor detects a change in the status (or state) of a NetExpert process, this information is passed to the nxprocmon.sh shell script. This script can be set to

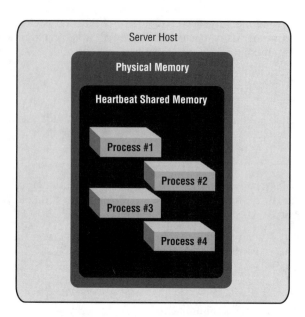

FIGURE 5.8   Shared memory segment.

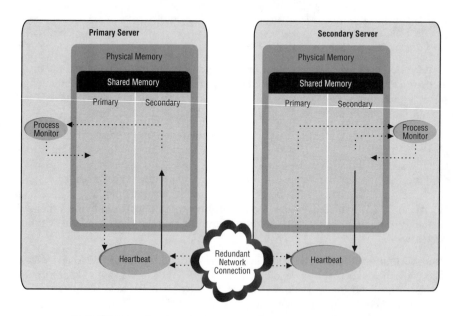

FIGURE 5.9   Synchronized memory segments (multiple servers).

process a status change in a specific way, depending on user requirements and the severity of the failure. For example, the script can be used to notify NetExpert operators, restart a specific NetExpert process, or switch over to a secondary system. If a "switchover" is initiated, notifications are sent to all remote gateways and operator workstation processes to redirect their messages to the secondary server. Figure 5.9 shows multiple servers with memory synchronized segments.

The **database replication process, Replica**, allows one NetExpert system, known as the secondary, to serve as a hot standby (or backup) for the primary NetExpert system. Although the secondary remains passive until needed, it performs the functions listed below to guarantee the database integrity of both the primary and secondary systems:

- **Database updates**   Operation requests are forwarded to the secondary by the primary IDEAS system.
- **Control message processing**   Control messages received from the primary system are processed.

Database replication does not require intervention by the database management system. Instead, Replica assumes responsibility for synchronization of NetExpert application transactions.

Replica also creates a shared memory segment to communicate with IDEAS.

If the primary system fails, the secondary assumes normal event processing duties. Upon switchover, gateways are automatically rehomed and external event processing is resumed on the secondary system.

If the primary system recovers and switchover again occurs, all data-replication requests generated by the secondary system while acting as the primary are processed in sequence by the primary system prior to normal external event processing.

The **database synchronization process (*DBSync*)** is used to synchronize the primary and secondary databases for updates that occur outside the scope of the runtime system, such as those performed through NetExpert editors, Package Administration, or the *cleandb* process.

Transactions are recorded in the NXTRANLOG table. When DBSync is run, the recorded transactions are "forced" to the secondary database, thereby synchronizing it with all nonruntime updates. Note that the replica process is responsible for propagating all runtime updates to the secondary system.

DBSync can be used to perform comparisons of selected tables and report the differences between the primary and secondary tables. In addition, DBSync can be instructed to selectively update one or more secondary tables based upon reported differences or to bulk-copy data from one or more primary tables to their corresponding secondary counterparts.

5.3.3   OPERATOR WORKSTATION

The Operator Workstation module of the NetExpert framework contains processes for controlling gateway activities, monitoring system performance, interacting with specific network elements, reviewing MIB data, and managing paging operations.

The Client Manager window is the control center of the NetExpert system. All client programs are started from this window.

The **Gateway Control** window controls the activities of the various gateways. From this window, gateways can be started and stopped and various logs and archives created. The operator can review this window to see which managed objects are currently sending data to NetExpert.

Gateways must be started to initiate data flow into NetExpert. Gateways can be set up to start automatically when NetExpert is started. Gateways that are not autostarted can be started from the Gateway Control window. In order for a gateway to start, there must be an available port. Ports are the physical connection through which the data passes from the managed object to the gateway.

VisualAgent Client Windows   The heart of VisualAgent is the ability to open windows (Figure 5.10) through which any part of the network environment can be

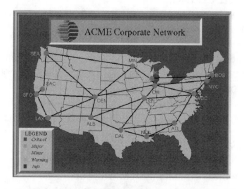

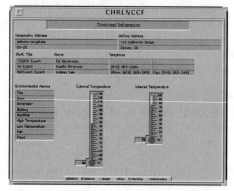

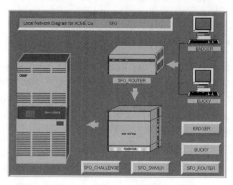

FIGURE 5.10   VisualAgent windows.

viewed. VisualAgent enables users to navigate back and forth between windows that display new or more detailed views of the network. Clicking the mouse on a graphic object can display new models in separate windows or replace the current model in the active window. Windows can "stack" on each other to make it possible to burrow in or out of a detailed network view.

The detail and definition in graphic windows can range from maps of the world to logical displays or schematics of the smallest reporting device in the managed environment. Status and alert information is propagated through all graphic layers.

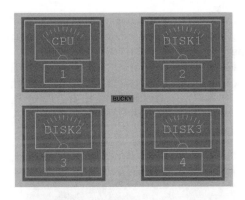

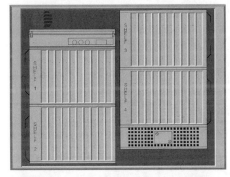

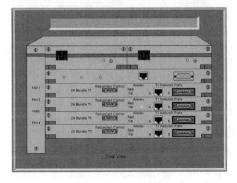

FIGURE 5.10 (continued)

Users can get a top level view, zoom to detail views, or inclusively review multi-level components.

NetExpert VisualAgent windows portray the status of the network using graphic objects. The graphic objects represent actual managed objects found in the network. The color of the graphic object changes when the status of the managed object changes.

The real-time status of any network element can be displayed in graphics, such as dials and meters. For example, the CPU usage of a server on the network can be shown in a bar graph or meter. VisualAgent also allows managed objects to be shown as icons, geographic or topographic maps, logical views, floor plans, equipment racks, scrolling lists, etc.

Input widgets such as sliders, toggles, buttons, and fill-in forms ease operator actions. For example, a slider can be used to set a threshold or a form can be used to fill in provisioning information.

Pop-up menus (Figure 5.11) can be configured to allow operators to perform certain actions by clicking mouse buttons. Clicking the right mouse button anywhere in a graphic window displays a context-sensitive menu. The menu displayed depends upon the graphic it is selected from.

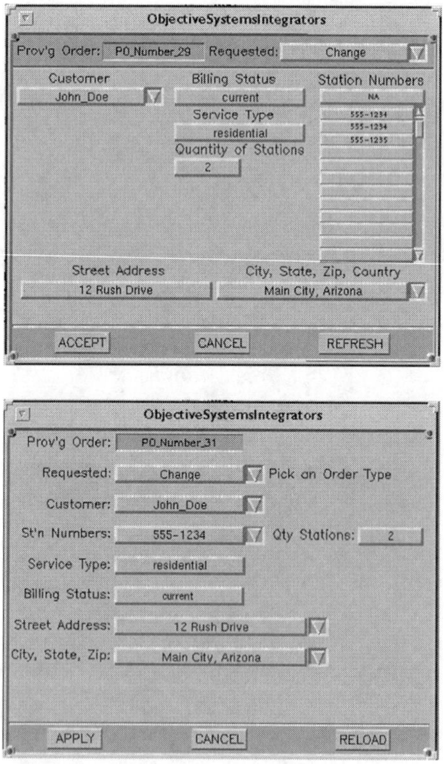

FIGURE 5.11   VisualAgent pop-up menus.

Among the menu items that can be configured are:

- **Zoom to Alert Display**   Open an Alert Display window to view alerts for the device or location
- **Zoom to Managed Objects Configuration System (MOCS)**   Open the Managed Object window to view attributes of the selected graphical object
- **Generate Event**   Generate a NetExpert event and pass it to IDEAS
- **Initiate Dialog**   Generate a NetExpert event that initiates a dialog
- **Send Email Notification**   Execute a shell script to send an email notification
- **Open Window**   Open another VisualAgent window
- **Save Workspace**   Save the current arrangement of graphic windows on the screen
- **Quit**   Exit from VisualAgent

VisualAgent also can establish a direct connection to a gateway — bypassing IDEAS — and display device attributes in a dialog box (Figure 5.12).

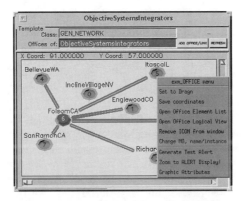

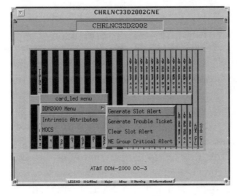

FIGURE 5.12   VisualAgent dialog forms with scrolling lists and toggle buttons.

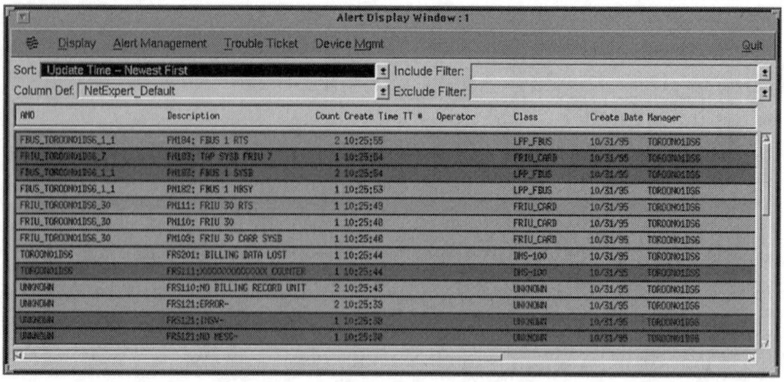

FIGURE 5.13   Alert display window.

The **Alert Display** windows (Figure 5.13) serve as the foundation for viewing and working with the alerts generated by the NetExpert system.

The Alert Summary window provides the user with a summary of the number of alerts in the system and allows the user to create filters and column definitions to customize the display of alerts and to open additional Alert Display windows.

The Alert Display window displays active alerts and notifications and allows the user to acknowledge and clear open alerts; review online advice, device data, and database information for a selected alert; zoom to the graphical view of an alert; create a new trouble ticket or edit an existing one; and start the Command and Response tool. Up to 10 Alert Display windows can be open at the same time.

With the Alert Command feature, users can define multiple commands to appear in the Alert Management menu. These commands may be UNIX executables or shell scripts. The menu label, command to be executed, and any alert values (such as alert ID, number, or severity) to be passed with the command are user-definable.

The Alert Display window-filtering system can be used to control which alerts appear in each window. The filtering feature enables operators to create focused views of current network alerts. Alerts can be filtered by Manager Class, Manager, Affected Managed Object Class, Affected Managed Object, Relationship, Operator, Acknowledged State, Severity, Count, and Alert Type — or by any desired combination of these. Filters can be named and are reusable.

Operators can customize their display windows by using preference settings to control the display order of alerts (by severity or creation or update time) and window columns, font size of alert text, pen style, background color of the display window, window title, and alert-severity blinking rates. These preferences, as well as filter selection, notification preferences, and window size, position, and layout can be saved for later use.

Operators can select an alert and open the Online Advisor to display additional information about the alert. In addition to status and general alert data, Online Advisor displays information about when the alert was created, acknowledged, and cleared; lists all the events that led to the generation of the alert; and shows the raw data associated with each event. One section of the window includes documentation on probable causes, recovery steps, and escalation and test procedures. This documentation, entered in the Rule Editor when the alert is created, leverages expert advice from proficient users and makes it available to all users. This allows novice operators to handle more difficult problems quickly and leaves experts free to focus on unusual and severe failures.

The **Inhibit Alert** feature allows an operator to suppress alert creation for specified devices for a defined period of time. Alerts can be inhibited for a specified time period (up to 24 hours) on one day, for several days, or for an indefinite period. Each alert can be inhibited for a network element or for all elements reported on by a particular managed object.

The Inhibit Alert feature also allows operators to change the list of inhibited alerts without stopping NetExpert. If an inhibited alert occurs within the specified time/date range, the alert is not generated and does not appear on the Alert Display Summary or Alert Display window. Also, when an alert is inhibited, the occurrence of the alert is not recorded in the database.

Inhibit Alert actions are recorded to the system log file (*syslog*) to maintain audit-trail accountability.

The **Command and Response System (CARS)** allows operators to initiate a dialog with or generate an event for a selected network device. The Command and Response window has two panels, Dialog and Event, that allow the operator to send different commands for specialized tasks. Only one Command and Response window can be open at any time.

With the Dialog Panel selected, the operator can initiate a dialog or cut-through session. The dialog to be initiated can be selected from a menu of dialogs created in the Dialog Editor. These dialogs contain the device-specific commands and responses necessary to complete a specific task. A cut-through session enables an operator to interact directly with a device, to log into and out of the device, query its status, determine problem causes, send device commands, and initiate test functions.

With the Event Panel selected, the operator can send an event to IDEAS. The event to be initiated can be selected from a menu of events created in the Event Editor. The operator can specify values for any required event attributes. This enables users to perform database updates and generate alerts.

The **Managed Objects Configuration System (MOCS)** window (Figure 5.14) provides information about the affected managed object of the selected alert, including containment and management relationships, and the attribute values and reported-as ("alias") names that have been assigned to it.

The **Trouble Ticket** windows allow operators to create or edit information relating to an alert and track the status of a trouble ticket used in resolving the problem causing the alert. Operators can select alerts and assign them to trouble tickets. Multiple alerts can be assigned to one trouble ticket. There are two windows associated with trouble tickets.

The Trouble Ticket Tracker is a summary window that displays a list of all existing Trouble Tickets and identifies their number, status, and a short description (either entered by the user or by an alert).

The Trouble Ticket window provides information related to an alert. Operators can track trouble tickets with alerts, keep the tickets updated, and make comments concerning the alert associated with the trouble ticket. A date/time-stamped log is available to record the steps taken in problem resolution.

Report Maker allows users to run a variety of predesigned reports against their NetExpert database. Available reports include Alerts and Correlated Events, Alert

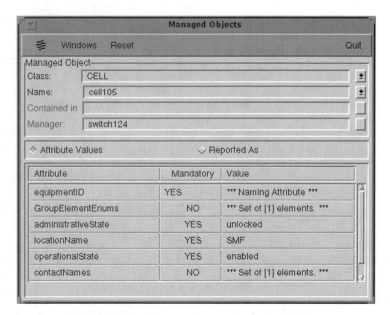

FIGURE 5.14 Managed objects window.

History, Event History, Polling Schedules, and Trouble Ticket History. Reports can be customized by selecting the desired parameters and date/time ranges. All users have access to all reports.

Data Browser allows operators to search data in runtime and archived data files. It also allows operators to monitor data from a specific device. The Data Browser feature provides the following functions:

- Search patterns can be predefined.
- Search patterns can be compared against saved data as well as live data.
- Search patterns can be constrained by time and date.
- Multiple search conditions can be applied at one time using both message definitions and message filters.
- The count of messages that meet search conditions is displayed.

Access to the browser is restricted to authorized NetExpert users as defined in the Authorization Editor. Operators can only view data for managed objects that are included in that operator's authorization.

The **Paging Editor** allows operators to define a list of contacts that are automatically paged when specified alerts are generated for selected network elements. The alert paging feature provides the ability to perform the following functions:

- Create a list of contact names with associated pager numbers and messages
- Associate a sequence of pager contacts to be called when a specific alert or class of alerts is generated
- Set the status of paging by managed object manually using the Page State field

- Initiate automatic scheduling of paging for managed objects at defined times using the Start and End time fields
- Set the status of paging for the entire system manually using the All On, All Off, or All Scheduled buttons

Access to the Paging Editors is restricted to authorized NetExpert users as defined in the Authorization Editor. Operators can only view data for managed objects that are included in the operator's authorization.

WebOperator Client   The WebOperator applet is served up to Internet browsers using web-server software. Any Internet browser that supports Java version 1.0.2 can be used to load and execute the WebOperator applet.

WebOperator functionality includes the Alert Display and Summary windows, but will be expanded in 1998 to include web-based support for VisualAgent and other operator workstation applications.

WebOperator Client displays (Figure 5.15) can be customized to suit user preferences. Users can choose from a variety of alert-notification methods, including beep, flash, and auto-scroll. Users can specify the font size (12 to 32 points) and style (normal or bold) used in alert rows, as well as the color of the window heading and alert row boundary lines. Alert severity can be indicated by coloring the entire row or by using a colored dot at the beginning of the row. The status bar at the bottom of the window can be toggled on and off. All these selections — along with column definitions, sorting, and filtering options — may be saved in a preferences file for reuse.

5.4   FRAMEWORK ENABLERS

There are a number of other framework applications, called enablers, that can be embedded in other functions and management applications. In case of the NetExpert framework, four enablers are important:

DataArchiver for maintaining data
VisualAgent for presenting data
Peer-to-Peer to build server connections
Package Administrator to export/import from/to databases

5.4.1   DATAARCHIVER

DataArchiver is a message-management system that enables users to collect and store large quantities of network element data in non-NetExpert relational databases for future retrieval and analysis.

DataArchiver is integrated with NetExpert's intelligent gateway processes through Event Forwarding Discriminators (EFDs). EFDs are objects that filter events identified by a gateway process and pass the events on to a specified archiver process. Events can be identified and forwarded to one or more DataArchiver instances and/or to IDEAS. DataArchiver instances map the event data to user-defined database tables. A NetExpert managed object class determines the set of events, while an EFD specifies which events should be forwarded to a particular destination. NetExpert reviews existing EFDs to identify executables and dependencies, and ensures destination dependencies are operational.

FIGURE 5.15   WebOperator client windows.

Multiple DataArchivers can be associated with a NetExpert system. This provides the ability to fine tune the performance characteristics of each DataArchiver instance. Conversely, multiple network elements can forward events to one DataArchiver. The data-storage system can be designed for the most efficient use of network resources — to decrease network traffic and consolidate data in readily accessible locations.

DataArchiver facilitates fault-tolerance of network nodes and databases through the ability to forward an event identified by a gateway to multiple DataArchivers that update replicated databases. If the network node or the archiver database goes down, valuable data will still be captured by the replicated system.

DataArchiver can be configured to log events that are not successfully processed, so that crucial data from a network element is not lost due to transient or permanent failures in the environment.

By default, all events that contain analysis rules are sent to IDEAS for processing after they are identified by a gateway process. However, EFDs can be created to determine which events are sent to IDEAS. Selective forwarding of events to IDEAS can reduce processing overhead and improve system performance.

Runtime processing also is simple. DataArchiver processes are managed by NetExAdmin, the process that manages the NetExpert system. When gateways that

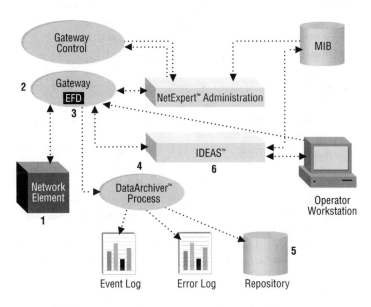

FIGURE 5.16   NetExpert system with DataArchiver process.

interface to the network elements are started in NetExpert, the DataArchivers that process events from those network elements are started automatically. At startup, DataArchiver verifies configuration data and confirms the accessibility of destination databases. The gateway then loads active EFDs and message forwarding to the archiver process is initiated.

When DataArchiver is added to a standard NetExpert system, as in Figure 5.16, event data flows from am "EMOA **1**" through a "gateway **2**" and is filtered through "EFDs **3**." Data is then forwarded to a "DataArchiver process **4**" for mapping to "repository Tables 5." Event data can still be sent to "IDEAS **6**" or held in the gateway for additional processing.

NetExAdmin acts as the administrator for the system, including the archiver process, and interacts with Gateway Control and the MIB to control gateways in the system.

5.4.2   VISUALAGENT

VisualAgent is a system for creating custom Motif-compliant graphic and forms interfaces within a NetExpert system. It provides individual users with a customized, dynamic view of practical, operative information. VisualAgent is the focal point for operator and administrative interaction, allowing users to manage the network environment from the graphics displays.

VisualAgent processes provide sophisticated development tools for displays that are more meaningful than raw data. VisualAgent enhances NetExpert's functionality through real-time graphic displays of NetExpert object attribute values. Changes to object attributes can automatically drive graphic dynamics, representing a managed object's condition by transforming animations and color changes. Graphics enhance the immediacy of data that represents the status of the managed environment. VisualAgent enables at-a-glance recognition and acknowledgment of the status of the network, allowing operators and developers to directly alter data concerning a managed network and see immediate results. Intuitive features, such as automatically updated graphic data and interactive controls, such as user-defined pull-down and pop-up menus and forms, endow management with unparalleled precision. Menus also can initiate functions with third-party applications such as trouble ticketing, spreadsheets, and data-analysis tools.

VisualAgent's main purpose is to provide visual access and control — display and control — of NetExpert objects and states. VisualAgent displays managed object information in a runtime environment and is distributable in a client/server environment — separate from the NetExpert server. Operators or graphic developers can graphically define managed-object information the same way that class attributes and behaviors are defined in the MIB.

Using the SL-DRAW2 object-oriented graphical modeling system, VisualAgent developers can depict any aspect of the managed environment. Graphics can come from any .DXF file (AutoCAD) or in any file format usable by Corel Draw. VisualAgent also can use geographic location for items on its displays using files like those from MapInfo and the USGS.

VisualAgent uses templates and retrieves unique information from the MIB. Easy-to-use template models simplify graphic development and allow users to quickly generate model views. The same picture can be displayed with different data, reducing graphics development time. VisualAgent developers produce customized graphic objects that become a set of templates. Once a dynamic template is developed, changes or additions to objects or models are replicated through the templates. Template configuration is performed using basic definitions at the class level. New instances of a template can be automatically configured as the basic template is altered.

Users also have access to a full range of dynamic behavior displays. Event and performance data reported to NetExpert can be represented dynamically in three-dimensional graphs, dials, bar charts, pie charts, or in many other useful forms within the VisualAgent user interface. For example, users can rotate graphic objects, change bitmap representation, and display or hide objects in response to dynamic changes. Both numeric or textual data can be displayed, enabling operators and other users of the NetExpert system to see and interpret that data as it is collected.

Through the propagation of status and alert information, combined with graphic dynamics and animation, the status of alerts is evident without bringing up the Alert Display. A few clicks of the mouse button narrow the view to the specific network element that caused the alert while specific information about the nature of the alert is available in detail through NetExpert's Alert Display window. Conversely, if an alert in the Alert Display window catches the operator's eye, a click of the mouse button can display in a graphic window the most descriptive and detailed available view of the problem. Operators also can zoom from graphic models to the Alert Display.

NetExpert actions can be initiated via VisualAgent. Users can create pull-down or pop-up menus and selectable scrolling lists using simple resource definitions to trigger events, request or change attribute values, or execute an event or dialog to perform an action in the network, such as to provision a device.

VisualAgent supports the use of NetExpert's DialogForm operator. DialogForm allows an event to kick off a dialog box that appears on the operator's screen. When the operator clicks the appropriate button in the dialog box, a predefined command is implemented. DialogForm enables the operator to fill in variable parameters to be used in the event. DialogForm facilitates network functions such as provisioning, traffic monitoring and control, and device testing.

Users also can create displays that list or highlight specific information. This helps operators better understand what is happening with a particular device, system, or application. Displays can show anything from "pictures" of equipment, or schematic diagrams of the entire management environment, to graphic views of performance data.

VisualAgent consists of three main processes:

1. *vaserver* is a central process that establishes a connection with IDEAS, receives information from IDEAS about the state of managed objects, and passes that information on to client processes at the workstation level. It can run on the server computer or on another server to assist resource balancing.
2. The *vaclient* process runs on the workstation and provides the graphical interface to the operator. Any number of *vaclient* processes can be running at any given time on different workstations, each communicating with the *vaserver* process. Operator access to VisualAgent and to specific objects and actions available through VisualAgent is controlled through the Authorization Editor.
3. *vablinker* manages the blinking of colors in graphic objects. It enables the graphic representation of an unacknowledged alert to blink by alternating the alert color with a neutral shade of gray. The process also regulates the rate of blinking, which varies depending on the severity of the alert.

VisualAgent uses the *vaserver*, *vaclient*, and *vablinker* processes at startup and as alert and update information from the management environment comes in. NetExpert runs the *vaserver* startup sequence each time NetExpert is brought up. This process identifies the operator to the system, gets his or her authorizations, brings up the proper root model, and starts the flow of information from IDEAS and the NetExpert MIB.

When *vaclient* initiates, the registration sequence process identifies the operator to *vaserver*, clearing security issues and preparing the display for the operator. Once this happens, the flow of runtime data from IDEAS through *vaserver* reaches the operator via *vaclient*.

When *vaserver* receives an alert from IDEAS, the alert sequence process tells *vaclient* which screens on the operator's terminal to open and channels the information from IDEAS through *vaserver* and *vaclient* to the operator.

FIGURE 5.17   Peer-to-peer Server simplified overview.

vaTool, included with VisualAgent, is used to debug network models. It provides reports to help track system-wide modeling problems at the server level.

5.4.3 PEER-TO-PEER SERVER

The **Peer-to-Peer Server** provides loosely-coupled data sharing among NetExpert systems in a distributed-management environment (Figure 5.17). Loosely-coupled sharing supports inter-provider/carrier collaboration. For example, two service providers who independently implement end-to-end TMN solutions based on NetExpert may "bond" their NetExpert solutions easily to provide inter-provider/carrier services. Peer NetExpert systems on the network forward event, alert, and object messages to a Peer-to-Peer Server, which employs filters from its own MIB file to determine which messages to forward as events to other peer NetExpert systems. Messages can be passed between domains in a network or hierarchically between the lower layers of a network to higher layers of management.

Key features of the Peer-to-Peer Server include the following:

- Omnidirectional data sharing with many peer NetExpert systems
- Rule-triggered sharing of alert, object, and event notifications
- EFD filtering of shared notifications, providing forwarding of selected notifications while discarding irrelevant notifications
- Many applications, including regional distribution, intercorporate, and interdepartmental transaction sharing, and function isolation
- GUI Editor for quick and easy system creation

Peer-to-Peer Server performs the task of receiving, filtering, translating, and forwarding notifications from peer NetExpert systems to other peer NetExpert systems via a Peer-to-Peer Server. When a Peer-to-Peer Server registers with IDEAS, it receives messages from IDEAS and determines whether these received notifications should be forwarded to one or more peer systems.

Notifications that peer NetExpert systems can send to a Peer-to-Peer Server are:

- Alerts that have been created, deleted, or changed
- Objects that have been created, deleted, or changed
- Events that have occurred or been cleared

EFDs determine which notifications are to be forwarded to which peer system(s). The attributes associated with an EFD govern the disposition of a message, causing some messages to be discarded and others to be forwarded to a specified peer system. They also determine which event attribute values are forwarded with the event.

FIGURE 5.18   Package Administration simplified overview.

The Peer-to-Peer Server supports the use of NetExpert's PeerEvent operator. PeerEvent enables IDEAS to forward event notifications to a defined destination without using EFDs.

Log objects are definitions in the Peer-to-Peer MIB file that provide logging of notifications and other information to any of several destinations. Filters can be included in the log objects to log certain types of notifications while others are discarded. Several destinations can be specified to receive log data, including *stdout*, *stderr*, *syslog*, and two alternating log files whose names can be specified. It is possible to have more than two log files by creating additional log objects (two per log object).

A Peer-to-Peer system can coordinate the operation of several local NetExpert systems used for different purposes, or it can coordinate data sharing among NetExpert systems performing similar functions in different geographical regions. For complex hierarchical networks having multiple levels of NetExpert systems, multiple Peer-to-Peer Servers can be installed. Peer-to-Peer Server installations also can be designed to minimize network traffic.

5.4.4   PACKAGE ADMINISTRATION

Package Administration provides a GUI for exporting and importing NetExpert rulesets from or to NetExpert databases. The application works with Informix, Oracle, and Sybase databases and enables migration of ruleset components from one type of database to another. For example, classes can be exported from an Oracle database and imported into an Informix database (Figure 5.18).

Exported rulesets, consisting of classes, associated class-specific objects, and global nonclass-specific objects, are converted into UNIX directories containing ASCII files.

To import these files to a database, the Package Administration application translates the flat files into a format appropriate for the destination database. The imported ruleset components replace any existing components in the database that have the same names.

Package Administration's user interface includes the following main windows (Figure 5.19):

- **Database Workspace** — Used to select and register databases to work with
- **Class Hierarchy window** — Used to view the containment tree of database classes and export selected ones to UNIX directories
- **UNIX Directory Workspace** — Used to view the UNIX hierarchy tree of database directories and exported files. It also allows users to select and import them into NetExpert databases.

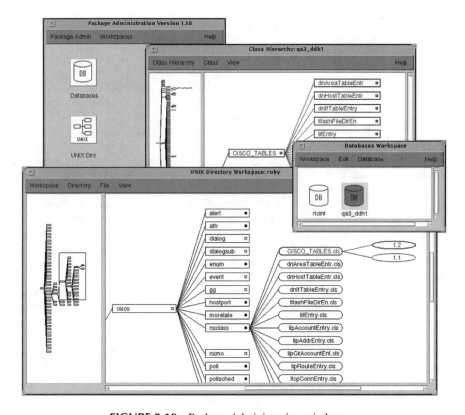

FIGURE 5.19   Package Administration windows.

Package Administration consists of two primary components: the GUI application and the *pautil* import/export utility. Non-GUI utilities also are provided to create class hierarchy lists, generate export scripts, and perform configuration management tasks, such as file creation and check-in/check-out. These utilities can be run from the UNIX command line.

Package Administration can be enabled to use configuration management systems such as SCCS and RCS. When configuration management is enabled, *pautil* will add exported files to the selected configuration management system. Revision histories can be maintained, and all previous revisions of a file can be viewed and imported to a database.

5.5   IMPLEMENTATION/DEVELOPMENT TOOLS

The NetExpert framework includes a variety of implementation/development tools that enable users to customize the system and integrate other OSSs with NetExpert or develop custom in-house applications to enhance the network management solution. These tools include a suite of specification editors, several application program interfaces, and additional system management tools.

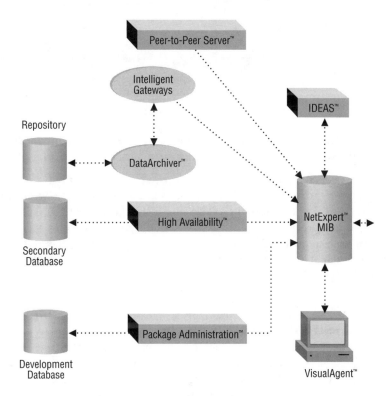

FIGURE 5.20   NetExpert MIB data flows.

Editors   The NetExpert specification editors are used to enter specific network data into the NetExpert MIB (Figure 5.20).

Rule Editor   NetExpert receives information, called raw data, from objects to which it is connected. The incoming messages are reviewed to determine which are important and which can be ignored. The criteria used to determine which messages are important are referred to as Identification Rules. Relevant information from the identified messages can be saved into variables called "attributes." Saving message information into attributes is called "parsing" and is specified by defining parse rules.

When a particular message from a device is identified as important, this information is packaged as an event. Any relevant information that is parsed from the message is also saved as an attribute of the event.

Once an event has been identified and the relevant data parsed, it is then passed on for analysis. This analysis can be performed at the gateway or in IDEAS. During analysis, NetExpert determines what action should be taken in response to the incoming message. Should an alert be generated? Should an alert be cleared? Should a parsed value be compared to a threshold level and, if the level is exceeded, should an alert be generated? The Gateway Analysis and Event Analysis panels are used to build intelligence into the message evaluation and action process.

FIGURE 5.21   NetExpert data flow.

Specific alerts to be generated by the various events are created in the Alert Edit window of the Rule Editor.

Rule Editor and NetExpert Data Flow   Raw data flows from network devices through a gateway, to the main NetExpert processor, and on to operator workstations. Rules, developed in the Rule Editor, determine which data is passed on from the gateway, how it is processed, and what information is displayed to operators. Rules also can be used to create dialogs, polls, and events to be generated by operators and sent through the gateway to connected network devices. This data flow is illustrated in Figure 5.21.

Defining Alerts   The Alert Edit window is used to specify what the operator sees in the runtime Alert Display window. New alerts are created and existing alert definitions are modified in the Alert Edit window. The alert description, severity, alert type, problem type, time out, and detailed description are specified in the Alert Fields panel. Problem-solving information specific to the alert can be entered in the Alert Advice panel. Events to be generated when the alert is acknowledged or cleared can be specified in the Alert Events panel. Alerts must be defined in the Alert Edit window before they can be referenced in the Analysis panel of the Event Edit window.

Managed Objects Editor   NetExpert is object-oriented, in that all parts of a network are treated as objects. Every object in the network has specific attributes or properties associated with it. All objects in NetExpert are organized into classes that share common attributes. For example, the class "equipment" is defined as having the attributes "equipmentID," "locationName," and "operationalState." Within the class "equipment", the managed object "modem" might have the attribute values "smartmod_2743," "main_office," and "enabled." Thus, classes represent collections of similar types of objects and managed objects represent specific instances of a class. If an order is received for a new circuit and a set of rules exists to provision a SONET switch, in order to establish that circuit then, the specific switch to be affected by the generic provisioning rule is the "affected managed object."

The network is defined by creating the classes of objects in it, defining the attributes those objects possess, defining specific network objects within the classes and the characteristic values of their attributes, and specifying relationships between classes of objects and between specific network objects. All the steps in this process are performed using windows in the Managed Objects Editor.

Classes are created in the Class Definitions window and attributes are defined in the Attribute window. Specific managed objects within the classes are defined and their attribute values specified in the Managed Object window. Rules for generating events and alerts for the various managed objects are defined in the Rule Editor.

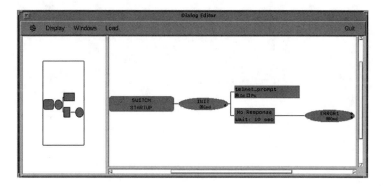

FIGURE 5.22   Dialog editor.

Class Relationships   The establishment of "relationships" between managed objects enables generic rules to be implemented which then execute only on the specific objects related to an affected managed object at runtime. This is one of the most important features of NetExpert because it significantly reduces rules. Establishing "relationships" allows one to generate generic rules. For example, the rules can address the ADM associated with a SONET switch. Then, when those rules are executed for a specific SONET switch (the affected managed object), NetExpert knows which ADM is affected because of "relationships." (See Section 7.3.1.) NetExpert conforms to the Network Management Forum (NMF) OMNI Point 1.0 standard regarding network specification of classes and objects. Standard NMF classes and associated attributes are "seeded" — built into the NetExpert database at startup. Seeded NMF classes are "read only" and cannot be modified.

All classes of objects inherit attributes from their parent class(es). A child class inherits all the attributes associated with its parent, with its parent's parent, etc. Many of the information/selection panels in the NetExpert windows show the inherited parent data. All the classes created in the NetExpert database inherit from the class of "top", which is seeded from the NMF standards. The top class has 13 children that are also seeded from the NMF standards and cannot be changed. Normally, classes are defined from one of the 13 seeded child classes.

Dialog Editor   A dialog is an automated script used to communicate directly with a managed object (Figure 5.22). Dialogs can replace routine and repetitive interactions such as user logins, device restarts, and paging of operators. They can query the status of a device within the network and replace a command/response interaction between an operator and a device. Dialog substitution groups can be created to enable a dialog to send the same command to several different managed objects within the managed-object class. These groups are defined in the Dialog Substitutions window.

Process Overview   When a dialog is executed, several processes are involved. Figure 5.23 shows a dialog process and the two files that handle data entering and leaving the gateway. The SNMP gateway does not send or receive data through a FIFO file.

Dialog commands are sent from the gateway to the *FIFO.MIN* file. The protocol agent receives the commands, converts them to the appropriate device protocol, and then sends the commands to the device.

The device responds to the protocol agent. Data is converted to ASCII format and passed via the *FIFO.MOUT* file to the gateway for identification and parsing.

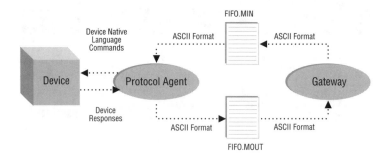

FIGURE 5.23 NetExpert dialog processes.

Depending on the response received from the device, the dialog moves to a different command state and the command/response cycle can be repeated.

Command states execute a command or series of commands and are unique to each dialog. Each dialog must include a starting state, an error state, and an ending state. The command set used in dialogs is a subset of the command set available in IDEAS. Once a command has been issued to a managed object, a response is expected. Within the command state, a wait time is defined to indicate how long the system should wait for a response to be returned. A "no response" state also is defined to specify what action should be taken if a response is not received within the defined wait time.

Responses define the information expected back from the managed object after it receives a command. They are similar to events in that they can have identification and parse rules associated with them. These rules identify the data returned from the managed object and enable the dialog to determine the next command state to process.

The **Administration Editor** is used to define the physical connection between a NetExpert gateway server and a managed object. A gateway server is not the same as a network server and can reside on a server or a client in a networked system.

The Gateway Configuration windows are used to define each gateway; specify the protocol agent that processes data from the managed object, host, and port that the data is received on; and determine which NetExpert functions are available to the gateway.

The Protocol windows are used to identify specific device characteristics and associate them with a protocol agent. Protocol agents supplied with the framework cannot be changed. However, users can purchase additional protocol agents or create their own, and enter specifications for these additional protocol agents in the Protocol windows.

The Host & Port window is used to specify the ports available on each host system in the network. Ports are later allocated to specific gateways in the Gateway window.

The Polling window and Polling Schedule window are used to define a variety of time schedules and apply these schedules to specific managed objects to generate dialogs and events at predetermined times. Polling schedules are not class-or managed-object specific, so a single polling schedule can be applied to several different managed objects.

Name	Alert Auth	Cut Through Auth	Partition	Application List	Dialog Group	Event Group	Password
carrie	ALL	ALL	<ALL>	<ALL>	<ALL>	<ALL>	******
celeste	EXCL_SMF	ALL	<ALL>	<ALL>	<ALL>	<ALL>	******
karl	ALL	ALL	DFW	DFW_OPERATIO	XC	<ALL>	******
rick	INCL_SMF	ALL	<ALL>	<ALL>	<ALL>	<ALL>	******
terry	ALL	ALL	SMF	SMF_OPERATIO	<ALL>	<ALL>	******

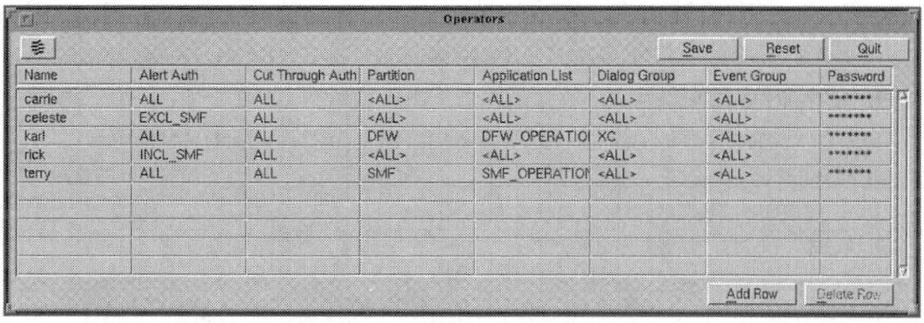

FIGURE 5.24   Operators window.

The **Authorization Editor** allows users to design a secure system by creating a variety of authorization levels for all NetExpert applications available from the Client Manager.

System operators can be restricted from certain areas of the network. Some operators may need access to only one class of managed objects, while others may need access to all managed objects. These restrictions, which can be specified down to the managed object level, are set up in the Authorizations window.

Group elements can be created and attached to network elements in the Managed Objects Editor. Collections of elements can be organized into groups and groups can be organized into partitions. Operators are associated with partitions and are only able to perform NetExpert functions such as accessing alerts, generating dialogs or events, or paging on managed objects that reference group elements contained within the authorized partition. These restrictions, which can be specified down to the managed object level, are set up in the Authorization Partitions window.

In addition, lists of applications and operator-initiated actions can be created and associated with operators. Operators are only able to access the applications or actions contained in the lists with which they are associated (Figure 5.24). These lists are set up in the Application List Editor window.

Operators also can be associated with lists of Dialogs and of Events and are only able to initiate/generate dialogs/events that are elements of the associated list. These restrictions are set up in the Dialog Group Editor and Event List Editor windows.

Finally, these authorization types can be assigned to an operator in the Operators window. This window also records the operator's password. Once an operator is logged into the system, runtime network-management duties can be performed at the operator workstation within the constraints of that individual's assigned authorization level.

The **High Availability Editor** is used to define the specific processes and operating parameters for a High Availability installation. The editor is used to:

- Create resource files (*hrt.rc* and *nxp.rc*) to set the monitored processes and define operating parameters for both primary and secondary servers.
- Modify the database subsystem table to ensure that ProcessMonitor starts on the correct host(s).
- Create UNIX shell script files to start Heartbeat and Replica, using the information contained in the resource files.

The **Peer-to-Peer Editor** is a graphical interface that provides control of the Peer-to-Peer Server MIB file. The main Peer-to-Peer Editor window enables users to specify peer-to-peer servers and peer NetExpert systems, establish links between them, and define event forwarding discriminators and log objects.

5.6 APPLICATION PROGRAMMING INTERFACES

The NetExpert framework includes several application programming interfaces (APIs) that enable users to integrate other management applications with NetExpert.

The **Protocol Agent Toolkit** consists of a library of C-language object modules that are linked together with a set of user-written C functions. Included with the C-language object library is a main function that includes all of the necessary IPMH interactions. This function uses a select-system call to monitor I/O access within the protocol agent.

The **CMIP Protocol Agent** can be extended to include application-specific programs that are coupled with the CMIP Protocol Agent. This logic is provided through various handlers. The CMIP Protocol Agent Toolkit enables users to create their own versions of the special handlers for Distinguished Name, Affected Managed Object mapping, or foreign language translations.

The **SNMP Client API** is a series of C++ classes designed and coded to facilitate communication between the SNMP Server and the Intelligent SNMP Gateway and Trap Daemon. It can be used to create additional SNMP management applications and link them to NetExpert through the Intelligent SNMP Gateway or to the MIB through the SNMP Server.

CORBA Access is an interface that provides software developers with programmatic access to NetExpert. CORBA Access provides a CORBA interface definition language (IDL) that allows a software developer to write applications in C++, Smalltalk, C, and other languages with mappings to CORBA IDL. CORBA Access also provides notifications and the ability to manipulate managed objects.

CORBA Access enables users to create programs to act as mediation devices, provide OSS to OSS interoperation, and create custom user interfaces.

The CORBA Access interface includes access to several functions and notifications. Functions include create and delete managed objects; Get and Set managed object attribute values; create, acknowledge, and close alerts; and create events. Notifications include Alert Creation, Alert Change (acknowledgment), Alert Clear, Managed Object Creation, Managed Object Deletion, and Managed Object Attribute Value Change. Notifications can be filtered so only the specified information is sent.

The **Trouble Ticket Agent API** consists of a library of C-language object modules that allow users to link third-party trouble ticket management applications with the NetExpert Alert Display process.

The **IPMH Toolkit** consists of a library of C-language object modules that allow user-written applications to intercept and process all classes of NetExpert messages. This enables developers to integrate custom applications with NetExpert.

Protocol Agent Toolkit consists of a library of C-language object modules that are linked together with a set of user-written C functions. Included with the C-language object library is a main function that includes all of the necessary IPMH interactions. This function uses a select-system call to monitor I/O access within the protocol agent. The **CMIP Protocol Agent Toolkit** can be extended to include application-specific programs that are coupled with the CMIP Protocol Agent. This logic is provided through various handlers. The CMIP Protocol Agent Toolkit enables users to create their own versions of the special handlers for Distinguished Name, Affected Managed Object mapping, or foreign language translations.

5.7 INTEGRATED SOLUTION ARCHITECTURE

A functional overview of OSI's proposed long-range solution is shown in Figure 5.25 — NetExpert-Based, Long-Range Operations Support System (OSS) Architecture. This architecture supports the integration of new services and technologies as they are added. It provides central network control, while still enabling domain autonomy. The advantages of this solution allows the deployment of operations capabilities for individual domains as they are installed, with minimal impact to centralized operations functions.

The centralized network management layer (NML) functions include:

- Configuration and service activation control (NetExpert iSAC)
- Inventory control [NetExpert intelligent Inventory Management System (iIMS)]
- Network-wide traffic/performance management (NetExpert trafficMASTER)
- Network-wide alarm-correlation and root-cause analysis (NetExpert Fault Manager)
- Centralized trouble ticket management
- CDR collection and normalization (NetExpert amaMANAGER)
- Centralized work force management

Each Domain Manager provides the following capabilities:

- Network element provisioning
- Alarm gathering, filtering, and forwarding to NML
- Performance data collection and forwarding to NML
- CDR/AMA data collection and forwarding to NML/SML

The Domain Managers may be remotely located, or centralized, and may even support more than one domain. This flexibility is intended to accommodate a variety of domain sizes, and to scale with network growth. Depending on how customers choose to build their long-term operations, one may choose to distribute operations in the future and create regional first-line operations centers that could manage across the local elements, that is, switches, BTSs, DACS, etc., for local correlation and filtering. This type of layered architecture would reduce the amount of data communications network traffic to the National Operations center, and allow local or regional personnel to manage across their whole "domain."

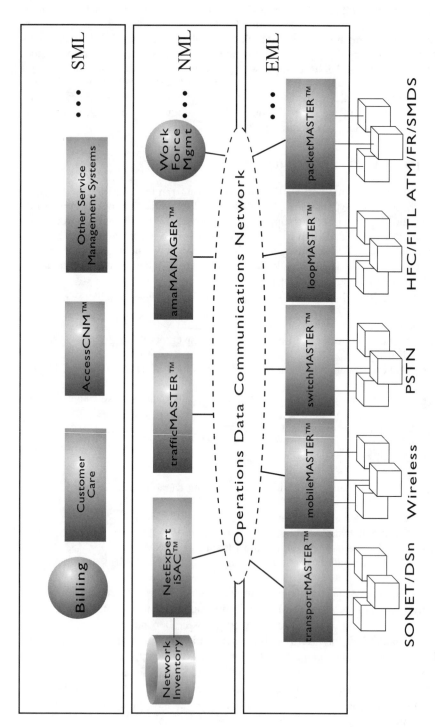

FIGURE 5.25 NetExpert-based, long-range operations support system (OSS) architecture.

However customers choose to evolve their operations, NetExpert's flexible architecture will protect their investment and support their ability to change directions.

5.8 SUMMARY

The NetExpert framework provides the basis to develop and run rule packages. Rule packages are tailored to individual customer environments. The rulesets are subdivided into point, domain, and corporate rules.

More examples of rulesets are shown in subsequent chapters.

6 Supporting TMN with NetExpert

CONTENTS

FIGURES

This chapter describes how OSI's NetExpert framework supports TMN standards-based solutions. NetExpert provides an extensive array of operations solutions that fit well within Telecommunications Management Network (TMN) principles and recommendations supported by recognized standards organizations. OSI's solutions are intended to meet critical business needs while supporting practical standards and industry evolution.

The NetExpert framework supports a broad array of operations solutions that are in accord with the Telecommunications Management Network (TMN) principles and standards recommended by various organizations, including the Telecommunications Standardization Sector of the International Telecommunications Union (ITU-T). The primary TMN principles supported by NetExpert are included in these solutions:

NetExpert's layered operations solutions which meet TMN standards for functional areas and layers.

OSI's standard TMN interfaces, such as NetExpert's Common Management Information Protocol (CMIP) Gateway for TMN Q3 interfaces and its electronic bonding TMN X interface.

Other non-TMN interfaces such as Simple Network Management Protocol (SNMP) which supports legacy and non-TMN systems.

It is always important that new technologies and standards be employed in accordance with a company's business goals. Since the telecommunications industry is changing much faster than standards are being developed, business needs must drive the implementation of standards and technologies, not the reverse. OSI believes in this approach and provides NetExpert as a flexible framework for meeting a broad range of operations business needs.

OSI understands that many companies depend on an embedded base of non-standard legacy systems which cannot be replaced overnight. In addition, even standards experts agree that new technologies and network elements (NEs) are often introduced prior to their standardization. Therefore, NetExpert's ability to coordinate with existing legacy Operations Support Systems (OSSs) while simultaneously providing management capabilities for new network technologies and services makes NetExpert one of the most practical solutions available.

Since TMN will not become ubiquitous overnight, and nonstandard systems and interfaces are likely to persist for some time, telecommunications management must live in a hybrid world. NetExpert allows this by supporting a wide variety of standard and nonstandard interfaces.

6.1 SUPPORTING TMN FUNCTIONALITY

NetExpert's flexibility allows management functionality to be partitioned easily among TMN functional areas and logical layers as needed. It also provides for the development of applications that support the main TMN management areas in the following list. Since these represent functional categories and not physical boundaries, NetExpert can naturally be used to implement management applications covering multiple areas:

- *Fault Management*, including alarm surveillance, fault localization, fault correlating fault correction, testing, and trouble management
- *Configuration Management*, including network and service planning, engineering, installation, inventory, and provisioning
- *Performance Management*, including performance monitoring, control, analysis, and quality assurance
- *Accounting Management*, including usage measurement collection
- *Security Management*, including prevention, detection, containment, recovery, and administration

NetExpert's framework also supports the implementation of layered management architectures. Basic principles for layering management functionality include:

- Distributing technology-dependent and supplier-specific functions to lower layers which are logically "close" to the individual elements being managed
- Performing network-wide management, independent of device-specific operations

FUNCTIONAL	FUNCTIONAL AREAS				
LAYERS	Configuration	Fault	Performance	Accounting	Security
Business	X	X	X	X	X
Service	X	X	X	X	X
Network	X	X	X	X	X
Element	X	X	X	X	X
NE					

FIGURE 6.1   TMN functional areas and layers supported by NetExpert.

- Providing end-to-end management of services, independent of the network infrastructure
- Allowing corporate-wide business practices to impact service and network management without such practices having to be embedded into each process and each layer
- Minimizing the dependencies between layers, and making service management resilient to changes in the network and network management resilient to changes in suppliers and elements

When implemented with a NetExpert solution, these principles enable the rapid introduction of new services and network technologies without requiring major changes in all levels of the operations architecture. This is a current limitation of many legacy OSSs that can significantly constrain a telecommunication provider's ability to compete in a changing industry. NetExpert solves this problem.

The combination of TMN functional areas and layers provides a complete matrix for the definition and description of the management capabilities offered by any OSS in any TMN. As shown in Figure 6.1, the NetExpert framework allows layered management solutions to be implemented for all functional areas and layers above the NE Layer.

6.2. SUPPORTING TMN ARCHITECTURES

TMN architecture, as described in the ITU-T Standardization Sector Recommendation M.3010, Principles for a Telecommunications Management Network, is comprised of three perspectives: TMN functional architecture, TMN physical architecture, and TMN informational architecture.

Functional Architecture   The TMN functional architecture encompasses the basic components shown in Figure 6.2.

OSI's NetExpert framework supports OSS functions with TMN Q3 and X interfaces (proprietary protocols), workstation functions, and adaptation for meeting non-TMN interfaces (illustrated in Figure 6.3).

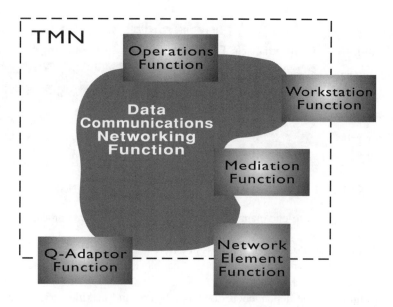

FIGURE 6.2   Components of TMN functional architecture.

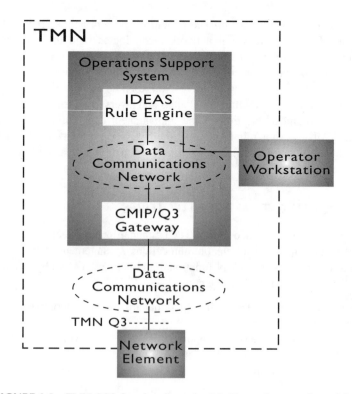

FIGURE 6.3   TMN OSS functionality using NetExpert framework modules.

NetExpert's framework architecture is comprised of functional modules of smaller granularity than those in the TMN functional architecture. Therefore, most TMN functions shown in Figure 6.2 may be created through different configurations of NetExpert's modular components. The individual NetExpert modules that support these TMN functions include:

- *IDEAS* rule engine provides flexible, event-based operations functionality.
- *Operator Workstation* allows GUI-based monitoring and control functionality.
- *CMIP Gateway* provides CMIP/Q3 adaptation and interfacing between the NetExpert framework and TMN Agents.
- *EB Engine/EB Gateway* supplies electronic bonding operations, interface, and mediation functionality for trouble administration.
- *Generic Gateway* adapts and interfaces between the NetExpert framework and non-TMN systems.
- *SNMP Gateway* provides adaptation and interfacing between the NetExpert framework and network elements or systems using SNMP.

Although the SNMP Gateway and Generic Gateway do not support TMN interfaces, they allow NetExpert-based solutions to integrate a variety of nonstandard, standard, legacy, and proprietary systems through adaptation and mediation. This facilitates the practical migration to standard management solutions without necessitating the immediate abandonment of embedded systems and solutions.

Physical Architecture   The NetExpert framework is flexible, allowing combinations of modular building blocks for a wide range of functionality. A basic physical architecture providing TMN operations support system functionality with a Q3 interface is shown in Figure 6.3. Since, as mentioned, NetExpert's modules are of smaller granularity than functions in the TMN functional architecture, it is possible to separate the OSS functionality into two components which remain interconnected through a data communications network: the IDEAS rule engine performs the management application functionality, while the CMIP manager gateway provides the TMN Q3 interface to the OSS.

A more comprehensive physical architecture that provides operations functionality with interfaces to both TMN and non-TMN systems is shown in Figure 6.4. External interfaces supported by NetExpert gateway modules include CMIP (TMN Q3 and X), SNMP, and other non-TMN interfaces.

Figure 6.4 illustrates the important ability of the NetExpert framework to support multiple interface protocols (both TMN and non-TMN) with the same management application. In other words, since the OSS management application is separate from the interface gateways, the same application can meet several different interface protocols. The NetExpert gateways (CMIP, SNMP, and Generic) provide adaptation and mediation functionality for each different interface.

This separation of functionality between the application and the interface within the OSS also allows the interface gateways to incorporate information models and behaviors that are specific to the network elements or systems they manage. Addi-

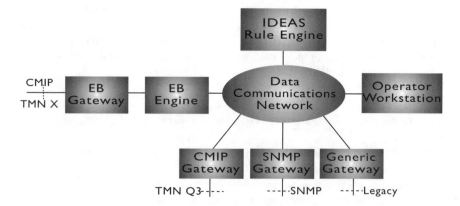

FIGURE 6.4   NetExpert framework supporting TMN and non-TMN interfaces.

tionally, the OSS application itself makes use of a more generic behavior and information model. This enables the OSS application to easily manage different systems with different information models or interface protocols.

Informational Architecture   The TMN informational architecture defines object-oriented management with manager/agent interfaces. NetExpert's framework supports both of these aspects. The NetExpert CMIP Gateway supports TMN Q3 manager interfaces. OSI's electronic bonding modules support both CMIP agent interfaces with electronic bonding engine and gateway functions. All of OSI's TMN interfaces support standard information models.

The NetExpert framework relies heavily on object-oriented principles to allow for rapid development and deployment of management applications. This increases the degree of software reuse, and decreases the cost of implementing operations solutions. Object-oriented information modeling and manager/agent interfaces are key NetExpert components that support the TMN informational architecture.

6.3   SUPPORTING TMN INTERFACES

The **NetExpert CMIP Gateway** supports TMN Q3 manager interfaces. As shown in Figure 6.5, this gateway includes the following main components:

- Management Information Model (MIM) Loader
- CMIP Protocol Agent
- Multiplexing Gateway (Mux)

The MIM Loader checks the syntax and structure of the interface information model and loads it into the CMIP gateway. It can validate and load Guidelines for the Definition of Managed Objects (GDMO) and Abstract Syntax Notation.1 (ASN.1) modules. The outputs include a MIM file, which is used by the CMIP Protocol Agent, and a log file. The log file contains the results of the processing and loading of the information models, including errors.

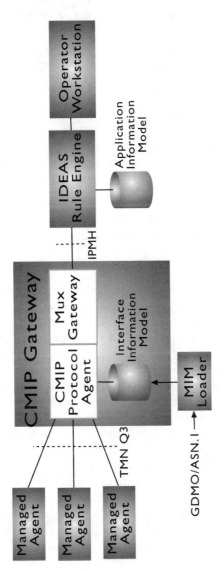

FIGURE 6.5 NetExpert CMIP Gateway.

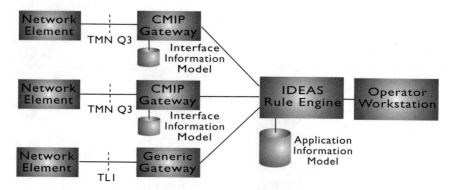

FIGURE 6.6   Mediation between interface and application information models.

The CMIP Protocol Agent allows NetExpert to communicate with CMIP Agents. It supports multiple managed CMIP agents, as well as multiple associations per agent. It also enables a rich support of CMIP, including:

- Full association handling
- Services for all operations
- Scoping, filtering, and synchronization
- Linked replies
- Access control

The CMIP Protocol Agent converts CMIP ASN.1 protocol data units into internal formats for mediation and additional processing by the Multiplexing Gateway. Runtime tracing can also be enabled and disabled for this component.

The primary functions of the Multiplexing Gateway are to extract information from incoming CMIP messages, mediate between the CMIP interface information model and the operations application information model, and execute bidirectional communication "dialogs" between operations applications and CMIP-managed agents. Multiple, nonblocking dialogs are concurrently supported.

CMIP gateway functions are defined using familiar NetExpert rule and dialog editors. Mediation between operations applications and interface information models is supported through NetExpert ID, parse, and dialog rules that define the construction and decomposition of CMIP messages, respectively. This flexibility allows multiple CMIP interface information models to be supported by a common operations application, potentially using a more generic information model. This is an important capability, particularly for multivendor environments where network elements use different, and often nonstandard, implementations of standard information models. Figure 6.6 shows the mediation between interface and application information models.

6.4 SUMMARY

TMN is an evolving standard which is gradually changing the telecommunications industry. The NetExpert framework and its applications support multiple TMN layers. *Each specific application will be individually addressed in the following chapters.*

Part II

Management
of Voice Related Services

The entire communications industry is in the midst of unprecedented upheaval. Competition, technology, and market demand have combined to present the industry with a set of problems and opportunities that have never been experienced before.

Service Diversification   Providing dial tone or routing network television is no longer the key to a successful telephone or cable company. The market is demanding advanced services for retail, entertainment, medical, conference calls/views, banking, and the home office, to name a few. These are features that obscure the traditional telephone/cable company boundaries, both organizationally and technically.

Increased Service Velocity   Delivery cycles for development, deployment, and activation of services are now driven by the competition, the technology, and ultimately, the consumer. Deliverables are now measured in minutes, seconds, and sometimes, even milliseconds, rather than in the day, week, and month measurements of the past.

Heterogeneous Networks and Operations   Niche-market technology providers, and business mergers and partnerships, have increased the diversity in both network equipment and operational systems within an organization. At the same time, the need to manage these elements has become increasingly critical to these organizations.

Protection of Installed Base   The appropriate expenditure of limited funds within the communications industry is often the difference between success and failure. The allure of new service introduction often overrides significant operational support spending. It is imperative to be able to reuse the existing infrastructure to its highest capacity and yet ensure the flexibility and effectiveness of any new support systems.

Uncertainty The following suppositions and assumptions are presented: It is apparent that this market will continue to grow and diversify. That increased bandwidth and control will breed new applications and services by the marketplace itself. That support systems will be in a constant state of change. That perhaps no ultimate standards solution may evolve. And, finally, that the successful organization will be the one that can adapt more quickly than its competition.

These criteria have forced the communications industry to reevaluate the way they have engineered and procured their operation support infrastructure. Framework tools and expert systems that allow maximum flexibility to change, integrate to existing systems, and which are usable by business experts will keep their organizations on the competitive edge of the marketplace.

This part concentrates first on service provisioning and activation processes supported by the NetExpert framework and its key integrated, packaged rulesets, such as the integrated Service Activation Controller (iSAC).

Physical provisioning and operations control are supported by domain integration rules packaged into switchMASTER. Performance evaluation, traffic management and optimization are supported by another set of domain integration rules that are packaged into trafficMASTER. All the prepackaged rulesets work with a large selection of network elements, such as exchanges, central offices, and private branch exchanges.

In order to evaluate call records and support accounting, marketing, and capacity planning functions, the amaMANAGER — another package of domain integration rulesets — is provided. The amaMANAGER works with DataArchiver to manage and process large amounts of call detail records.

7 Service Management and Service Activation

CONTENTS

FIGURES

TABLES

Due to the growing demand on services in general, and service changes and extensions in particular, operations support systems are expected to offer flexibility, scalability, and good performance. This chapter concentrates on service management and activation. Supporting applications and packaged rulesets will be addressed in the following chapters of the same part.

7.1 CUSTOMER REQUIREMENTS

Objective Systems Integrators has worked continually with numerous clients on requirements for provisioning, service activation, and service management systems (SMS).

These requirements are fairly consistent in content, although scope has varied substantially. Some SMS projects have been mostly service activation with minimal data management. Some have been "local" SMS dealing with particular service sets like ATM or AIN. Others have been very broad in scope addressing the majority of functions performed within a telco. Finally, some projects are visionary, combining new network and operational features into the customer facing, service creation, and the provisioning aspects of SMS.

Central to these varied agendas was a set of core functionality essential to service management. This was not to imply these were static functions. On the contrary, these base functions had to be delivered in the most open and flexible manner possible to allow the individual users to tailor their service management systems to their particular environments and markets.

Figure 7.1 shows the core functions that are the building blocks for service management. The functions have been categorized in two ways: as a set of core functions, that is, SMS request validation; and by application area, that is, service activation. This document will describe individual application areas, the required system capabilities, and NetExpert's ability to deliver those.

7.2 PRINCIPAL PROCESSES OF SERVICE MANAGEMENT

Service Activation Fundamentally, service activation is the process by which customer requests become functioning services in the network. The success criteria are that the network completes calls associated with the service as prescribed, that the service can be maintained, and most importantly, that the service can be billed. A service activation system is effective if these criteria are successfully met in a time frame appropriate to the specific service.

The process, however, is more complex than the definition.

Although all network elements can be provisioned, the mechanics of provisioning can range from rudimentary updating of cryptic tables in older network elements, to relational database updates in modern platforms, to standards-based interfaces in state-of-the-art technology.

Varying combinations of network elements may have to be activated depending on the service and geographic network configuration.

Client facing systems may also vary by service or across an organization. Services may require input from more than one source to complete a customer

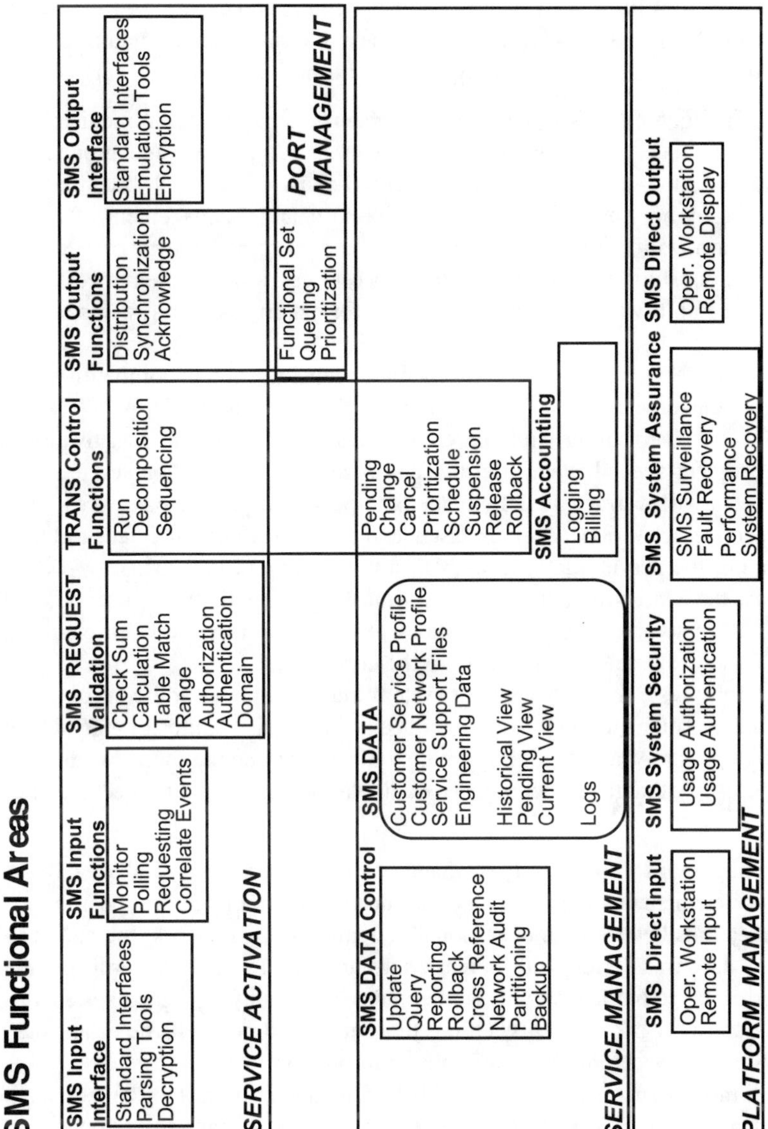

SMS Functional Areas

FIGURE 7.1 Processes and functions of service management.

request. The service may need to extract data from other OSSs such as facility and number assignment. Finally, the successful activation of a service will require the updating of numerous support systems such as billing, inventory, and directory.

In an increasingly competitive environment, the rate at which new services are being introduced to the activation process has greatly accelerated.

Therefore, a service activation method must have the capability to interface to a variety of OSSs and network elements. These interfaces must remain as flexible and maintainable as possible to allow for rapid introduction of new services.

Service Management   Service activation systems generally act immediately upon an incoming request, send acknowledgment upon completion, and produce audit and/or log information. Service management systems take a much broader view of a service. They are concerned with pending and change management of a service request, prioritization of request processing and historical views of a customer service, relationships between the customer's service view and the network service view, and the ability to provide this service information to other systems and organizations.

Timing of service activation will be a major consideration in competitive service offerings. "On demand" levels of response will be required for video services, emergency service routing, and conference bridging. Specific time activation will be necessary for special event offerings and other specific duration services. These requirements will warrant flexible methods to pend and prioritize services.

As service activation becomes faster and more customer-directed, the task of coordinating the service data, from both a customer and network view, becomes critical. As the SMS will be the conduit for most service activity, be it service orders going to the network or customer updates returned from the network elements, it becomes a key source of information to the corporation.

Therefore, a service management system must be able to add control and data management features to the flexible interface and analysis features of a good service activation system. Again, flexibility to allow the user to define their own prioritization, reporting, validation, queuing, and data relationship is vital to meet the clients' distinct needs.

Port Management is the ability to manage a set of ports to an individual system or network element. As the SMS takes an increasingly more central position in the communications between the operational systems and the network, the amount and variety of data transferred will increase substantially. In the near term, the major SMS performance bottleneck will be the interface capacity of legacy systems, both on OSSs and network elements. Often, more than one port will be established to these systems to accommodate the required throughput.

Once more than one port is established to an individual system, then port management becomes a key component of a service management system. If all managed transactions to a system are identical, then simple queuing may suffice. However, in most systems transactions may have different priorities, ports may have different protocols, or may have different functional characteristics, such as native UNIX port, applications port, SQLPLUS port, or a cut-through port.

The management of these states both at system initialization and during runtime can greatly enhance an organization's ability to manage the costs and effectiveness of these interfaces. Effective management of these ports also allows the SMS to reach optimum performance without reenginering the clients' existing systems.

Platform Management consists of the features of the SMS that concern the operations of the system itself and the underlying hardware and databases that the system is built on.

A service management system is an integral part of an organization's infrastructure and must have a full suite of administrative and control processes. These functions must include the ability to interact with the system on a real-time basis for both administrative and operational purposes; mechanisms both manual and automated to discern and affect the state of the system; processes to restrict access and select authority; and tools for backup and recovery.

7.3   THE ROLE OF THE NetExpert FRAMEWORK

There are four principal functions of the NetExpert framework that may be utilized in the service management process. Each of them will be addressed in depth.

7.3.1   INTEROPERABILITY

Interoperability is the process by which NetExpert exchanges information with other systems. The gateway processes have made NetExpert the industry leader in providing this ability. Two methods of delivering interoperability exist: standard and generic gateway facilities.

Standard gateways adhere to existing industry standards and allow data to be applied directly between interfacing systems using recognized data and facilities definitions. Examples of standard gateways are SNMP, CMIP, and COBRA.

The Generic Gateway provides two layers of interoperability: protocol agents and parsing/dialoging abilities. The protocol agents provide interoperability at the facilities level; that is, X.25, TCP/IP, RS 232, and Shell levels while the parse/dialog functions define the types of data sessions to be performed.

The ID and Parsing tools are the state-of-the-art in data acquisition. With GUI-based definition supported by integrated raw data and online testing abilities, these tools allow subject area experts to quickly identify and capture incoming data events. Features such as context sensitive parsing allow even the most complex data structures to be processed via these tools.

Example   This example will be used throughout this chapter to illustrate the features of NetExpert.

The following is a new service order type that needs to be provisioned for a simple AIN Do Not Disturb service. The assumptions are that an SMS exists and AIN service orders are already being processed, that gateways to the service order system, service control point (SCP) and switch are available and that standard routines for repetitive action already exist.

The following graphics illustrate the data acquisition rules required to capture the new service type from the service order system.

ASCII data structure received from service order gateway:

```
USOC 401 416-232-9879
ST - 17:00 ET - 19:00 PIN - 1225
```

The following rules identify the ASCII stream as being an "event" to NetExpert and position the parsing tools for data capture. A successful ID process will generate an event to the Server for rule processing.

This command states:

find the ASCII string USOC 401 anywhere in the input line, consider this line 1 of the record.

Test	Line	Col	ID String	Col
TRUE	1		USOC	

Now data can be parsed out of the input stream into attributes.
The fifth command states:

find the ASCII string PIN — in line 2, put all following data until end of that line into attribute $PIN.

If the Col fields are used, data can be extracted by actual location in the line.

Line	Col	Start String	Attribute	Col End	String
1		USOC .	$USOC_NUM	.	
1		`USOC ....`	$TEL_NUM	^$	
2		ST.-.	$START_TIME	.	
2		ET.-.	$END_TIME	.	
2		PIN.-.	$PIN	.	

When all data has been captured, NetExpert will associate the incoming event to a managed object (a request transaction).

This command states:

associate this event with the managed class SO_401 with the instance being the value of attribute $TEL_NUM.

If the affected object associated is a new instance, the process will create a new managed object data structure.

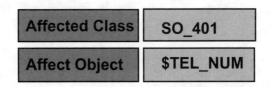

The *Dialog (Command and Response)* tool allows the user to conduct complex interactions with remote systems. Using Command scripts to insert data into the remote systems and Response (ID/Parse features) to capture complex reply scenarios, this facility can manage everything from outputting simple data streams to detailed interactive sessions. The GUI nature of the tool allows a graphic representation of the dialog session that simplifies development and maintenance activities.

Example   Continuing our previous AIN example, the following graphics represent the dialog session required to provision this service on an SCP. (Each SCP would differ. This format is a rough representation of an ISCP v2.2 interface.)

Note: The rule example later in this chapter would actually run this dialog.

This is the dialog tree associated with the USOC 401 service. The ovals are command lines and the rectangles are response processes.

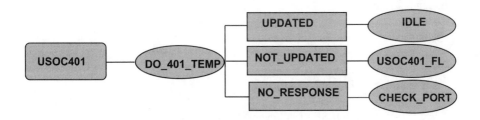

In this particular case, the gateway expects a specific set instruction to be received. It waits for the final "}" character to interpret the input. The DO_401_TEMP command is expanded below. As no intermediate responses are generated by the SCP, these commands can be sent as one.

Three potential responses can be received. In this scenario, these responses are consistent for different service so these response routines could be reused from previous service dialogs.

Each response generates its own subsequent command. The IDLE and CHECK_PORT commands are standard for this type of SCP session. The specific USOC401_FL routine would generate an event to initiate error processing for any failed 410 service order and could be developed using NetExpert analysis rules and alerts.

Template: USOC_401;
Key= (${TEL_NUM},TEL,10);
Status=Active;
Parameters
{
TOD_Start = ${START_TIME},Time};
EOD_time = (${END_TIME},Time)
PIN_Override = (${PIN}, I, 4)
}

This suite of tools has proven very successful in developing interfaces as diverse as monitoring printer ports on switches, providing SNMP provisioning data to routers, and capturing service order data from a legacy customer facing system. Ultimately these tools have allowed the clients to interface with their varied operational and network interfaces, allowing them to build the relationships and processes necessary to manage their services.

7.3.2 PROCESS RULES

Process Rules provide the functionality of a programming language through a user-friendly, GUI-interface. The following are samples of some of the analysis commands available:

Control	IF
	ELSE
	DO
Comparative	EQUAL
	NOTEQUAL
	DOESEXIST (event)
Logical	AND
	OR
Extended	CORRELATE (event)
	CLEAR (event)
	GENERATE (event)
	DIALOG
	CREATE (managed object)
	SET (attribute)

There are many more commands and new ones are introduced with each new release. The Extended Commands are primarily used to manage the object-oriented nature of the product by controlling events, dialogs, and managing object-oriented data.

These commands are maintained in a GUI form as follows:

Example   These are the set of rules that would be triggered by the ID event for the Do Not Disturb AIN service described above.

As you can see in line 2, it also initiates the USOC 401 dialog described earlier.

This routine acts immediately on the service order, activating the service on the SCP and switch.

The switch dialog is a Destination Trigger and would be the same for all AIN services using that feature.

The USOC 401 dialog was defined earlier. If these updates are successful (USOC401_FL DoesNotExist), then the POST (standardize for all services) event is fired to update the customer subscription current values.

The Set command in this example changes the Affect Manager for the event and the port the event is associated with so that the dialog is presented to the correct device (DMS_SEA003).

Operator	Operand1	Operand2	Relate	Event	Alert
Set	@AMANGER	$SCP			
Dialog	USOC401				
Set	@AMANGER	$SWITCH			
Dialog	DN_TRIG				
If					
DoesNotExist					$USOC401_FL
Then					
Set	@AMANGER	SQL_MAN			
Generate				POST_REQ	
EndIf					

The Relate column allows clients to create and use object oriented relationships between managed objects in the database. The Alert column allows the user to issue predefined information displays to the Operator Workstation and check for or clear existing alerts.

The combination of the three examples above and the definition of the SO_401 managed object and the USOC_401_FL event would complete the coding for the Do Not Disturb Service.

Should a client wish to augment the rules based process, they can incorporate C++ routines into their process flows via the NOW API product from OSI. However, a vast majority of the clients have found the rule-based processes have provided them with all the power they need.

Figure 7.2 shows the service activation process using NetExpert.

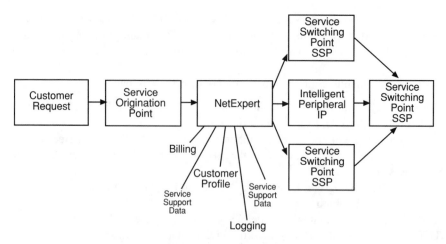

FIGURE 7.2   Service activation with NetExpert.

7.3.3 DATA MANAGEMENT

Data Management provides the core for service management. The object-oriented nature of NetExpert lends itself to establishing relationships between managed objects. In service management, the core objects will be a customer, their subscription (past, present, and future views), service support data, and network elements that supply the service. Supporting data for billing and traffic can also be captured and maintained.

NetExpert's *Managed Object Editor* allows the client to define Classes and establish relationship via either the "Relates" feature or via "Attributes." This editor can also create individual instances of these classes. Supporting processes such as "SNMP MIB loader" and "managed object creations on-the-fly," allow managing the instances of the Classes either individually or in batch loads.

The following is a partial example of a standard TMN-based Managed Class Model (Table 7.1) that is supplied with each NetExpert installation plus the additional classes to support service management.

NetExpert's interfacing ability also allows it to manage this information in other databases and/or on other systems.

These abilities give NetExpert clients the flexibility to manage their service data to the level of detail required by their organization.

7.3.4   ONLINE ACCESS TO INFORMATION

Access   The operator workstation is NetExpert's optimal point of access. Most control functions can be initiated from here. In a management environment, the prime functions for this operator will be monitoring for abnormal events, monitoring and managing gateways, and issuing recovery or maintenance events. This operator also has direct access to the MIB data where immediate changes can be implemented, if necessary.

TABLE 7.1
Classes Hierarchy of Service Management

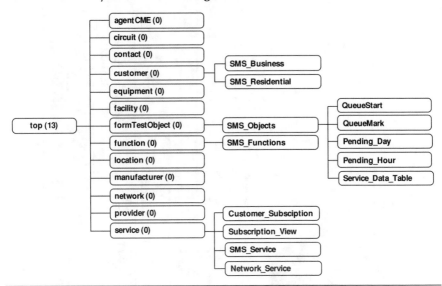

The following is an example of some of the information that could be displayed to the operator via the operator workstation (Figure 7.3).

A powerful feature of this product is the ability to open direct cut-through sessions across the provisioned gateways. This allows the operator to intervene at the NetExpert level, but also to intervene directly to other systems if he has appropriate clearance.

The operator workstation can be configured by the user so that different users can have different authorizations.

Some clients have needed more specific online access to NetExpert. To support these needs, two alternative access mechanisms exist.

Remote cut through is the process whereby NetExpert can, upon remote request, attach a cut-through session to a remote (nonoperator workstation) x_terminal. This feature has allowed clients to integrate their port management facilities and offer flexible system connection across an enterprise network.

Forms windows were driven primarily from AIN clients who needed to capture supplementary data beyond their current Service Order Systems ability. In this case, NetExpert can bring up remote forms-based x_windows sessions and allow remote operators to access and update data directly with NetExpert.

7.4 SERVICE CREATION FUNCTIONS

NetExpert's Operation Creation Environment (OCE) is a combination of NetExpert's base functionality described above, and a set of prepackaged, managed objects and processing rules that supply the user with all the building blocks needed to rapidly deploy full-function Service Activation and Management Systems. Figure 7.4 shows the positioning of OCE.

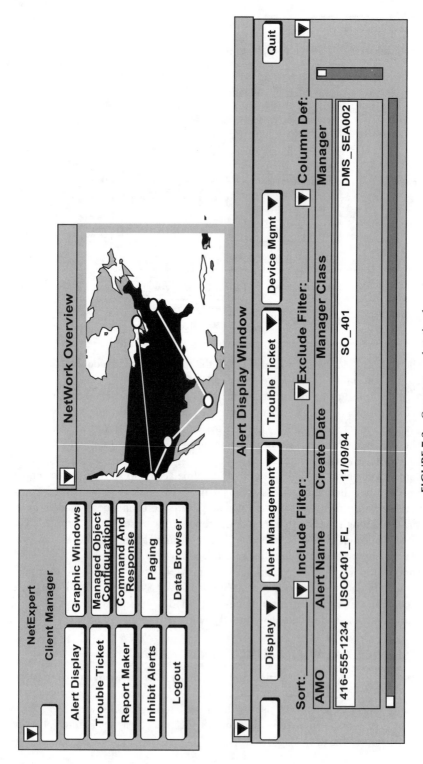

FIGURE 7.3   Operator workstation layout.

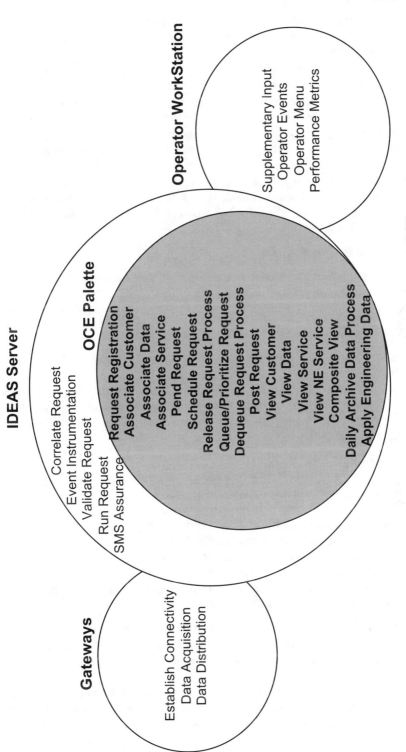

FIGURE 7.4 Positioning the Operation Creation Environment (OCE).

The prepackaged component is referred to as the OCE palette. This palette provides a core set of features that provide a full function SMS while maintaining the flexibility of NetExpert.

The purpose of the OCE is to provide NetExpert clients the most comprehensive set of functions to manage the service management requirements. An SMS should be a facilitation device, not a predefined, static system that needs to be accommodated by an organization at tremendous cost. OSI has gone to great lengths to provide flexible productivity. The goal is to deliver a solution that can be put in production in a matter of weeks but still grow to meet the most complex and unique service management requirements.

The following describes the suite of tools provided with NetExpert's Operation Creation Environment. Each section states the function, the principal application area where it is used, and whether the function is part of the base functionality of part of the OCE palette.

Service Creation Functions are:

Establish Connectivity   (Service Activation, Port Management)

Base Functionality   GUI, rule-based tools that allow quick establishment of connectivity to major communication protocols either via standard interface gateways to CMIP and SNMP, or with the generic gateway and predefined protocol agents such as TCP/IP, X.25, Serial, and Telnet.

Protocol Agents also can be developed to support encryption and decryption capabilities.

Baud rates and other configuration criteria are entered via these high-level tools.

This function allows data to be presented to the Data Acquisition and Distribution functions. Appropriate "managers" are assigned to each "port" to facilitate both input and output data definition.

Data Acquisition   (Service Activation)

Base Functionality   GUI, rule-based tools (ID/Parse) that allow rapid development of routines that will identify incoming data, parse pertinent data into user-defined attributes, and initiate specified processing.

These tools provide the principal input connectivity flexibility of the NetExpert framework. Users, or their agents, can generate rules to perform complex data capture via parsing and identification rules in a matter of hours.

This facility is where a large part of the "user definability" will be recognized. Here, the user will be able to model input interfaces to both legacy and current systems with minimal or no effect on those applications.

Data Distribution   (Service Activation, Port Management)

Base Functionality   GUI, rule-based tools (Dialog) that allow rapid development of routines that will output data, perform interactive processing, and initiate data retrieval.

These tools provide the principal output connectivity flexibility of the NetExpert Platform (see Section 3). Users, or their agents, can generate rules to perform complex data distribution and emulation in minimal time.

This facility is where a large part of the "user definability" will be recognized. Here, the user will be able to model output interfaces to both legacy and current systems with minimal or no effect on those applications.

The functions (or command and response dialogs) created here define the richness of the SMS solution. The methods and procedures required to manage individual network elements are maintained via these tools. Not only the initial service addition and change routines should be considered, but also service suspension, rollback, and network audit features can be developed to provide comprehensive network elements management layer. These routines are mostly unique to the particular network elements and often NetExpert clients apply specific business practices to their management. OSI has packaged suites of rules for certain network elements, and constantly is developing new ones. NetExpert's tools allow the maximum flexibility for this customization while greatly reducing the development time.

Request Registration   (Service Activation Service Management)

OCE Supplied Function/Data Structure   Process that registers a request for action in the Service Management platform. At this point the SMS owns the request and may return acknowledgment to the requesting system. The SMS will now process this request to some form of logical conclusion and maintain a record of the activity.

The OCE Palette contains a Request data structure that provides the building block for future OCE function processing. This structure contains base request attributes (Requester Name, Pending Date, etc.), a tag oriented current values array that allows the structure to be service independent, a set of relationships (Service Type, QueueUp, etc.) allows the request to associate itself to other objects in the MIB for ease of processing.

This object also maintains a relationship to any associated corrective passes. The subscription structure is shown in Table 7.2.

The request type attribute designates the particular processing for the request. Supported functions include:

- Add
- Change
- Cancel
- Suspend
- Release

The request data structure is both service and data independent. The same request structure can support new service class information, customer service information and service support or engineering data. This allows functions like pending to be applied to a request regardless of the data content.

Associate Customer   (Service Management)

OCE Supplied Function/Data Structure   Process associates request with the corresponding subscription and customer to be initiated. If these objects do not exist and the request has appropriate privileges, they will be created. The request then is

TABLE 7.2
OCE Subscription_View Object

Class	views	
Manage Object	req_010982	
Reported As	so-893456	
Attributes	**Type**	**Value**

Subscription_key	String	4162350990	
Request_type	Enum	change	
Request_date	Date	94/12/22	**Request Info**
Pending_date	Date	94/29/22 10:00:00	
Priority	Int	3	
Requestor	String	Bill Smith	

Service_type	Link	DoNotDisturb	
Queue_link	Link	Queue_1_Pri_3	**Relationships**
Composite_View	Link	Composite_View_001	

Current_values	Setof Str	17:00	
		19:00	
		1234	**Tag Values**
Table_rows	Int	0	

Corrective_Passes	Link	req_011038	**Corrective Pass**
Corr_Pass_Cnt	Int	1	

Next_Event	String	DND_Run_001	**Next Event**

linked to the corresponding subscription, service, and previous requests. If this request is a corrective pass, then a special association to the original pended request is established.

The OCE supplies a data configuration for Customer Services. Objects for Customer, Customer/Subscription, and Service all relate and reference the individual requests. This combination of objects and relationship make up the Customer Service Profile. The relationship is shown in Table 7.3.

Associate Data   (Service Management)

OCE Supplied Function/Data Structure   Process similar to Associate Customer except the request concerns service independent table data, such as Service Support Data (ANI/ZIP) or Engineering Data (Feature Interaction and network Resource).

Sample data structures for this type of data will be provided with the OCE palette. However, by nature, this data is very application specific and tailoring will be needed. NetExpert's managed object editor and analysis rule editor make customization simple. The OCE customer and subscription objects are shown in Table 7.4. The OCE service object is shown in Table 7.5

TABLE 7.3
OCE Customer Profile Relationships

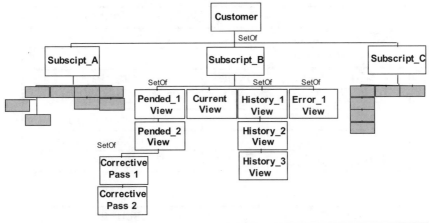

Associate Service   (Service Management)

   OCE Supplied Function/Data Structure   Process similar to Associate Customer except the request concerns service classes.

   This process allows for the management of the transaction concerned with service classes. Specific processing rules for services deployment and maintenance can be developed to automate these tasks within an organization.

   The SMS_Service object in the OCE MIB is affected by these transactions.

Correlate Request   (Service Activation)

   Base Functionality   The ability to associate two or more distinct requests into one logical request (such as combining a request from a service order system with supplementary data input from a PC) are accomplished by using core NetExpert rule operators, such as DoesExist and WatchDog. The user, or their agents, can quickly develop relationships between incoming events to match their business needs.

Validate Request   (Service Management)

   Base Functionality   Data comparisons can be managed via the analysis rules. Table lookup processing has been packaged (see Apply Engineering Data) and the use of other OSI offerings and the electronic bonding package can also assist in complex validation.

Pend Request   (Service Management)

   OCE Supplied Function/Data Structure   This process delays action on a request until a predefined time in the future. Granularity of time is from a specific day down to a specific minute.

   The OCE palette supplies both the managed object structures and base rules to perform these functions. (See Figure 7.5.)

TABLE 7.4
OCE Customer and Subscription Objects

Class	SMS_Residential	
Manage Object	4162350990	
Reported As		
Attributes	**Type**	**Value**
Subcriptions	Setof MO	DoNotDIsturb_Sub
		FindMe_Sub
		CallForward_Sub
Service_Status	Enum	Active

Class	Cust_Subscription	
Manage Object	4162350990_DoNotDisturb	
Reported As		
Attributes	**Type**	**Value**
Current_Values	Setof Str	17:00
Table_rows	Int	0
Views	Setof MO	so-59346
		so-30981
Service_Status	Enum	Active
Service_Type	Link	DoNotDisturb
Network_Link	Link	DMS_SEA003

Schedule Request   (Service Management)
 OCE Supplied Function/Data Structure   Similar to pending, but includes the ability to discern previous pended actions for the same time period that would affect the delivery of the current pended request. Processing to facilitate request scheduling contention include priority placement and best schedule notification.

Release Request Process   (Service Management)
 OCE Supplied Function/Data Structure   A background process that releases all pended or scheduled events at the predefined time.

Queue/Prioritize Request   (Service Management, Port Management)
 OCE Supplied Function/Data Structure   The process where requests are queued for sequential processing. This queuing is often necessary to manage slow network element interfaces or to support distinct business practices. Priority levels are usually applied to allow specified requests to have precedence over other requests.

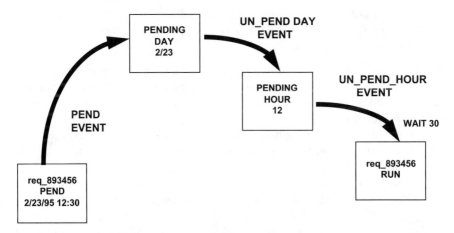

FIGURE 7.5   OCE pending request process flow.

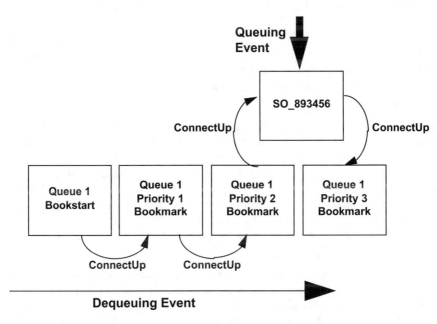

FIGURE 7.6   OCE priority queue process flow.

The OCE palette supplies managed objects, queue rules and also a new queue generation routine to allow the user to manage their queuing requirements. The queuing objects are available for modification to allow further queue refinement (that is, the use of counters to make sure that 1 priority 2 request is processed for every 5 priority 1 request). (See Figure 7.6.) Note that the queuing and requests structures of the OCE allow dissimilar request types to the same queue. This facilitates using one queue to manage all traffic to a particular network element interface port

Dequeue Request Process   (Service Management)

OCE Supplied Function/Data Structure   A background process that dequeues the next request from a queue.

Run Request   (Service Activation, Service Management, Port Management)

Base Functionality   GUI, rule-based tools (Analysis) that allow rapid development of routines that allow the user, or their agents, to implement their unique business practices for request/service management.

These tools provide the principal processing flexibility of the NetExpert platform. Users, or their agents, quickly generate rules to perform functions such as Synchronization and Decomposition.

As each service will require different processing rules, this function will be developed for each service and will call on the other functions defined in this document.

There will be more than one "Run" process for each service request.

The "receive" processing is closely coupled with the Data Acquisition Function and identifies the particular request type (Add, Change, Suspend, Release, Cancel, etc.) and will define which processes are to be applied to the incoming request (Pend, Queue, etc.).

The "action" process will be the actual request processing which will handle the request decomposition into distinct network operations, synchronization of external processes, the application of engineering data, error processing and the posting processes.

It is conceivable that there may be more run routines if multiple queuing operations are needed.

Note: Given the OCE base functions and NetExpert's high-level tools, these "run" processes can be developed in a matter of days, if not hours, while still giving the user full control of their service processing requirements.

As processing rules are developed, many may become building blocks for subsequent services. For example, an SCP service suspension request may be to interject an SCP service that routes an originating call to the organization's collection department for a suspended customer. This function could be applied to any suspension requests for all service subscription types associated to that particular SCP class.

Post Request   (Service Management)

OCE Supplied Function/Data Structure   This process posts the successfully completed request to the current values of the customer's subscription record for the affected customer. The Subscription object is updated to reflect the status change of the request from pending to historical. Network references are updated as applicable.

Uses the composite View function to update current values.

View Customer   (Service Management)

OCE Supplied Function/Data Structure   Function returns to the requester details on all subscriptions associated with the specified customer. Current View values are returned unless a date is specified, in which case the Composite View function is used to build a pending (future) or historical view of the customer's subscription.

View Data   (Service Management)

OCE Supplied Function/Data Structure   Display or query contents of a support data structure.

TABLE 7.5
OCE Service Object

Class	SMS_Service	
Manage Object	donotdisturb	
Reported As		
Attributes	**Type**	**Value**
Subscription_type	String	DoNotDisturb
Tag_names	Setof Str	Start Time
		End Time
		PIN Number
Last_value	Int	3

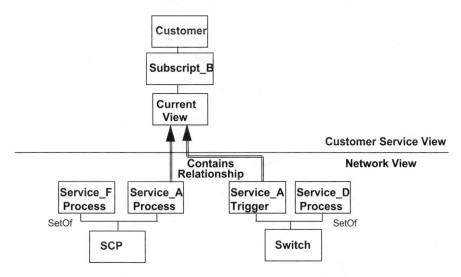

FIGURE 7.7   OCE network/customer profile relationship.

View Service   (Service Management)
   OCE Supplied Function/Data Structure   Display the parameters of a specific service type and all current subscribers.

View Network Element Service   (Service Management)
   OCE Supplied Function/Data Structure   Display current service parameters for a particular network element instance and all active subscriptions. (See Figure 7.7.)

Composite View   (Service Management)
   OCE Supplied Function/Data Structure   Uses the tag value nature of the request and subscription objects to build a view that represents a specific view updated by all subsequent changes until a designated time.

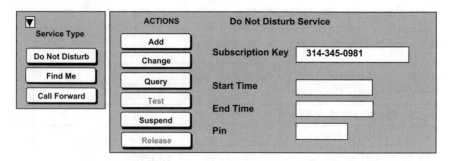

FIGURE 7.8   NetExpert graphics forms usage.

Apply Engineering Data   (Service Activation, Service Management)
OCE Supplied Function/Data Structure   Table lookup functions to apply network or service criteria to request processing; that is, lookup appropriate network elements to implement a particular service for a specified telephone number or to check feature interaction table for service compatibility.

These lookup processes can be applied to validation, network resource, authorization, feature interaction, and domain verification.

Supplementary Input   (Service Activation, Service Management, Platform Management)
Base Functionality   The ability to quickly produce tailored forms that allow input of specified data directly to the SMS platform.

These routines can be used as an auxiliary data input function, for supplementary data, or for specific control functions. (See Figure 7.8.)

Operator Events   (Platform Management)
Base Functionality   Specific events can be created that can only be generated from the operator workstation. These events can perform backup, system or fault recovery processes, query data, or perform any function definable within the ruleset.

Alerts can be established that prompt the SMS operator to perform specific function as the need requires. This combination allows the developers to automate and document many of their procedures within the tool itself.

Operator Menu   (Platform Management)
Base Functionality   This is the base NetExpert Menu that allows a variety of function to the operator, notably:

- Backup database
- Reinitialize rules (new software routines)
- Authorization and authentication

SMS Assurance   (Platform Management)
Base Functionality   NetExpert can use its network management facilities to monitor both the hardware and processes that are critical to the SMS operation. Alerts and recovery routines can be implemented to ensure the availability of the SMS platform.

Event Instrumentation   (Service Management, Platform Management)

Base Functionality   NetExpert can apply threshold criteria to any event. This allows service requests, errors, and queues to be monitored for user-defined load criteria.

Archive Daily Data Process   (Service Management, Platform Management)

OCE Supplied Function/Data Structure   Daily process that moves user-selected data classes to an SQL database. This data can be maintained for ongoing reporting of service management data without directly affecting the SMS performance.

Performance Metrics   (Service Management, Platform Management)

OCE Supplied Function/Data Structure   Using NetExpert advanced graphics abilities, thresholds established with the Event Instrumentation function can be displayed at the operator workstation or at a remote site.

7.5   SUMMARY

The NetExpert framework offers two unique abilities to its users:

1. Interfacing to almost any system
2. Rapid deployment of customer requests into services

Service management based on a flexible framework allows the full implementation of customer requirements and not only the best fit adoption of a traditionally developed system with little opportunity for customization and long enhancement times. The ability to develop unique, value-added solutions to service management requirements is part of a company's competitive advantage.

The confidence of an organization in its service management solutions to handle changes rapidly and with a high completion ratio could be the differential in a competitive environment. A tool that allows an organization to integrate its existing systems into a living, service management solution that will grow as standards and market demands evolve. The NetExpert framework offers a simple, rule-based interface that allows an organization's subject matter experts to develop and modify rules directly.

The combination of the benefits listed coupled with the ability to implement functionality incrementally, with confidence that long-term planning objectives are not being risked, allows users to apply scarce financial and human resources in the most cost effective manner.

8 The Integrated Service Activation Controller (iSAC)

CONTENTS

FIGURES

This chapter addresses the service activation segment of service management in greater detail. OSI provides an enterprise level component package of integrated rulesets and extended operators, called iSAC, that is the basis which can be modified to meet service provider needs.

As convergence in the telecommunications industry becomes more evident, so does the need for automated service activation. Service providers in the new competitive age will be offering both mass market residential and business services. These services will be increasingly robust and complex as they evolve.

8.1   WORKFLOW MANAGEMENT

Key to the success of this process is the ability to integrate the critical business functions such as order entry, trouble management, network engineering and workforce with activation at the network and element layers. In Figure 8.1 the activation flow occurs from the left, where a Customer Service Representative performs the customer negotiation process, to right where service is activated on the network. Automated activation both simplifies and reduces the cost of activation. To complete a service request, various manual processes and computerized interactions with

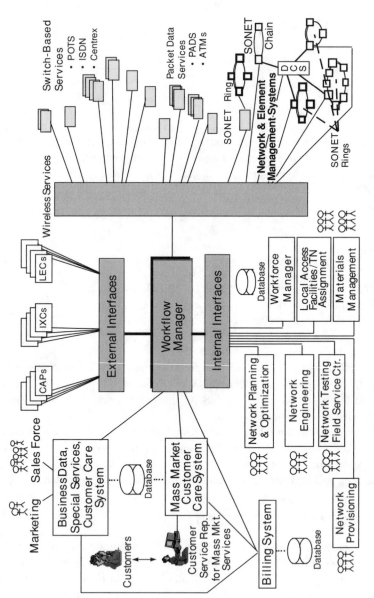

FIGURE 8.1 Service activation and management — an overview.

internal and external systems must occur in a well-defined sequence. This is Workflow Management and its coupling with automated flow-through activation is made possible through:

- an increase in the intelligence of the network elements
- the introduction of flexible and extensible network management software solutions

Central to an integrated business solution is the introduction of a workflow management capability that will utilize business rules to manage the "end-to-end" flow of events associated with a specific service offering.

The requirements of a workflow-based service activation solution include:

- an intuitive and flexible suite of tools for the new services and changes to existing services
- a single source of data and order management for all lines of business.
- the increased automation of routine and unimportant tasks allowing resources to concentrate more on a customer's complex needs
- the ability to integrate internal legacy systems and support a migration to modern systems
- activation capabilities go directly to the devices or their management systems
- the support for interacting with external organizations

An automated workflow/service activation solution will provide the following benefits to a service provider (Figure 8.2):

- a single source of data and order management for all lines of business
- the reduced need for additional staffing by automation
- the guarantee of a higher quality of service to customers
- a mechanism for growth and change
- a more effective, customer-focused use of human and network resources
- an overall reduction in the cost for service activation

Building on the intelligent Service Activation Controller (iSAC) base, OSI uses its NetExpert framework to provide a workflow capability. NetExpert is suited for workflow because the generic gateway provides a robust mechanism for interfacing with internal and external systems. It is expected that a significant amount of change in the external interface areas will occur as convergence is realized. The Workflow/Service Activation engine will evolve as new standards are implemented.

The IDEAS engine provides the analysis capabilities necessary to execute and manage process flows.

The graphical rules creation environment is ideal for representing business processes and the individual activities that comprise them.

In Figure 8.2, the grayed area depicts the OSI focus. Note that existing OSI solutions (transportMASTER, switchMASTER, mobileMASTER, and packetMAS-

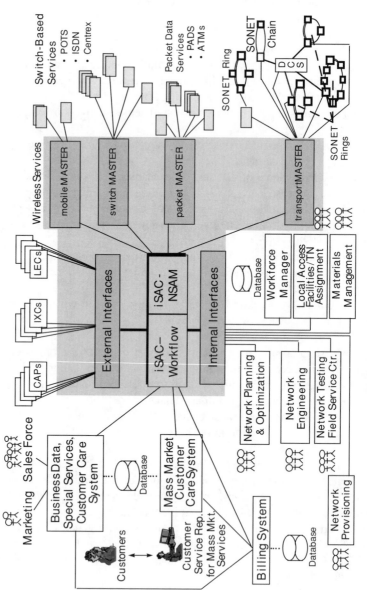

FIGURE 8.2   OSI workflow solution.

TER) are leveraged. There are two components of iSAC that support end-to-end service provisioning.

iSAC (Workflow Manager)   The Workflow Manager provides the ability to define the processes necessary to provision services and, when requests are received from Customer Care, a process definition is assigned to manage the activities required to complete the customer request.

iSAC (Network Service Activation Manager (iSAC–NSAM)   The Network Service Activation Manager coordinates and manages automated requests to the network. At the network and element layers, current OSI capabilities are utilized to enact the activation primitives.

8.2   iSAC — NETWORK SERVICE ACTIVATION MANAGER (iSAC–NSAM)

The introduction of iSAC–NSAM is the essential first stage to the final delivery of a workflow solution. It is a module that will increase reuse and decrease the need for custom development through:

- leveraging existing provisioning and activation rulesets with minimal customization
- providing a well-defined interface (interface B) to incorporate new Element Management Layer rulesets that support the integration of upstream systems and introduce consistency across all telecommunication domains
- providing a well-defined interface (interface A) to upstream systems such as Customer Care, Order Entry, and Workflow using either a general order format that will require minimal customization (0% in most cases) with the upstream systems (if systems do not enforce a proprietary format)
- proprietary format: Rule developers will use NetExpert ID/Parse rules to decompose the service order and invoke iSAC object and events over a fully-documented interface

The iSAC environment consists of a palette of objects, events, and graphics that perform the following:

Pending   Service orders can be "pended" indefinitely for later processing. Pending allows for the advanced scheduling of requests. Pended orders are persistent within iSAC until complete. Pends can be set up as leading or lagging for the prescribed interval.

Critical date management   Service orders that are in the active state (INPROGRESS) are managed within a work queue. A NetExpert PendEvent will fire, in the event of a risk for service orders and their components, to ensure that service levels are met. Service Orders, that have not yet completed and have exceeded the critical date specified, will alarm. These alarms are displayed in the alert display. If notification is required by an upstream system then this functionality can be rapidly introduced through Interface A.

Intraservice order dependency resolution   Component object dependencies support preconditions, postconditions, parallel, and serial event execution. Dependencies are defined for rollback as well.

Manual and automatic rollback policies are supported. If the policy is manual, iSAC does not send CANCEL requests to the Element Management System (EMS). If the policy is automatic, CANCEL requests are sent to the device IDs of all INPROGRESS components in the order.

Interservice order reference   Supported as a related service order, iSAC can support through custom events, requests to update existing orders. A future serviceManager palette will support different order updates, such as scheduled date, priority, or component change.

Service order state management   Service order state changes are automatically provided in real-time to the ISAC_OrderMgr for propagation to upstream systems.

Time-stamping   For auditing purposes, this data can be used to determine inefficiencies in the service activation process. Eliminating the bottlenecks supports ongoing process improvements. Also, where external interfaces persist, the performance operating efficiencies of partners and competitors can be monitored and reported.

The iSAC domain also includes the following functions.

Service order decomposition   Custom order formats can be supported over the A interface using Core iSAC. This interface is detailed in Figure 8.3. If the upstream system adheres to the specified order format and event interfaces, no customization is required.

Registered domain manager specifications   An interface class, ISAC_PeerMgr, has been provided to support the interface to local (using the same NetExpert system) or remote (using peer NetExpert systems) systems. As part of this interface, EMSs register the devices they can provision and their support services. This registration process allows iSAC to use the Network ID of components to dynamically route them to the appropriate EMS.

8.3   EXTERNAL SYSTEM INTERFACES — FLOW CONTROL BETWEEN iSAC AND THE DOMAIN MANAGERS

Figure 8.3 shows the relationship between iSAC and the serviceMANAGER. Functions of a serviceMANAGER were detailed in Chapter 7.

In this figure, the iSAC scope is all the contents of the dotted box.

A — Generic Gateway   This is the entry point for service orders into NetExpert. The data must arrive in ACSII format.

B — Service Order Decomposition   Decomposes the input stream into a service order and components. A generic service order format is provided in the body of the specification and the management of this format is defined in the iSAC_OrderMgr class. (Customization: Using iSAC_OrderMgr (0%), otherwise it is implementation dependent (20–80%))

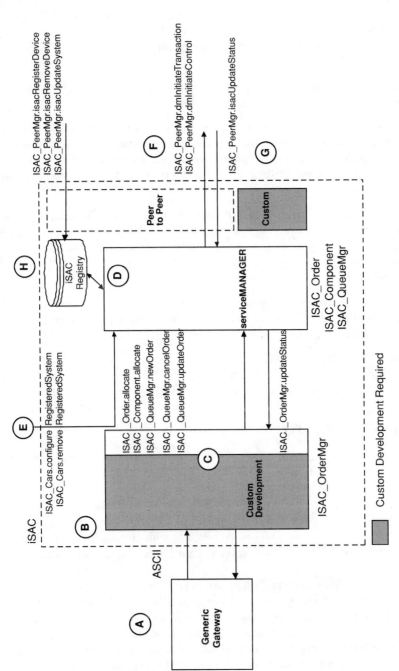

FIGURE 8.3 Overall iSAC architecture.

C — Input specification to iSAC   Orders and Components are allocated and populated with the decomposed data. Once populated, generation of service order events will place the service order in the control of the serviceMANAGER. Service order updates are sent out of the serviceMANAGER. Customization is minimal and well scoped: 0% (if OSI standard order format using iSAC_OrderMgr is adhered to) to 20–80% (for custom interfaces).

D — serviceMANAGER   Manages the service order through its lifecycle (pending to complete, error, or cancelled. (Customization: Versioned)

E — CARS Interface   Actions that must be invoked manually yet still require automation are invoked through the CARS graphic interface. For example, adding a registered system (a known NetExpert system accessible via peer-to-peer) must be done manually, but the CARS event automates the bulk of the process. (Customization: Versioned)

F — Interface to NetExpert-based Domain Managers   This interface is defined by mandatory component attributes that identify the peer system and action. Statuses are returned from the peer system upon completion, error, or acknowledgment.

Transaction routing is performed through the use of the serviceMANAGER registry. (Customization: Versioned)

G — Custom Interfaces   If the network element or Element Management System is not managed by an OSI solution, then custom development is required to support the iSAC interface. (Customization: Implementation dependent [20–80%])

H — iSAC Registry   Network elements that can be provisioned through iSAC must register with iSAC Registry. Registration will facilitate component routing at interface F. (Customization: Versioned)

8.4   iSAC — WORKFLOW

The iSAC integrated solution consists of three main components which follow the standard naming nomenclature of the Workflow Management Forum. They are:

1. *iSAC/Workflow Engine (iSAC/WE)* is a rule-based application using the NetExpert Framework. The engine utilizes workflow templates created by the iSAC/PDT and populates the template with customer order data collected via the iSAC/CA to create an internal order and its component work tasks.
2. *iSAC/Workflow Process Definition Tool (iSAC/PDT)* consists of two components. The iSAC graphical user interface (iSAC/GUI) provides a graphical representation of expert system rules that define work tasks comprising a workflow and allows users the ability to construct new workflows or modify existing workflows from a predefined set of work tasks. The iSAC Task Definition Tool (iSAC/TDT) allows selected users to define new work task types which are added to the palette of work tasks residing in the iSAC/GUI.

3. *iSAC/Client Application (iSAC/CA)* is a series of windows applications that provide the user interface for the iSAC. Users are able to view and enter data into internal orders and their component work tasks.

8.5 SUMMARY

The service provisioning process is the heart of operations at telecommunications service providers. Its speed, flexibility, and reliability determine success or failure. Due to high complexity and volume, manual provisioning must be replaced with solutions based on intelligent, expert systems. This chapter has shown how the NetExpert framework and its pre-packaged domain management application components can help to deploy a highly effective service provisioning system.

9 switchMASTER

CONTENTS

FIGURES

switchMASTER is a domain level integration ruleset to manage switches and legacy service-support infrastructures.

switchMASTER provides consolidated fault, performance, configuration, accounting, and security management for heterogeneous, narrowband networks. Its standards-based, object-oriented platform minimizes development and maintenance efforts while maximizing interoperability and extensibility. (See Figure 9.1.)

Flexible interfaces, prepackaged network element interfaces, and a distributed architecture allow NetExpert-based switchMASTER applications to quickly and cost-effectively accommodate an organization's current and future needs.

9.1   BENEFITS OF USING switchMASTER

switchMASTER has the following attributes:

- The ability to integrate proprietary or standards-based interfaces with network and operational systems
- Network management capabilities to maximize automating problem resolution and minimize support requirements
- Can implement custom functions to facilitate future services such as AIN, ISDN, xDSL, and Internet access
- Distributes functionality in order to maximize network and computer resources
- Minimizes operator training through use of standard interfaces and intuitive GUI displays

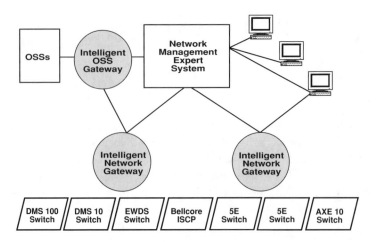

FIGURE 9.1   Generic structure of switchMASTER.

Its most recognizable advantages are:

- *Profitability*   Deployment of a NetExpert-based solution will enable the customer to minimize capital expenditures, administration, and training costs by providing a single management platform that uses existing hardware within an organization. NetExpert runs on a variety of UNIX platforms including SUN, HP, and IBM. Database support across these platforms includes that offered by Oracle, Sybase and Informix.
- *Internal process improvements*   OSI's application packages are designed with the flexibility to conform to customer's current and future process flows and business practices. The power of the NetExpert framework allows for dynamic business change of not only the network, but the organization which must manage it.
- *Increased Service Responsiveness*   The rapid provisioning capabilities of NetExpert-based solutions allow faster revenue recognition and improved customer service.
- *Effective Use Of Existing Staff*   The automation of repetitive tasks enables more efficient use of personnel resources and decreases the chances of human error.
- *Rapid Deployment of New Services*   The flexibility and ease-of-use of the NetExpert tool facilitates the rapid introduction of new products and services to customers ahead of the competition. Emerging technologies can be readily supported.
- *Service Assurance*   NetExpert-based solutions provide proactive management of the customer's network. Correlation of faults and automated testing results in improved mean-time-to-repair. The customer could provide service-level guarantees to their own, internal or external, customers.

9.2   INTELLIGENT GATEWAY IN THE ROLE
OF ELEMENT MANAGERS

Switched Services   The role of the access switch is changing dramatically. The addition of ISDN, AIN, xDSL, and the general demand for rapid delivery of customizable services is redefining telecommunications. The competitive edge will belong to the organization that can cost-effectively transition to the advanced, flexible infrastructure necessary to meet the market's changing service requirements.

Flexible Integration   switchMASTER simplifies the formidable task of migrating from today's network and operational infrastructures to the competitive new model needed to get services to market first. switchMASTER helps customers modernize existing operational infrastructures while positioning for rapid service introduction.

Regardless of current network components or legacy systems, switchMASTER provides a full suite of integrated network management functions that are customozed to meet each organization's needs. No more cumbersome systems, crowded with functions that could not be trimmed from an inflexible solution.

switchMASTER combines high-level, rule-based integration tools with distributable building blocks that provide speed to market and flexibility.

Intelligent gateways provide today's switched networks with many element management capabilities that can manage NEs and OSSs:

- *Data Normalization* allows the expert systems to remain independent of the proprietary nature of specific interfaces. The data from these interfaces is standardized into object-oriented instances.
- *Port Management* allows the specific management of the intricacies of a particular interface to control flow, rollback, error checking, and reconfiguration to maximize use.
- *Local Processing* provides local filtering, correlation, decomposition, and analysis of data between the upstream systems and the downstream managed elements.
- *Local Access* gives local staff manual access to the managed elements and local data.
- *Local Data Repository* provides a store of data for local access, translation verification, and data normalization.
- *Data Distribution* to any location is provided, allowing data to be routed to the expert server, other intelligent gateways, remote data repositories, or directly to OSSs.
- *Network Conservation* is achieved by performing all appropriate filtering and analysis at the local level, greatly reducing network data flow.

Figure 9.2 shows the structure of the intelligent gateways. They can function as front-ends, or concentrators, for many different types of switches.

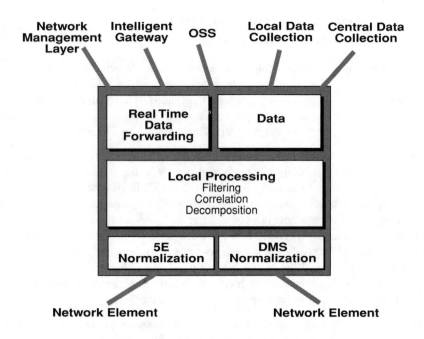

FIGURE 9.2   Structure of intelligent gateways.

9.3   NetExpert IN THE ROLE OF EXPERT SERVERS

The work of the intelligent gateways is coordinated by the expert server or servers.

Figure 9.3 shows the internal structure and the interactions with intelligent gateways and operator workstations.

Expert Servers provide comprehensive network management functions, including managing the data and its interaction with the Intelligent Gateways.

Functional Distribution is provided by the portability of the rule-based applications. Carriers can customize functional distribution along client processing guidelines, organizational requirements, or other specific procedures.

Global Correlation is achieved by managing information collected from the Intelligent Gateways and determining root cause on a network basis.

Global Data Repository provides network information to the entire organization.

Global Access is provided via the client operator workstations.

Global Problem Resolution is achieved by using powerful root-cause analysis capabilities, integrated management processes, and interfaces. The easy-to-use nature of 4GL editors allows expert servers to capture subject matter expertise, eliminating manual intervention and speeding problem resolution.

9.4   switchMASTER FEATURES AND FUNCTIONS

switchMASTER is built on and takes advantage of the benefits of the NetExpert framework. Like NetExpert, switchMASTER has open interfaces and is standards-

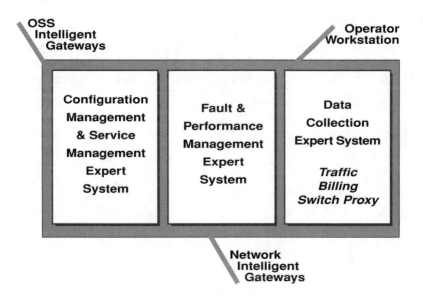

FIGURE 9.3 Internal structure and interactions of expert servers.

based and object-oriented. It easily integrates with other ruleset packages like transportMASTER and mobileMASTER.

switchMASTER's principal functions include:

Configuration Management — Switch Configuration managment includes:

- The *Shadow Database* is a snapshot of the switch data required for configuration. Engineering and modeling function can be performed without affecting switch operations or provisioning. The shadow database provides guarantees that requested resources are available when provisioned because the information is synchronized with the switch.
- *Automatic Service Provisioning/Transaction Management* Complex switch transactions are managed throughout the provisioning process to the network element.
- *Rollback* is provided at the transaction level. Failed transactions are returned to the initial state.
- *User interfaces* are provided for engineering staff. These multiplatform distributed applications simplify the provisioning process for engineers and service personnel.

Fault and Test Management — Switch Fault and test management include:

- *Cross-Vendor Fault* — by a single interaction
- *Fault Localization*
- *Root Cause Analysis*
- *Physical and Logical Monitoring* The hardware and software inplementations are monitored and correlated.

9.5 SUMMARY

switchMASTER represents the example of an application component. It integrates with with other application components, such as transportMASTER, mobileMAS-TER, and iSAC. Its rulesets can be easily changed to meet special customization needs of customers.

10 Traffic Management with trafficMASTER

CONTENTS

FIGURES

Traffic management has emerged as a strategic tool to keep pace with increasing competition between communications providers. In today's environment, it is essential to design a network that maximizes resources. This chapter provides a discussion of issues related to traffic management and an introduction to a traffic management system based on OSI's NetExpert framework.

10.1 TRAFFIC MANAGEMENT

In today's telephony environment, the only constant is change. In mature markets, competition is causing market fragmentation and spawning new kinds of telephony services and service providers. In emerging markets, the demand for the installation of new lines is causing growth at unprecedented rates. In all markets, Grade Of Service (GOS) levels must be maintained and even increased using the same or fewer resources.

More and more, network management is seen as a strategic tool to meet these challenges. One particularly important area of network management is traffic management of switched telephone services.

This chapter addresses traffic management issues in three main areas:

1. Traffic engineering and planning
2. Real-time traffic management
3. Traffic analysis

It describes the motivation for traffic management, the requirements for a traffic management system, and typical situations where traffic management is a benefit. Finally, an implementation of a traffic management system is presented.

From a business perspective, the benefits of traffic management include:

Increased call completions   The large volume of calls processed by most networks means even small increases in completions can have a significant impact on revenue.

Decreased capital cost expenditures   Without proper management of network facilities and resources, networks tend to be over-engineered in some areas and under-engineered in others. Traffic management ensures that capacity is allocated correctly to each area of the network.

Increased GOS levels   Even when a network is engineered correctly, sudden increases in network traffic and inevitable equipment failures can cause sharp decreases in GOS levels. Dynamic traffic management can ensure that network diversity is fully used and that vital network services are protected.

10.2   THE TRAFFIC MANAGEMENT PROCESS

Network Structure   The majority of the world's telephone networks are based on Fixed Hierarchical Routing (FHR). Switches are connected to each other in a routing hierarchy. Local switches (end-office switches) have direct connections to subscriber equipment such as telephone and PBX. End offices typically have direct connections to other end offices in the same geographical area. Tandem switches have connections to other switches, not to customer premises.

Switching is done hierarchically to provide alternate routing for traffic overloads and facility failures. Each switch is given a class designation depending on its location in the hierarchy.

In the case of failure or overload, lower class switches can route calls to higher class switches. Figure 10.1 represents a portion of a typical public voice network.

A typical routing hierarchy is shown in more detail in Figure 10.2. Traffic between two local switches is normally routed though a High Usage (HU) trunk group. High usage trunk groups are engineered so that some traffic during the busy hours will overflow to an Intermediate High Usage (IH) trunk group. Traffic from

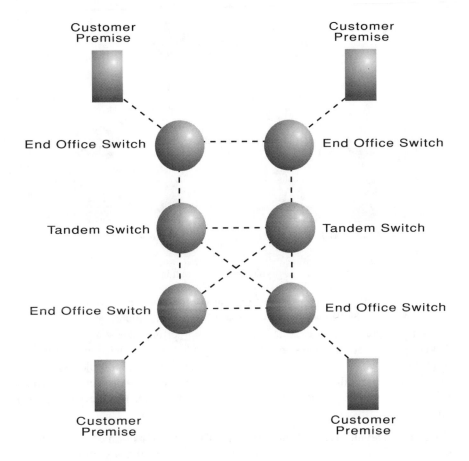

FIGURE 10.1 Simple view of a public voice network structure.

IH trunk groups overflow to Alternate Final (AF) trunk groups. Overflow traffic from AF trunk groups goes to a "No Circuits Available" announcement.

In practice, networks, switches, and routing appear in many configurations. For example, some switches perform both tandem and end-office functions and some networks do not have any end offices or tandem switches at all.

Traffic Engineering and Planning A telephone network is a dynamic entity. Traffic volumes, calling patterns, and the introduction of new services all create the need to constantly "reengineer" the network.

The ultimate goal of traffic engineering and planning is to cost effectively provide customers with a target GOS. Simply put, GOS is expressed as the percentage of unsuccessful call attempts because of insufficient network capacity. Many service providers engineer their network for a certain GOS level measured during the "busy hour." The busy hour is that hour of day in which there is the greatest amount of

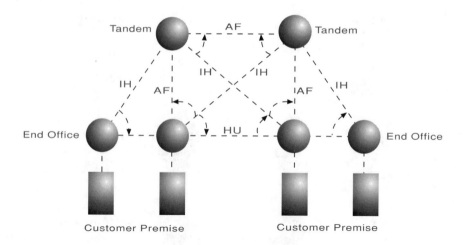

FIGURE 10.2   Typical alternate routing hierarchy.

traffic. A GOS of two percent means that during the busy hour, no more than two call attempts out of 100 will be unsuccessful because all circuits are occupied.

Network engineers analyze facility usage and performance statistics to ensure that current network resources are appropriately allocated and functioning correctly. Network planners use the same data to allocate new resources for projected growth. In order to engineer the network, several factors are taken into consideration, including previous growth, planned introduction of new services, and projected customer growth.

Goals of Real-Time Traffic Management   The goals of real-time traffic management are to ensure maximum use of all equipment and facilities at all times and to complete as many messages (calls) as possible.

Network events, such as mass callings, equipment failures, and focused overloads, can seriously affect customer service levels and operating revenues. In situations of extreme network overload, real-time traffic management maximizes network resource usage and protects vital network services. In cases of localized network overload, traffic can be routed to those parts of the network having resources available.

Some examples of network problems include:

- General network overload, when the entire network is saturated with calls. For instance, this can happen on a national holiday.
- Focused overload, when an unusually high number of calls are directed at a single telephone number or location. Examples of this include radio show call-ins and ticket sales.
- Equipment failures, including minor failure of equipment, such as a line card or multiplexer, or events as serious as one or more switches failing to operate.

In order to maximize call completions and preserve network resources, the following principles apply when considering network management actions:

- Keep all circuits filled with messages
- Use all available circuits
- Give priority to single-link connections when all available trunks are exhausted
- Inhibit switching congestion

Network Management Controls   The most effective way to respond to network events is through the use of network management (switch) controls. There are two major categories:

1. Protective controls
2. Expansive controls

Protective controls limit the amount of traffic that can come into the network. Traffic offered to a switch may need to be limited if that switch is receiving traffic beyond its capacity. When a network is overloaded, not all traffic can complete, so preference should be given to traffic with the highest chance of completing.

Expansive controls reroute traffic from those areas of the network experiencing overload to areas with capacity available. Network overload may occur because of above-average calling levels or network equipment failures.

The most common switching controls are:

Cancel To/From   These controls prevent traffic from alternate routing to/from a trunk group.

Skip   A skip control prevents traffic from accessing a trunk group.

Reroute   The reroute control diverts traffic from one area of the network to another.

Code Block/Call Gap   These controls are applied to a telephone number or code (like an area code). Traffic destined for the telephone number is prevented from completing.

Directional Reservation   A directional reservation reserves trunk groups for incoming or outgoing traffic.

Most switches allow a certain level (or percentage) of traffic to be affected. For example, a code block that blocks only 50% of the calls destined for an area code could be applied.

Using Reroutes   By using reroutes, a network manager can take advantage of time-zone and busy-hour differences. On a peak-traffic day, all circuits are often fully occupied and network managers can take advantage of calling patterns that are time sensitive. For example, 8 p.m. on the west coast is 11 p.m. on the east coast. Typically, calling drops off after 10:30 p.m. Therefore, calls destined for the central region can

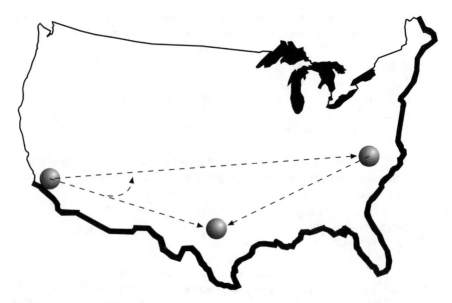

FIGURE 10.3   Rerouting example. The arrow indicates traffic that is rerouted to switches in the eastern region and then advances to switches in the central region.

be rerouted through the eastern region to take advantage of spare capacity. This is shown in Figure 10.3.

Focused Overloads   Here is an example of Focused Overload. It is Monday morning in a major city and the phone lines have just opened up to ticket sales for the concert of a popular rock group. Suddenly, the downtown network is flooded with call attempts, most of which will reach a busy signal. Within minutes, people experience delays in getting a dial-tone. In this situation, there are two major concerns:

1. To ensure that the number of attempts in the network is kept within the switching capacity. (If too many attempts are allowed in the network, switches will spend an inordinate amount of time processing call attempts, with queues becoming longer and longer. If left unchecked, switches may stop processing altogether and restart themselves, preventing any calls from completing.)
2. To give priority to traffic that has the greatest chance of completing. (Assume that 80 percent of all calls to the ticket office will receive a busy signal. If these call attempts are limited by the use of switching controls, other call attempts will have a better chance of completing.)

The focused overload situation is depicted in Figure 10.4. The response to a focused overload is to block calls that are destined for the ticket office as close to

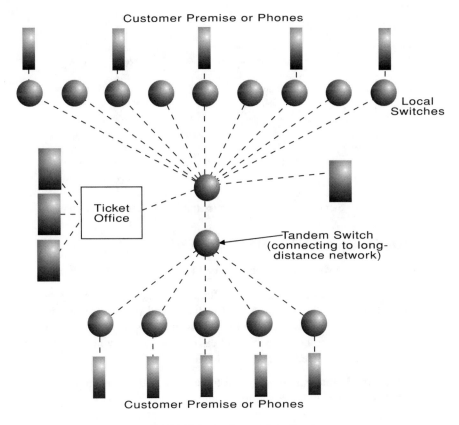

FIGURE 10.4 Focused overload.

the source as possible. In this case, Code Block or Call Gap controls should be put in place to block a percentage of calls destined for the ticket office. The controls should be put into every switch that is processing calls attempting to reach the ticket office. This includes local switches and remote switches accessing the local network via tandems.

Traffic analysis is a broad category concerned with the analysis of traffic and network problems. These may include intermittent traffic problems, equipment-related problems, or volatile traffic patterns. This function is often performed in conjunction with other groups, such as facility or transmission management.

Traffic analysis requires diverse network information that allows network managers to track the source of network problems. Often, traffic data may reveal that a problem exists. However, transmission alarms or facility information may be needed to determine the root cause of the problem.

The network analysis group must be able to generate ad hoc reports and analyze network information interactively to determine patterns and recognize unusual network events.

Network Element	Parameter	Detail Level
Switch	Call Completions	Busy Hour, Daily,
Remote Switching Unit	Call Failures by Reason	Weekly, Monthly,
	Projected Call Increase	Yearly
Trunk Group	Usage	Busy Hour, Daily,
	Overflows	Weekly, Monthly,
	Attempts	Yearly
	Answer to Seizure Ratio	
Network	Call Completions	Busy Hour, Daily,
	Call Failures by Reason	Weekly, Monthly,
		Yearly

FIGURE 10.5   Sample data requirements for traffic engineering and planning.

10.3   TOOLS FOR TRAFFIC MANAGEMENT

This section discusses some of the tools needed to manage network traffic.

Traffic Engineering and Planning   The traffic engineering and planning functions require tools that can collect traffic-related statistics and produce reports in both detailed and summary formats. Because the data collected is so varied, as shown in Figure 10.5, tools must be both flexible and easy to use. Some typical traffic management reports are discussed later in this chapter.

Real-Time Traffic Management   In order to diagnose and respond to traffic problems, the network manager needs the following:

- Access to real-time traffic statistics
- Easy to use applications for issuing controls to any switches in the network
- Applications that enable predefining and applying groups of controls, called control preplans
- *Traffic Statistics*   Traffic managers need access to real-time traffic information in order to diagnose and manage network problems. Trunk group statistics are obtained from switches in five- or fifteen-minute intervals.

Examples of useful statistics include:

Attempts	*Percent occupancy*
Seizures	*Answer to Seizure Ratio (ASR)*
Answers	*Out-of-service circuits*
Overflows	*Attempts/circuit/hour*
Usage	*Average holding time*

Because of the number of trunk groups in most networks, important traffic statistics should be thresholded and used to drive alarms that indicate traffic problems.

Switching Controls   Switching controls are the network manager's most effective method for responding to network events. In order to be effective, a network management system should provide methods for rapid implementation of network controls. It also should insulate the user from differences between switch models and the effects of software load changes. Controls should be tracked in real time so the network manager always knows what controls are in effect.

Tools for Traffic Analysis   The traffic analysis function requires a flexible set of tools that allow a user to detect and analyze network traffic problems. Questions faced include:

> Why are we losing calls on Trunk Group A in the middle of the night?
> Why is traffic decreasing in this part of the network?
> What was the cause of a severe drop in GOS levels in the western region last
>   night?
> Did we receive the amount of traffic from Carrier X that was promised?

In order to determine the cause of problems in the network, a user needs access to detailed and summary traffic data, switch configuration data, and other network alarm information. The ability to interactively generate ad hoc reports and queries is essential for determining the source of problems in the network.

Calls may be lost because of sudden increases in traffic, faulty trunk groups (minor equipment failures), or incorrect configuration within the switch itself. As networks increase in complexity, access to information becomes more and more important for analysis of network problems.

Standards for Traffic Management   The main body of standards that apply to switched telephone networks are those from the ITU-CCITT1. While overall standards — such as the Telecommunications Management Network (TMN) standards — for network management exist, standards for traffic management still are being defined. Even when traffic management standards are fully defined, it may be a long time before switch manufacturers comply. In any case, a service provider should protect current investment in equipment that has proprietary protocols and interfaces while preparing for a standards-based future.

Constantly Changing Networks   Networks are constantly changing. New equipment is being brought in, new switches are being introduced, current switches are being upgraded, the structure of the network is evolving, and new services have to be managed.

In this environment, it is essential to have a network management system that facilitates the rapid deployment of new network management features. It also should remain flexible enough to accommodate changes in new input and output formats without requiring changes to software programs.

10.4 TRAFFIC MANAGEMENT IN PRACTICE

OSI provides an integrated traffic management solution combining its NetExpert framework with trafficMASTER, an application component that addresses the three main areas of traffic management:

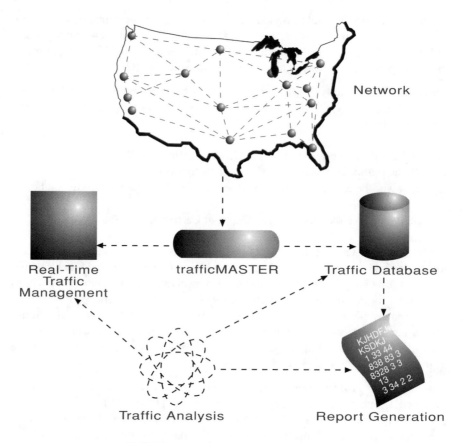

FIGURE 10.6   trafficMASTER overview.

1. Traffic engineering and planning
2. Real-time traffic management
3. Traffic analysis

This solution is depicted in Figure 10.6.

International Telecommunications Union (ITU)   Consultative Committee on International Telegraphy and Telephony (CCITT). The CCITT was set up within the ITU's jurisdiction in order to adequately manage expansion in the rapidly growing evolving worldwide telecommuncations industry.

NetExpert Architecture   NetExpert supports centralized data collection for both real-time network management and population of the traffic management database. Data is collected by gateways that handle all communication between the switch and NetExpert.

From the gateways, data is forwarded to the traffic database for storage. It is simultaneously sent to IDEAS (Intelligent Dynamic Events Analysis Subsystem), NetExpert's rule-based expert system. IDEAS provides the following:

- Event correlation with suppression of sympathetic events
- Translation between information models to present a switch-independent view of traffic
- Filtering, thresholding, and concentration of traffic data and events
- Automated recognition and response to traffic events
- Standards Compliance

NetExpert complies with the architectural goals of the ITU-CCITT TMN specifications. TMN defines a distributed network management architecture that is needed to support the large amounts of data required for efficient traffic management. NetExpert supports standards-based and nonstandards-based protocols, such as Common Management Information Protocol (CMIP) and Simple Network Management Protocol (SNMP), as well as proprietary switch protocols. This allows NetExpert to communicate with current network elements while providing a solution that meets future requirements.

NetExpert's Management Information Base (MIB) conforms to the object model specifications set by the Network Management Forum.

Development Environment   The specification of the IDEAS and gateway functions are accomplished without programming. Defnining rules in this manner means that solutions can be put in place in a fraction of the time required by other technologies. Changes in switching controls, traffic-report formats, or switch messages are handled by rule changes that the user completely controls. This translates into:

- Lower implementation costs
- Faster implementation
- Less reliance on the network management system supplier for new functions and/or changes

Traffic Engineering and Planning Support   trafficMASTER includes a flexible traffic report generation application. This application allows the user to define reports and to execute them both interactively and in batch mode. While the underlying data is stored in a relational database, the user interacts with a point-and-click interface without having to write complex SQL queries.

Generated reports go beyond just data. They contain information valuable for engineering, marketing, provisioning, and network analysis groups.

Following are descriptions of sample traffic reports:

The **Historical Busy Hour Report** summarizes traffic data for a number of months on switching elements such as trunk groups or line units. (See Figure 10.7) The example that follows shows data for trunk groups. This report may be customized in these ways:

1. Use data within a start date and end date.
2. Exclude any chosen day, including holidays, odd days, and weekends.
3. Exclude days without enough data.

		Avg BH	Avg BH		Avg Hold	CCS			Ckts		
Month/	Study	Usage	Maint.	Avg BH	Time	Per	%	Overflow	in	Ckts
Year	Days	(CCS)	(CCS)	Attempts	(Seconds)	A/L	Cap	Attempts	Svc	Req.

Historical Busy Hour

Exchange: East 1   Type: Bouncing Busy Hour

Office Type: 5ESS   Report Period: From MMYY- TO MMYY

Trunk Group: toll5 — **Access Lines: 1200Target GOS B.03**

Month/Year	Study Days	Avg BH Usage (CCS)	Avg BH Maint. (CCS)	Avg BH Attempts	Avg Hold Time (Seconds)	CCS Per A/L	% Cap	Overflow Attempts	Ckts in Svc	Ckts Req.
Jan. 98	20	xx	xx	xx	xx	xx	xx	xx	xx	xx
Feb. 98	22	xx	xx	xx	xx	xx	xx	xx	xx	xx
March 98	18	xx	xx	xx	xx	xx	xx	xx	xx	xx
April 98	10	xx	xx	xx	xx	xx	xx	xx	xx	xx
May 98	22	xx	xx	xx	xx	xx	xx	xx	xx	xx
. . .										
Dec. 98	22	xx	xx	xx	xx	xx	xx	xx	xx	xx
Average Busy Seas.	xxx	xxx	xxx	xxx	xxx	xxx	xx	xxx	xxx	xxx

FIGURE 10.7   Historical busy hour report.

Report Generated On: May 1, 1998	Monthly Trunk Group Exception Report Switch: East1				Report interval: April 1998	
Trunk Group	Date	Hour	Circuits	Usage	%Ovfl	%Util
TrkGrp1	April 4	18:00	100	100	10	100
TrkGrp2	April 4	20:00	100	95	0	95
TrkGrp10	April 3	17:00	50	45	0	90
TrkGrp34	April 7	16:00	50	47	0	94

FIGURE 10.8   Monthly trunk group exception report.

4. Exclude low days of each week. (For example, exclude the two lowest days of the week.)
5. Use Bouncing or Time Consistent Busy Hour.
6. Select busy hour from certain hours of the day.
7. Use engineering tables.

The **Monthly Trunk Group Exception Report** includes statistics for those trunk groups that have exceeded a given threshold. (See Figure 10.8.) Thresholds are defined in the report-generation database and may be changed at any time. Threshold values are based on hourly statistics. If a traffic value for any hour during the month has exceeded the defined threshold, it is included in the report.

The threshold value appears in the last column of the report. Other trunk group information is included for comparative purposes.

Report Generated On: May 1, 1998	Weekly Quality of Service Report Switch: East1		Report Interval: April 1, 1998– April 28, 1998	
Week	April 1– April 7	April 8– April 14	April 15– April 21	April 22– April 28
No. of Attempts	1,234,456	1,398,294	1,345,091	1,401,782
Call Breakdown in Percent				
1) Call Connections	97.6	98.6	98.5	98.2
2) Ext. Tech. Irregularity	0.56	0.98	0.86	0.97
3) Unanswered Calls	10.2	11.1	10.7	10.8
4) Subscriber Busy	22.4	23.1	22.6	22.3
5) ATB DestinationBusy	0.56	0.79	0.81	0.62
6) ATB Home Exchange	0.21	0.34	0.23	0.27
7) Call to Non Working Number	0.02	0.04	0.03	0.04
8) Int. Tech Irregularity	1.07	0.98	0.80	0.81
9) Call Loss (2+5+8)	2.19	2.75	2.47	2.4

FIGURE 10.9   Quality of service report.

The **Weekly Quality of Service Report** shows calling statistics for a single switch broken down by call-completion categories. (See Figure 10.9.) At the bottom of the report is an indicator of the total percentage of calls lost due to insufficient capacity or technical problems. The percentages included in this total can be modified to include any combination of percentages in the report. This report also summarizes the originating and terminating traffic.

Real-Time Traffic Management Support   NetExpert provides graphical tools to support real-time network and traffic management. Traffic performance data thresholds are displayed graphically.

All switching controls are supported through a point-and-click interface that provides a common look and feel across different switch types. Controls are translated by trafficMASTER into switch-specific commands that insulate the user from having to learn different control protocols and switch interfaces. Predefined sets of controls, called preplans, also are supported to provide instant responses to network problems.

All responses to network events can be implemented manually by network managers or automatically through expert-system analysis of traffic and performance problems. In IDEAS, rules can be defined to recognize network problems and respond to them using switching controls. If desired, IDEAS can notify only the operator and suggest a course of action.

Traffic Management Applications   Examples of real-time applications are shown in Figure 10.10. In the background is a graphic display of elements in the network. Switches are represented by spheres, trunk groups by connecting lines. Each network object has associated dynamics that represent the status of the network element.

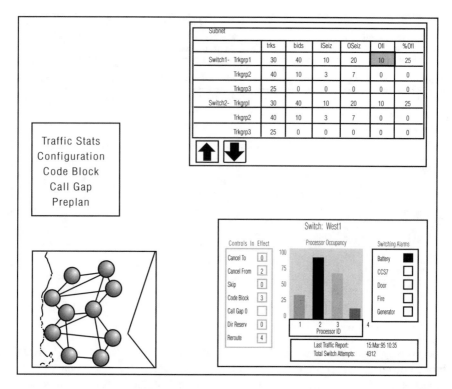

FIGURE 10.10   Traffic management display. At the top-right side of the display is a matrix containing traffic statistics for a user-defined set of trunk groups. This display updates automatically in 5-, 15-, or 30-minute increments, displaying the most recent traffic statistics. Cells change color to indicate that thresholds have been exceeded. At the bottom-right side of the display is a window containing switch status information updated in real time. At the bottom-left side of the display is a detailed graphical display for the area surrounding the selected switch. At middle left is a pull-down menu of choices to open a form or application.

Traffic Data Collection and Storage   trafficMASTER collects trunk group data from each switch in the network, then processes, thresholds, and displays it to the user in real time.

Supported trunk group parameters include:

- Attempts
- Seizures (OSeiz)
- Overflows (Ofl)
- % Overflows (% Ofl)
- Number of circuits
- Number of circuits out of service
- Usage
- Average holding time

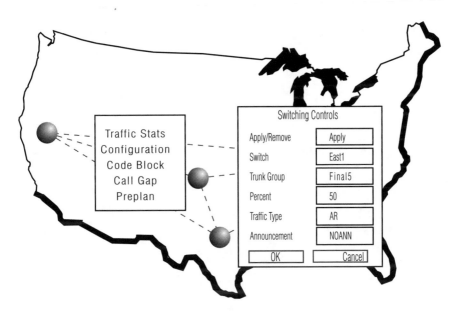

FIGURE 10.11   Issuing switching controls.

Applying Control with trafficMASTER   trafficMASTER supports the processing and tracking of controls through an intuitive, object-oriented graphical interface, as shown in Figure 10.11.

In issuing a control, when the user first selects an object (like a trunk group), a dialog box containing the set of parameters appropriate to the object appears. The user fills in the appropriate menu items and presses "OK." trafficMASTER translates the request into the control native to the switch and applies the control to the switch.

Regardless of switch type, the control interface is the same. This relieves the user from having to deal with different switch interfaces and protocols — an important feature when timing is critical.

10.5   SUMMARY

NetExpert provides strong support for the traffic analysis functions by providing integrated traffic reporting, real-time management, and fault management. Interactive report generation and ad hoc database queries are supported for single events, as is report generation for a number of traffic statistics. Report generation based on exceptions also is supported.

Traffic management can play a strategic role in ensuring an operating company's competitive role in the marketplace. Increased call completions and increased GOS can be achieved if the right information and applications are available. NetExpert's trafficMASTER provides the tools to achieve the goals of traffic management in a standards-based, easy-to-use environment.

11 Accounting Management with amaMANAGER

CONTENTS

FIGURES

Introduction Service usage is a principal component for the successful operation of a service supplier. There are a few powerful billing packages available.However, the raw data must be collected, compressed, and forwarded to the location where the billing packages are implemented and operating. In order to support multiple interfaces, standards are required. amaMANAGER provides the functions of interfacing multiple switching devices, collecting and consolidating data, and forwards them to the gateway required by the billing packages.

11.1 THE CONCEPT OF AUTOMATIC MESSAGE ACCOUNTING (AMA)

OSI's data collection products incorporate tools to rapidly address the integrated, multivendor environments typically associated with accounting management (management of information derived from the measurement of the use of network resources) applications. Centralized billing collection through DataArchiver allows real-time billing and quick capture of call record information for preprocessing, reformatting, and normalization of accounting management data. Users can extract database information via NetExpert's ReportMaker or use any standard SQL interface software. NetExpert's Data Message Handler (DMH) can be used for inter-carrier communication for billing and rating protocols. Now, with OSI's amaGATEWAY, an integral and efficient method of transmitting, processing, and managing a larger volume of Automatic Message Accounting (AMA) files becomes available.

New network services and comprehensive measurement strategies are creating a need for enhanced tools to manage this data. In addition to new data generating network elements, new applications are also being developed which require access to this data (for example, marketing, network and customer support). The value given to this data also varies widely, requiring the capability to treat different AMA data in a different manner. The amaGATEWAY provides an open and fully integrated architecture to support the special needs of multiple applications while retaining the high degree of quality, availability, and security expected for AMA data files. This provides the flexibility to meet future business needs on a timely basis and allows the user to create new business opportunities based on accounting management capabilities.

The increasing complexity of network services and the sophistication of customer needs are multiplying the degree of flexibility required of accounting management. The concept of "billing data" is changing as the network and operations systems infrastructures are enhanced to support increasingly complex products and services (data, broadband, xDSL, and AIN services) and new service platform technology (wireless access, SONET, and ATM). The amaGATEWAY provides function and process modularity in anticipation of new technological and service capabilities in order to respond to user needs and the increasingly distributed nature of the network.

11.2   GENERAL DESCRIPTION OF THE APPLICATION PACKAGE

The purpose of this implementation is to collect call detail record (CDR) files from various network elements and make them available for further downstream applications, such as billing. The files are polled from the network element via a Cisco or Dynatech router. The protocol between the switch and router is typically X.25 or BX.25, and the protocol between the router and amaGATEWAY is XOT (X.25 over TCP). XOT (a Cisco Systems specification defined in RFC 1613) is preferred because it allows networked Ethernet transport to the gateway. amaGATEWAY performs its tasks automatically, without human intervention, and has extensive administrator-controlled flexibility regarding how and when data is collected. The amaGATEWAY has the following high-level attributes:

- Collects CDR data from the network element
- Communicates with the element via a router using the XOT protocol
- Collects and stores all CDR data from the element onto gateway machine local hard disk accurately and reliably.
- Parses CDR data from its raw format into an ASCII format and stores this ASCII data on a local disk for remote access
- Provides a means for the billing system to attain access to ASCII format data

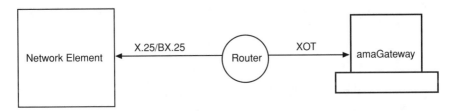

FIGURE 11.1   Architecture of the amaMANAGER.

The amaGATEWAY is a custom mediation device that will store CDR data on its local hard disk or disk array. (See Figure 11.1.) The typical system is a Sun Sparcstation running the Solaris 2.5 operating system. The data is stored in flat files in the native format of the network element and, if required, is parsed into ASCII-legible CDRs which can be read and understood easily by other processes or applications. The hardware platform of the amaGATEWAY is a stand-alone application running on a Sun workstation and runs separately from any other OSI/NetExpert applications.

Data obtained from the switch is stored on the Gateway as flat files and the billing system (or other application) will obtain the files via file transfer protocol (FTP). The application also can be configured to transmit the files automatically. The amaGATEWAY provides the central control for automatically collecting CDR data from the switch. To ensure integrity from the switch to the gateway, blocks of data are sequentially numbered at the switch and checkpointed as they teleprocess to the gateway. This sequential numbering scheme provides the ability to audit the data from the point of origination to the point of exit, thus ensuring end-to-end integrity. Polling operations are conducted on the basis of these data blocks and their sequence numbers.

11.3   FUNCTIONAL DESCRIPTION
OF THE APPLICATION PACKAGE

AMA data stored on the switch is classified as primary or secondary data. Primary is the raw data that has not been polled and acknowledged previously by the amaGATEWAY. Secondary data is information that has been transmitted, received, and acknowledged. Collecting raw data is the gateway's primary function. A schedule can be maintained which initiates a polling session for all available CDR blocks twice a day or whenever it is required by a customer.

Polled CDR blocks are stored internally on the amaGATEWAY. The blocks are stored in a flat-file structure and sized according to the estimated billing system transmittal frequency. Once the files have been polled (or automatically transmitted) and confirmed by the billing system, they are moved to a separate directory where they are made available to other applications.

The amaGATEWAY also provides monitoring of its disk storage. Once a files has been confirmed by the billing system, the name of the file is changed and it goes into the queue for automatic deletion after a configurable number of days. Only verified files are deleted. Alerts are presented when the file storage reaches 70% (minor), 80% (major), and 90% (critical) occupancy.

11.4   amaMANAGER

The functions of the amaGATEWAY are enhanced with the addition of DataArchiver to provide a complete management system for customer usage information. The amaMANAGER exists within the NetExpert framework which is a rule-based expert system. This means that changes in data formats, protocols, and applications can all be introduced rapidly, by the user, to meet evolving needs — something no other system offers. The amaMANAGER gives the user access to AMA data as a strategic source of information and enhances the billing process which adds value to customer products and services.

Within amaMANAGER, DataArchiver is used to store events and data flowing through the amaGATEWAY in relational databases. This allows the user to enhance the value of AMA data via specialized processing. Billing data is the most valuable information a company has about its customers. Once the data has been placed in the relational database, users can create reports using OSI supplied usage report definitions. These reports can be modified easily to suit specific requirements. Users can also define and generate their own reports as requirements change and mature. Reports created in amaMANAGER can provide real-time information on market penetration, financial performance, call patterns, fraud, service preferences, demographics, network efficiency and others.

11.5   DataArchiver

Network administrators often face a quandary between the need to create specialized reports and the limitations within proprietary data formats. The task of balancing these opposing forces can be bothersome at best and impossible at worst. OSI delivers a solution in with its NetExpert DataArchiver.

DataArchiver is a message-management system that meets diverse user requirements with a flexible interface. DataArchiver forwards information to relational tables, eliminating distortion from unusual formats, and allows high-volume collection and storage of event-specific information for later analysis. Data from NetExpert processes can be retrieved and normalized, simplifying access and eliminating dependence on legacy storage and reporting systems.

DataArchiver leverages NetExpert's ability to integrate multivendor, multiprotocol devices, storing the data from network elements into user-defined repositories. These may be existing relational databases used by other OSSs or user-defined databases designed to enhance data analysis through the use of third-party reporting tools. DataArchiver is the front-end message-management system for trafficMASTER, as well as for performance reporting.

Tailor Performance Management   DataArchiver gives service providers the ability to control computing and create detailed, historical analyses of performance events. Network administrators can pinpoint critical performance issues by collecting data on bit error rates, errored seconds, and other performance characteristics, based

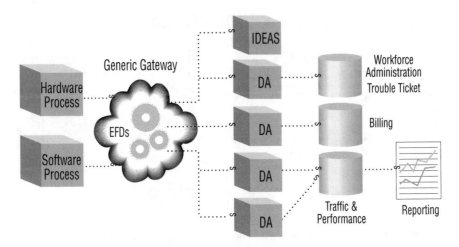

FIGURE 11.2   A sample DataArchiver application.

on attributes like circuit ID to identify weekly, monthly, and yearly performance values. Network elements can be polled for performance statistics, which can be browsed and sorted graphically or textually. Data from existing internal, remote, or external performance management systems can be received by DataArchiver and routed to log in, one or more, relational databases for use by reporting and analysis systems.

Extensible Billing Management   DataArchiver's flexible interface offers rapid development of many customized applications and numerous solutions for billing management. For example, billing data can be forwarded to multiple DataArchivers to guarantee that billing records are not lost because of host or database failure. For providers who need to maintain peak service levels, DataArchiver can maintain audit logs for historical analysis of service requests and provisioning actions. Incoming service-order information can automatically be distributed to activate service, modify customer billing records, and maintain audit trails. Additionally, DataArchiver can play a vital role in fraud detection by receiving call detail records from a switch and storing them for later processing by a fraud or rating system.

Versatile Traffic Management   DataArchiver offers a host of customizable options that broaden operator traffic-management abilities. Simple dialogs can request data from network elements, based on a polling schedule, to archive items such as traffic reports from network switches for volume statistics and trend analysis. Capacity requirements can be satisfied by correlating variables related to network overloads, equipment failures, and real-time traffic. Also, planning for future traffic may benefit by using the NetExpert trafficMASTER, a ruleset that addresses real-time traffic management and long-term traffic planning. A sample DataArchiver application is shown in Figure 11.2.

Dynamic Integration and Deployment   Based on NetExpert technology and driven by user-defined rules, DataArchiver meets management requirements with

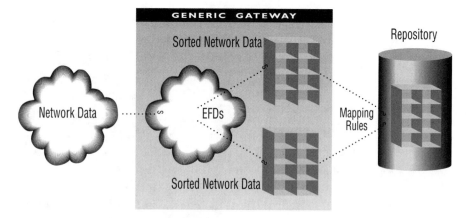

FIGURE 11.3   Raw data is transformed into useful, retrievable information through EFDs and user defined event mapping rules.

easy-to-use templates and user-defined data models. When specified data is identified and parsed by the Generic Gateway, Event Forwarding Discriminators (EFDs), which are NetExpert managed objects, filter and send the information to DataArchiver processes. Network data translates quickly and easily based on user-defined "event mapping" rules that determine how data should be mapped to tables or columns for storage and retrieval.

Distributed Environment   Network administrators always are concerned about bottlenecks in their systems. DataArchiver delivers additional benefit by allowing users to distribute processing, thus eliminating bottlenecks. DataArchiver processes can be colocated with Generic Gateways or distributed across the network, allowing flexible configuration of the DataArchiver application and optimal use of system resources. Administrators can partition DataArchiver activities between CPUs to optimize system performance and balance the processing load on units. DataArchiver is a fully extensible NetExpert component, designed for UNIX client-server distribution and plug-and-play management. Figure 11.3 shows how raw data is transformed into useful, retrievable information through event forwarding discriminators and user-defined event-mapping rules.

DataArchiver advantages are:

- Collect information from NetExpert processes and mediate between network data, NetExpert, and physical repositories.
- Prevent crucial data from being lost because of transient or permanent failures in the environment by logging indeterminate events.
- Multi-cast and/or partition events between IDEAS and DataArchiver as necessary, forwarding only events of interest to the destinations and clarifying performance-tuning requirements.
- Identify and forward events to multiple DataArchivers that update replicated databases, enabling fault tolerance by establishing dual data flows and confirming availability of data.

- Abolish time-consuming searches for information by writing rules with identifying discriminators that capture, filter, and store data in designated formats.
- Improve operator access and reporting abilities by developing customized relational data models to meet unique requirements.
- Achieve higher productivity by creating modules with seamless integration of network data sources to multiple operator, decision, and information support systems.
- Create audit trails to archive data for historical tracking, allowing users to take immediate corrective action and/or store the data.

11.6   SUMMARY

In order to support performance analysis, capacity planning, trending, and accounting, data is needed. The amaMANAGER extracts, transfers and processes data. Data can also be stored and maintained for further processing in DataArchiver. Customers may get reports generated by amaMANAGER or preprocessed data subject to further processing and analyzing by the customer.

Part III

Management
of Wireless Networks

The ability to communicate across geographic, economic, and corporate boundaries has never been more important than today. Privitization, deregulation, and advancing technology are fostering new competition in telecommunications.

Expanding opportunities and limited human resources in the wireless market are forcing providers to automate the way they manage and deliver cellular, digital, and personal communication services.

The concept of cellular radio networking goes as far back as 1947 when it first was invented by Bell Labs. Thirty-five years later, technology finally caught up with the cellular concept and allowed the first analog networks and terminals to be built. By the early 1980s, the Federal Communications Commission (FCC) had decided how it wanted to structure the industry and announced that it was dividing the country into well-defined market areas, designated by county. Under the FCC's plan, one license was granted to a company that was affiliated with an operating telephone company in a given market. The other license was awarded to an independent operator with no company affiliation in that area (this ruling has since been revised in the Telecommunications Reform Act of 1996). Two test systems were constructed in Washington, DC, and Chicago. At the conclusion of successful testing in 1983, these two cities became the first to offer commercial service. There are now more than 1,400 cellular systems operating across the U.S. and many cellular systems have been installed in Asia and Europe as well. The popularity of this technology is obvious.

Cellular technology uses element manager radio waves to transmit conversations. When the subscriber places a call on a cellular phone, the call is transferred via a computerized switch between operating areas known as cells. Each cell site is designed to provide coverage to specific geographic areas generally one mile to 20 miles in diameter, depending on the terrain and the capacity needs of the system. The coverage provided by a cell corresponds to the number of users that are likely to exist in that area. Hence, more densely populated areas demand smaller cells.

Each cell site has its own transmission tower linked to a Mobile Telephone Switching Office (MTSO), which connects the call to the public switched telephone network. Each tower is equipped with a radio transmitter/receiver and by adjusting the transmitter's power, the range of the radio frequencies can be shaped to fit a single cell. With proper engineering and maintenance, those same frequencies can be used in another cell not far away, with little chance of causing interference.

As the subscriber moves from one location to another, the call is handed off to the next cell site to provide optimal signal coverage and call clarity. This arrangement of multiple cells allows the subscriber to travel throughout a territory and maintain a quality conversation as the call is handed off from cell site to cell site. When a mobile phone begins to leave a cell, the network senses that the signal is becoming weaker and hands the call over to a tower that emits a stronger signal.

The technology most commonly used to transmit cellular calls is known as analog, where the voice is actually carried on the airwaves. The latest evolution in cellular, digital technology, is now being introduced in major metropolitan areas and will eventually be available in most markets. Digital transmission provides several benefits to cellular customers: It is more efficient than conventional analog transmission; it can handle three or more calls for every one made in an analog system. Its increased capacity also means fewer blocked calls, and offers a more private method of transmission as well as a platform for future wireless services, such as data transmission and interactive computers.

This segment of the book introduces GSM and its management challanges first. Management functions and the applicability of the TMN model are addressed in some depth. The following chapter investigates wireless technology as the key competing differentiator in tomorrow's market. The critical success factors of management are represented by proper support of automated configuration management, performance monitoring and management, expert fault management, fraud prevention through security management, and flow-through processing account management.

NetExpert's capabilities to manage wireless networks are evaluated in the following chapter. This chapter addresses all known wireless technologies, such as GSM, CDMA, TDMA, AMPS, TACS, JTACS, NMT, ESMR, and CT2.

Finally, the last chapter of this part of the book shows how mobileMASTER meets user expectations with its prepackaged, integrated domain rulesets.

12 Management of GSM Networks

CONTENTS

FIGURES

There currently are more than 50 million cellular subscribers in the United States alone. And in the rest of the world, mainly inspired by the relatively low cost and high capacity of the Global System for Mobile Communication (GSM), over 60 countries have selected this technology. The resulting explosion in mobile phones will push the world's population of cellular subscribers to more than 100 million by the turn of the century.

12.1 WHAT IS GLOBAL SYSTEM FOR MOBILE COMMUNICATION (GSM)?

In 1982, the European Conference of Posts and Telecommunications (CEPT), consisting of the postal and telecommunications administrations of 26 European nations, established Groupe Speciale Mobile. The team was put together to develop a set of common standards for what was then called the "Pan-European Digital Cellular Radio Network." This system, now called the "Global System for Mobile Communications" was to provide a common terminal network interface and enable roaming capabilities across Europe. The objective was to design a digital standard that would provide greater capacity, security, clarity, and services than was possible using the conventional analog technology.

In 1987 these countries drew up a Memorandum of Understanding which they invited any mobile telephone operator across the world to sign. The document provided the endorsement needed by manufacturers to allow them to pursue development of the system. The principles that apply to all former cellular technologies are equally relevant to GSM. The exception is that GSM uses digital technology which provides the following key benefits:

- Increased spectrum efficiency to provide greater network capacity
- Highly sophisticated subscriber authentication which reduces the possibility of fraud
- Sophisticated encryption techniques to prevent eavesdropping
- Error correction to provide more consistent speech quality and eliminate interference
- Simplification of data transmission to allow the connection of laptop and palmtop computers to future generations of GSM terminals, without need of a modem
- Standard network element interfaces that allow network operators to select network elements from multiple vendors

The functional architecture of the GSM Public Land Mobile Network (PLMN) is similar to most cellular radio networks. The main functional elements are (Figure 12.1):

- The mobile station (MS)
- The base station subsystem (BSS)
- The mobile switching center (MSC)

The mobile switching center is responsible for land line switching and some mobility control, while the base station subsystem includes all air interface control and additional mobility control.

The base substation is further broken down into the base station controller (BSC) and the base station transceiver station (BTS) to allow separation of control and radio aspects. Several registers (databases) are also part of the system:

- Home location register (HLR) containing subscriber information
- Visitor location register (VER) containing information on users in the local area
- Equipment identity register (EIR) containing information about good/bad equipment
- Authentication center (AuC) which assists with authentication operations
- Operations Systems (OS) to provide maintanence and administration

12.2 NETWORK AND CONFIGURATION MANAGEMENT

The concept behind GSM was to provide a common digital cellular system throughout Europe. This would provide suppliers with economies of scale not possible with

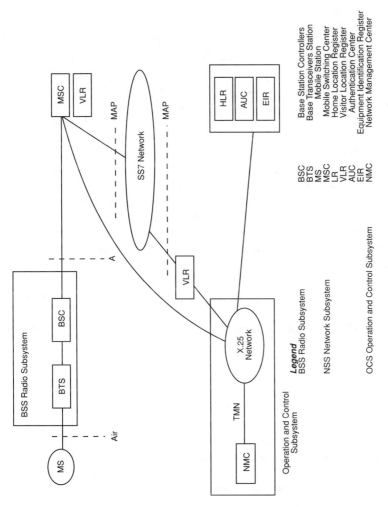

FIGURE 12.1 GSM structure.

prior, incompatible, analog systems. In order to provide benefits on the operator side, it was necessary to support a multivendor environment for the network infrastructure. This was done by standardizing on functionality and standardizing on various network interfaces. An additional benefit for the operator was the commonality in procedures that could be achieved by having a standardized system to operate. In the operations, administration, and maintenance areas, this meant standardizing on management interfaces and capabilities as well.

The entire management model defined for a GSM Public Land Mobile Network (PLMN) is extremely large and complex. Altogether there are more than 100 managed-object classes defined with almost 500 attributes.

The configuration-management component contains the provisioning service and the network element status and control services. These services are broken down into functions for growing, pruning, conditioning, and related functions. The service is defined only for the base station subsystem (BSS). The majority of the object classes are on the functional side of the model with a single-defined equipment object and three software/database objects. In addition, the model makes use of various ITU-defined objects and notifications for monitoring and logging. The focus on the functional side of the model was to assure maximum standardization and, thus, the greatest chance of interoperability.

Due to the nature of configuration management, some fictional objects were defined that correspond to elements more related to the physical aspects of the system (such as the BSC and BTS elements). Since it was not possible to restrict the implementation of these elements, they were defined on the functional side of the model but were provided with a GSM-defined attribute known as "relatedGSMequipment," which can identify the equipment and support the functionality. These same object classes were also provided with the package called EquipmentRelatedAlarms containing notifications that allow reporting of equipment alarms.

Several object classes were derived from the ITU M.3100 standard including:

- **The GSM Equipment attribute** To help alleviate need for additional subclassing.
- **The GSM Functional Objects attribute** To identify functional objects for the purpose of reporting equipment alarms through these objects.
- The **replaceable Software Unit** and **executable Software Unit attributes** To identify related files and software patch containment relationships.
- **The Operating Software Unit attribute** To provide for administrative control as well as identification of automatically restartable software units. Name bindings are defined for this class to the set of functional objects that could report equipment alarms.

12.3 PERFORMANCE MANAGEMENT OF GSM

Performance management is a very important function of cellular networks for several reasons. Cellular networks tend to be more complex than wired networks.

In addition, digital cellular networks tend to be more complicated than analog cellular networks. GSM is an all digital cellular standard so GSM networks are more complex. Another reason that performance mangement is so important in cellular networks is the rate of change. There are very high growth and churn rates as well. Some cellular networks are adding customers at the rate of 10,000 per week. Churn rates of 20% are not uncommon in the industry as well. Well-managed networks can accommodate the high growth rates and minimize the churn rates.

The lowest layer of a GSM cellular network, and any cellular network, is the radio transmission itself. The next higher layer is the data-link layer. GSM uses the integrated services network's (ISDN) link access procedure D (LAPD) protocol on this layer. The third functional layer is divided into three sublayers: The Radio Resources Management layer, the Mobility Management layer, and the Connection Management layer. The Radio Resources Management layer provides a reliable radio link between the mobile stations and the network infrastructure. It controls the setup, maintenance, and termination of radio and fixed channels. This sublayer contains the handover procedures. The Mobility Management layer is responsible for location management including updating and registration procedures as well as security and authentication. The Connection Management layer has three sublayers which manage call control, supplementary services, and short message services. The call control sublayer is closely related to ISDN call control (similar to the Q.931 standard). In fact, ISDN and TMN were both kept in mind when the GSM standards were written.

Cellular networks also are more complex because of the mobility of the customers. In a wired network the connection is kept for the duration of the call, but this is not true of cellular calls. Customers may be moving between cells while they are continuing a call. Calls are subject to what is called "handovers" in Europe or "handoffs" in North America.

Handovers, or handoffs, are the switching of an ongoing call to a different channel or cell. There are four types of handovers in GSM: Handovers between channels (time slots) in the same cell, cells under the control of the same BSC (Base Station Controller), cells of a different BSC but the same MSC (Mobile Services Switching Center), and between cells under the control of different MSCs. Handovers between channels in the same cell and between cells under the control of the same BSC are called "internal" handovers. Handovers between cells of different BSC but the same MSC and between cells under the control of different MSCs are called "external" handovers. Handovers are needed for two reasons: because customers are moving between cells, or as a means of load balancing by the MSCs.

Managing handovers is a very important part of cellular network management. When a handover is not successful it is called a "drop." With a drop the customer is basically cut off in the middle of their call. Understandably, customers get annoyed by too many drops. This is one of the reasons that churn rates can be high.

There are two commonly used handover management algorithms:

1. The minimum acceptable performance
2. The power budget method

The minimum acceptable performance algorithm is simpler and more common. The power budget method has better quality and eliminates the "smearing" that may occur near cell boundaries. Both algorithms are based on power control. When the power of the signal weakens to a certain level, then a handover takes place. Neither algorithm is specified in the GSM standards. In fact, no handover management algorithm is specified in the GSM standards.

12.4   FAULT MANAGEMENT

In cellular fault management, the supplier has to manage the land lines, which are usually leased from a separate provider, in addition to managing the wireless network elements. This added complexity makes fault management a very difficult task.

In the cellular market, customers must be satisfied with their service or they will look for another provider. This is evidenced by the high rate of customer turnover in the market,estimated as high as 20% annually for an average year.

This high rate of churn predicates the need for excellent fault management in cellular networks. The companies that maintain their networks with little fault activity will be the ones who will maintain, or even increase, their customer base once PCS begins to challenge the cellular market. A lack of network availability will be devastating to a cellular provider who is trying to increase market share.

Hardware redundancy was what most manufacturers employ as a way to provide almost seamless service from their equipment. Each of the components in the equipment has an identical piece to perform the same function. When a fault is detected in a component the control is passed over to the mirroring component and a notification is made to fix the faulty piece of equipment. This is a good method of backing up the hardware, although it can be very expensive because the supplier essentially has twice as much hardware as needed to perform the desired functions. The only problem with employing this method of fault management is that a single point of failure is introduced. The single point of failure being the software that is running on the redundant hardware. If this software fails, the redundancy in the hardware will not be able to help.

Software Redundancy   The latest trend in fault avoidance is software redundancy. With the high price of hardware and its ever increasing reliability, people are moving toward software redundancy schemes to replace the standard hardware redundancy that was formerly used. This could be a substantial undertaking based on the complexities of such a software system. However, this type of system has become a necessity in the cellular market where the supplier cannot afford to lose calls or, especially, a portion of the network. The supplier must provide a robust network for its customers that is always available. A good portion of cellular subscribers carry their phones for use in emergencies. If they cannot rely on the availability of one network at all times, they will seek another.

Software redundancy consists of having separately configured versions of the software running in each area of the different software areas. Each software instance will be monitoring the activity on the network and have access to all the critical data needed for transactions. One version of the software is designated the active version, the other is considered a "hot spare." The two software instances also monitor each other to verify the intended performance. When one instance sees that the other is not meeting certain thresholds and may be entering a fault condition, control is taken by the correctly functioning version. An alarm is sent to a network monitoring agent to specify the fault condition that has taken place and the need for someone to investigate the problem. This promotes the overall concept of fault avoidance through fault management. As stated before this is crucial to the highly competitive wireless industry.

Fault Monitoring   An OSS monitoring application can monitor network elements and alerts of any failures that occur on the cellular network. The operator who is receiving the alarm information is able to reroute traffic or change the software in any of the network elements from a workstation. The application also can identify faulty hardware units allowing the operator to dispatch technicians to the exact location of the network unit to replace it.

Fault Recovery   Companies employ remote digital cross connect switches on their networks to route traffic around a fault that may have occurred and they can restore facility paths at a remote site from their console. These remote capabilities greatly reduce the downtime of the network.

To be able to back up the wireless portion of their networks, cellular providers maintain mobile network units that can be dispatched to the area of a fault to get service back up as soon as possible without having to rely on fixing the faulty unit that created the network outage. One example is the use of microwave emergency response vehicles. These can be used to restore a microwave link that goes down and drops one of the facility paths of the network. The emergency response vehicle can put the link back online until the original unit is fixed. Another example of a mobile network element is the cell site on wheels. If a cell site is lost, the calls may be dropped when entering the area of the faulty cell or the signals become very weak. By dispatching a temporary cell site to the area, service to the the customer is restored while the problem is being investigated.

12.5   SECURITY MANAGEMENT

Security management is one of the hottest topics in the cellular community today. The cellular industry sees fraud costs of $1.3 Million per day. This adds up to an annual figure of over $500 million. These numbers are not limited to lost revenue, they also reflect out of pocket expense incurred when a provider is defrauded from another operating area. Providers still must pay for their use of the spectrum, even if they do not own that area. If fraud goes undetected, significant out of pocket expenses incurred by a small company can put it out of business.

Another instance of fraud falls directly on the customer. When a customer's cellular phone is cloned, he is confronted with the inconvenience of explaining to the provider that he is not making the calls that appear on his bill. This is a factor that can cause erosion of a provider's customer base. A company must provide an acceptable level of security if they intend to survive and maintain their revenue stream.

Customer Impacting Solutions   One of the temporary solutions provided by cellular providers was to brown-out any customer who went outside the local calling area. This allowed them to control the security within their own network without the worry of paying for use of the spectrum controlled by other entities. This limited fraud to lost revenue only, eliminating any out-of-pocket losses. The problem with this, however, was that it was a major inconvenience to the customer and not welcomed as a viable solution.

Another early solution was the use of Personal Identification Numbers (PIN). This was introduced to combat the phone cloning problem that exists in cellular today. Cloning is accomplished by using a scanner to capture a Mobile Identification Number (B)/Electronic Serial Number (ESN) combination. This combination is then installed on another cellular phone to produce a clone of the original phone. By allowing a user to enter a personal code that they had picked, added a degree of difficulty that successfully moved intruders away from these types of accounts. PIN numbers fall into two categories, Subscriber Personal Identification Number Intercept (SPINI) and Subscriber Personal Identification Number Access (SPINA). SPINI can require the user to enter the PIN before each call. SPINA requires the subscriber to enter a PIN to access the network. Access is granted for a specified period of time or until some network event occurs, such as the phone being turned off. Both types of PIN numbers are not supported by all switch manufacturers. This does not allow for universal use of PIN numbers across boundaries with unsupportive switching equipment. This limits PIN numbers to being a proprietary solution based on the switch vendors used by area providers. The effect PIN numbers had on fraud was to push thieves toward providers who did not use this technique.

Profiling is a system instituted by many providers as a way to be proactive in detecting fraud in order to reduce the amount of lost revenue and also to give customers a sense of security, that they were being protected by their provider. Profiling is the process of monitoring the calling patterns of subscribers in order to develop an overall customer profile.

The system is set up with certain thresholds based on a customer's calling profile and monitors the call to see if the patterns fit the customer's profile. Once the system determines that a pattern is irregular, a notification alarm is sent to a technician who will contact the customer to verify that he actually made the call. Providers monitor these systems 24 hours a day and have greatly reduced their losses by using this method. Providers must develop a method to eliminate fraud or they will constanty have their revenues undercut by technocrooks.

RF Fingerprinting is another technique that is being used to combat fraud in the cellular industry. RF Fingerprinting consists of comparing the radio signal information and characteristics of the customer's cellular phone with a copy of this information stored in the provider's database. This method severely limits the amount of fraud that can take place because the fingerprint of a given phone is unique and cannot be duplicated. This will eliminate the low-tech thief who scans numbers from the spectrum and produces clone phones quite easily.

However, this solution is very costly to implement. To secure the entire network, hardware, software, and communications equipment must be installed at each cell site. Any cell site that does not have the proper equipment will not be able to verify the fingerprint of the phone and will be susceptible to fraud. The problem is magnified by the fact that the equipment used to verify fingerprints is vendor specific. There are no industry standards set up for RF fingerprint verification at this time, which leads to different proprietary implementations. This limits fraud protection to a provider's network. If a customer leaves that 's network and roams, the fingerprint of the phone will not be known to another provider's network and cannot be verified.

Authentication is the industry's answer to the fraud problem. Authentication fills most of the holes in a provider's security network by severely limiting fraud to those with very sophisticated methods and special resources.

Authentication is clearly defined in GSM and is accomplished by having two functional units: a subscriber identity module (SIM) card which is held by the customer and can be inserted into any mobile unit, and the Authentication Center (AuC) which is maintained by the provider. A SIM card also is called a "smart card." Each subscriber has a secret key: one copy is stored on the SIM card while the other copy is stored in the AuC. When the subscriber accesses the network, the AuC validates the user ID by generating a random number and sending it to the subscriber's mobile device. The mobile device and AuC then use this random number, in conjunction with the subscriber's secret key and a ciphering program called A3, to generate a signed response (SRES) that is sent back to the AuC. The number received by the AuC is then compared with its own calculation. If the numbers agree, then the user is authenticated.

The same initial random number and key are also used to compute the ciphering key using an algorithm called A8. This ciphering key, together with the TDMA frame number, use the A5 algorithm to create a 114 bit sequence that is XORed with the 114 bits of a burst (the two 57 bit blocks). This optional enciphering creates another level of security beyond that of TDMA which codes and interleaves the message before transmission. This security feature eliminates the possibility of someone listening to a customer's conversation because of its encrypted nature.

Authentication is also performed on the mobile equipment used by the subscriber, in addition to the subscriber themselves. In GSM, a terminal is identified by a unique International Mobile Equipment Identity (IMEI) number. A list of IMEIs in the network is stored in the Equipment Identity Register (EIR). When an IMEI query is sent to the EIR, one of the following statuses will be returned:

- **White-listed**   The terminal is allowed to connect to the network.
- **Grey-listed**   The terminal is under observation from the network for possible problems.
- **Black-listed**   The terminal has either been reported stolen, or is not the type approved (the correct type of terminal for a GSM network). The terminal is not allowed to connect to the network.

Security management must provide for a solution that is transparent to the customer and requires the highest level of technical expertise in order to be successful. When this issue is dealt with, the cellular industry will see its profits increase with the reduction in lost revenue and the increase in customer confidence.

12.6 OPERATIONS, ADMINISTRATION, AND MAINTENANCE ASPECTS OF GSM

GSM is considered a special case in network management since it allowed the application of TMN standards to an entirely new network. In the process of GSM, certain elements of TMN were actively encompassed and others deliberately dropped. In addition, a model was devised to depict the management of a GSM network.

Specifications for GSM were developed by the Sub-Technical Committees of the Special Mobile Group. Of these groups STCSMG6 was responsible for OAM standards, the group's body of work is the application of TMN to GSM. As a system that was devised for a common digital cellular system, GSM was designed to take advantage of economies of scale and overcome the incompatibility issues of the older analog systems. The network management implications of this frame of mind appeared in such issues as support of a multivendor environment for network infrastructure and standardization of functionality and interfaces. Moreover, to achieve even greater functionality from the system, attention was given to standardizing management capabilities and interfaces.

STCSMG6 defined GSM TMN architecture using existing TMN standards. Q3 interfaces were defined to connect operations systems (OS) to network elements (NE). By definition, functional network building blocks could be combined in many forms. The specification was broken into five fictional management areas: Performance, Fault, Configuration, Accounting, and Security. In addition, specifications were made for general requirements, communication aspects, and message syntax.

Figure 12.2 shows GSM in a TMN structure (TOWL95).

Functionality was an important aspect in GSM and particularly in the application of OA&M. Diverse aspects that are not unique to GSM were not covered. For example, security management does not go into data transfers or access to a network element) because they are not specific to GSM. Further, only a portion of the standard was comprised so that development would be achievable in a reasonable time frame. An example of this is shown in the fact that fault and configuration management were different only in the Base Station Subsystem (BSS).

Three types of objects are defined in the GSM model:

1. Purely Functional objects
2. Functional objects that are generally related to equipment resources
3. Objects that represent equipment resources directly

   - Most of the objects that were defined support attribute value changes and create/delete notifications.
   - Objects with state support state change notifications.

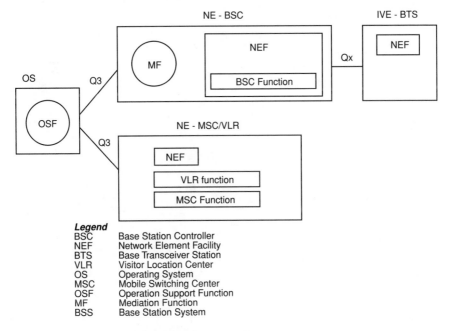

FIGURE 12.2 GSM in TMN stucture.

- Objects that are equipment related support user labeling attributes that allow name and location specification. In addition, these objects support reporting of equipment alarms.
- Only one object is defined to directly represent equipment resources. These general objects support user labeling, state control, and alarm notification capabilites. This was done to provide a standard equipment object that might be used to represent any arbitrary piece of equipment. Differentiation among equipment resources can be achieved through the type, name, and location attributes.

12.7 SUMMARY

GSM is a leading mobile technology. Management solutions are considered innovative. Even the recent management standards such as TMN are well supported. Many experiences from GSM management solutions, particularly in the field of security, can be ported to other areas of wireless networks.

13 Wireless Communications: Competing in Tomorrow's Market

CONTENTS

FIGURES

In the wireless market, massive business territories await carrier service, and a rapidly evolving market invites vendors to address current and future service opportunities. Cellular providers need powerful and flexible management tools to champion current and future services, improve operating costs, and generate new revenue streams. Access to vital information anytime, anywhere, and the ability to solve data management problems for geographically dispersed, complex networks is critical.

13.1 INTRODUCTION

Wireless communications is growing at unprecedented rates with current subscriber counts reaching almost 70 million worldwide, projected to hit 100 million by the turn of the century. The growth is fueling worldwide network construction that will spark significant competition between network operators. Network operators must keep pace with this rapid growth by properly positioning themselves for operational efficiency, relying on integrated and intelligent network management components in their operational support system arsenals.

The most important strategy a network operator can implement is an integrated expert network management solution capable of interfacing all network elements to streamline the work process flow of network and subscriber configuration tasks, performance and fault management, and resource inventory that will reflect the true cost of ownership.

Room to Grow with Standards and Open Architecture   An intelligent, integrated network management system (NMS) should be designed to support standard technologies and accommodate network growth and NMS upgrades. Based on open, scalable architecture, an intelligent NMS should be capable of supporting high message volumes and more network elements and applications as the network grows. It also should allow applications to be easily customized and new functionality to be added by network operators, system suppliers, or third parties. Support for industry standards is an essential component of open network management architecture. For service providers, adherence to the TMN standard is important to keep up with the latest telecommunications technologies. A TMN-compliant solution should address management issues of configuration, fault, performance, accounting, and security management. While the standardization of network management systems and solutions is progressing, it is important to note that for years to come wireless networks will continue to provide nonstandard network elements and business process challenges that network operators and network management systems must be flexible enough to deal with.

Integrated, Intelligent, Scalable, Network Management   The most efficient and functional network management systems should provide the operator an ability to grow and evolve applications and functionality. This growth capability must be coupled with an inherent integration that allows seamless scaling and facilitates transparent functional changes. It is important to note that many "mission critical" network functions are being turned over to the NMS and seamless integration into business operations and update capabilities is necessary. Figure 13.1 presents an example of these abilities.

The integrated NMS should provide the operator a consistent look and feel. It also should provide online assistance capabilities to incorporate equipment manufacturers or business process documentation and network element command simplification.

Simplified User Interface   The combination of object-oriented programming and an easy-to-use graphical user interface (GUI) such as Motif is a powerful contribution to effective network management. An object-oriented environment provides the necessary building blocks to implement network operation and management systems. Object-oriented programming tools not only simplify network management by categorizing devices as objects with similar behaviors, attributes, and dialogs, but also are intuitively helpful to users. Object-oriented programming lets customers define wireless telephony systems in terms of classes of objects and actions that the objects can perform. Using object-oriented programming tools, network administrators can build interfaces, rules to monitor, and dialogs with each class of object and then reuse the objects each time an element is added to that class. A common GUI would reduce operator training as well as training costs through straightforward, user-friendly func-

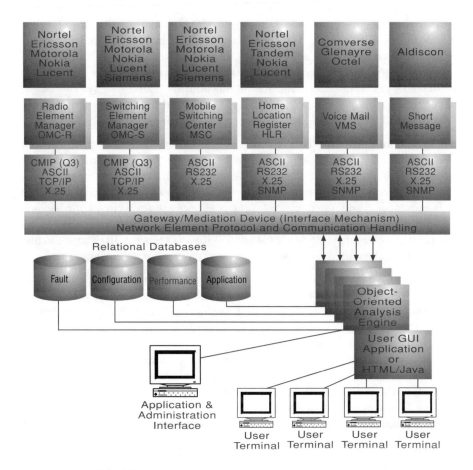

FIGURE 13.1   Integrated, intelligent network management.

tions. Intuitive features, such as automatic updates of graphic data and interactive controls like user-defined, pull-down menus and forms, enable network operators to manage with unparalleled precision. For example, an SNMP GUI can include applications that allow network operators to configure, initiate, and cancel polls, and control polling intervals. A GUI also provides the ability to create and modify rules, graphics, and alerts. Menus can be used to initiate functions with third-party applications, such as trouble ticketing, spreadsheet creation, and data analysis. Thresholds can be monitored and displayed graphically in tables, lists, and graphs.

13.2   AUTOMATED CONFIGURATION MANAGEMENT

The ideal NMS automates configuration tasks and is flexible enough to rapidly evolve to accommodate changes in business processes, customer needs, or network elements. Automated configuration tasks free the operator to handle more complex tasks and eliminate the need for the operator to know every language of every system for any network element. This benefit is particularly important in network and subscriber provisioning, which can become extremely complex in multivendor environments. It is important to also ensure that configuration management solutions are scalable. New networks may have the advantage of going online with a complete configuration management tool. Scalability is important to ensure existing networks can integrate small pieces of configuration management functionality until legacy systems can be disconnected. Configuration management can be viewed from two perspectives: the subscriber and the network. Each have significant, unique complexities to discuss.

Network Configuration This type of management can be applied to all network elements that rely on software to identify their unique functional character. In the wireless network, this includes most elements. Configuration management can include relatively simple, single-element manipulations such as switch trunk group adds, moves, or changes to complex tasks that allow end-to-end circuit provisioning across multiple switch, digital cross-connect and transport network elements. Wireless network configuration also should address Radio Frequency (RF) configuration management, allowing the operator a simplified method for cell manipulation.

Subscriber Configuration This type of configuration management can best be thought of as a mediation function between customer care systems and network elements. As wireless networks continue to expand, more advanced subscriber products and services based on enhancements to network elements will be made available. Such services will dictate changes in the network element provisioning process and dialogs. By employing a scalable, object-oriented configuration tool, network operators can economically continue to change and add network element interfaces without incurring high-cost modifications to existing customer care systems.

The following core functional components are critical to both customer and network configuration management, as shown in Figure 13.2:

Graphical User Interface The GUI provides the operator a synopsis of all required information in a menu-driven, table format written in plain English. Pull-down menus give the operator a variety of easily selectable process options and provide a standard view of functionally similar network elements from different vendors. HTML/Java web-based GUIs are currently the most flexible alternatives. Together, the two technologies let any security-cleared user — regardless of desktop platform — access the configuration system.

Network Element Provisioning Dialogs Dialogs are the command libraries that communicate with network elements to execute configuration changes. They are configured to respond only to appropriate input and identify when a command or series of commands has failed. Rollback functionality allows the operator to automatically return to previous configurations upon failure of command execution, improper network element response, or human intervention. Because there are almost as many interface possibilities as there are network elements, a flexible management tool — capable of handling both nonstandard interfaces at the ASCI level as well as the standards-based level with complex elements, such as SNMP- or CMIP-based ones — is desired.

Shadow Database The shadow database is an offline duplication of the network element configuration database. All data manipulations are executed to the shadow

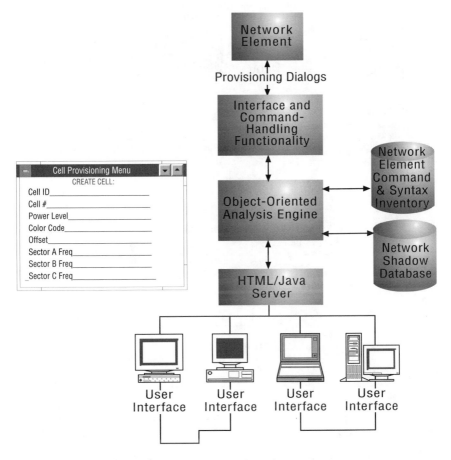

FIGURE 13.2   Wireless configuration management.

database, then scheduled and queued for maintenance-window network element download. Another benefit is the ability to review network element configuration data offline, rather than directly with the network element. This prevents overloading the network element payload with NMS dialog requests. Engineering and modeling functions can be performed without impacting switch operations or provisioning. Because information is synchronized with the switch, the shadow database guarantees that requested resources are available when provisioned. It also provides access points to link network inventory and financial applications.

13.3   PERFORMANCE MONITORING AND MANAGEMENT

There is more to performance management than just monitoring network performance data. It requires an integrated solution that lets one view and access all network-produced data sets, including billing reports, automatic message accounting (AMA), configuration database reports, and fault system-based reports. Performance management should bring harmony and clarity to the cacophony of traffic data

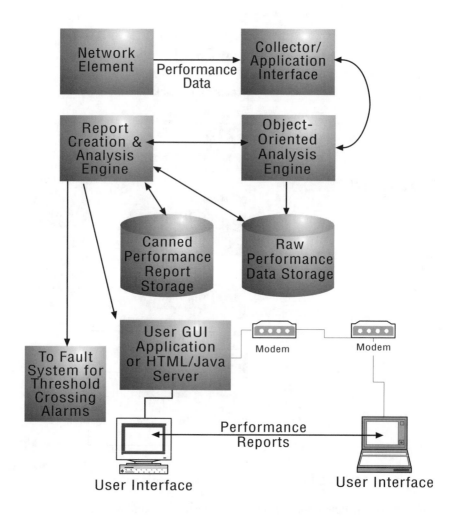

FIGURE 13.3   Integrated performance management.

generated by intelligent network elements. Wireless network performance management is rooted in the wireline network, since most wireless switches are modified versions of their land-based counterparts. However, the radio frequency (RF) environment adds a complex component to the performance management equation in volume of performance statistics alone. While wireline network engineering requires careful analysis to avoid overbuilding a network, the cost of an overbuilt trunk group does not compare to the cost of an unnecessary or misplaced cell site. For this reason, accuracy of performance management tools may mean the difference between profit or loss. A solid performance management solution consists of data collection, data storage, and report distribution systems, as illustrated in Figure 13.3.

Data Collection is the network element connection point. Low-level protocol and communication functions occur here, as well as a time stamp and normalization for forwarding to storage.

Data Storage should be as flexible as possible to let network operators search, access, and manipulate the raw data in scheduled and canned reports. There should not be any limit to the output or reuse of stored data.

Data/Report Distribution is critical to ensure the user community has access to information needed in a timely, reliable, easy-to-use format. Integration with configuration management is key to the accuracy of any performance reporting solution.

Performance management tools report more than just recent intervals. They generate reports on network trends, forecasts, queries, and data analysis — which are all accessible to unsophisticated users who lack programming skills. The performance reporting system should not constrain frequency of statistics transmission, nor should it impose restrictions on statistics reporting. It also should let topic experts define its content layout and the frequency of scheduled and ad hoc reports.

13.4   EXPERT FAULT MANAGEMENT

Fault management is the cornerstone of an integrated wireless network management solution. Fault management is the real-time collection, processing, and presentation of network events that can be as simple as a door opening, or as complex as a major network failure. The core components of fault systems are similar to performance management systems. They include the network element data communication interface, an analysis and forwarding mechanism, a storage system, and data presentation. Performance systems generally have fifteen minutes as the smallest window into their performance data stream, while fault systems must operate as near to real time as possible to match the network event data stream. Consequently, high throughput and fast analysis components are required to keep pace with a large network during high message volume event periods. This is illustrated in Figure 13.4.

An effective fault system provides information at various levels of complexity, based on the scale or scope of events. Better known as event correlation, this process identifies network events such as a failure of a cell site in a wireless network and combinations of network events. The expert system can see and digest incoming failure messages based on the sequence and type of events. Through event analysis, patterns can be matched with known event sequences, then conclusions drawn or decisions reached. An expert system can identify the end result of a failure sequence or it can perform analysis of all the sympathetic events as an aid to troubleshooting. The expert system greatly reduces the chances of misdiagnosis and allows the operator to view the network in a simplified steady state rather than a constant state of flux. Using predefined and user-definable rules, the expert system establishes relationships and identifies desired outcomes based upon known and unknown sequences of events. To understand the value of this process, consider the events involved in a cell failure. A cell failure is often preceded by frequent transport (or T-1) errors. These errors are symptomatic of the probable cause of the failure, and should be identified in the trouble-analysis process. However, the failure is the net effect, so the escalation and notification process associated with service affecting conditions should be initiated and the failed T-1 restored.

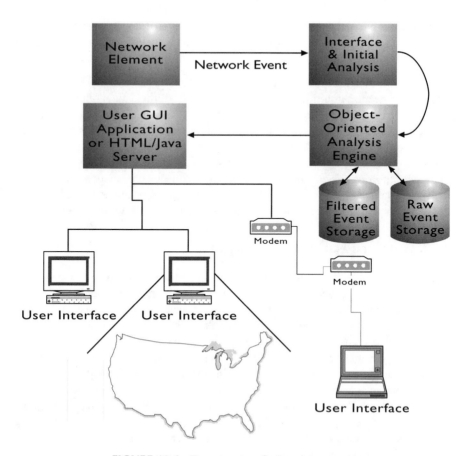

FIGURE 13.4 Expert system fault management.

Fraud Security management in wireless networks is divided into network management system security and customer network system security. Network management system security is best described as a sophisticated user authentication system that directs users through standard login and profile sequences that are authenticated using a controlled user-profile database. Each user should have access to all aspects of the network management system necessary for effective performance. This task should be defined in the system configuration tables. The network management system should not impede the network operator from employing any security management system dictated by business process. The customer network security system ensures fraud is effectively managed through round-the-clock monitoring. However, the network management system must forward particular data, such as configuration information or billing data, to the fraud system for it to function properly, and the network management system must be able to interface with the fraud system to receive real-time, fraud-related event information. Most importantly, a fraud system is much more effective if it is an integral part of full-time network management activities rather than a stand-alone entity.

13.5   ACCOUNTING MANAGEMENT

The ability to collect, process, and effectively employ billing information is essential to the financial livelihood of all wireless carriers. Yet, most carriers do not fully apply the billing data to other areas of business operation. Network operations and engineering can benefit from any information on calling patterns and loading behaviors throughout the daily cycle. Trending of usage patterns and growth rates can be useful for projected expansion plans. Accounting management at the network management level can best be thought of as a mediator between billing systems and network elements. As wireless networks continue to expand, more and more advanced subscriber products and services will continue to become available. The network must be flexible enough to add new data collection services to existing data collection efforts. By employing a scalable object-oriented accounting management tool, network operators can continue to change and add network element interfaces economically, without employing costly modifications to existing billing systems.

AMA data is the billing data that comes from network elements and contains customer call information valuable to many enterprise departments. State of the art NMS solutions should have an AMA data manifold to distribute information to appropriate applications or departments. Collection, normalization, time stamp, and storage will make up the main functional front-end components, with one data feed from the network element able to give support to many applications as the back-end output. All network operators are faced with a difficult challenge of dealing with the needs of the AMA data user community, which include fraud systems, performance management systems, and roaming systems, to name only a few. As advanced roaming and worldwide network connections occur, a standard, easily manipulated, scalable AMA data handler will be required in every successful wireless network.

13.6   SUMMARY

As today's wireless networks evolve, efficiency in network engineering and operations is key to the success of network operators. Sophisticated network management tools allow operators to operate more efficiently, and also provide the customer with a much more stable, uniform product. Managing network faults and properly anticipating problems results in the best possible wireless product. Managing the growth and inevitable changes tests the network operator's abilities to provide a quality product. Having state-of-the art tools that efficiently fill the gaps in today's technology is absolutely essential for anyone anticipating long-term success in wireless network operations.

14 Role of the NetExpert Framework to Manage Wireless

CONTENTS

FIGURES

Rather than the "best fit" adaptation of a traditionally developed legacy OSS with little time or opportunity for customization, providers choose NetExpert's flexible framework to achieve full implementation of unique technical and functional requirements. NetExpert's GUI-based, easy to use rule development tools are designed to let network analysts or engineers quickly and easily create operations support and management applications or modify existing ones to support new services. Monitoring cell sector output, keeping an up-to-date list of neighboring cells, and monitoring channel usage per cell are easily accomplished. Customers can smoothly integrate additions to new or existing cellular infrastructures, support load changes to OSS software, and easily modify rules to accommodate identification of different protocol and data formats from numerous device vendors.

All NetExpert applications are customizable through the use of NetExpert development tools for editing rules, dialogs, graphics, and authorizations.

The NetExpert framework can be used for nearly all major wireless technologies, such as GSM, CDMA, TDMA, AMPS, TACS, JTACS, NMT, ESMR, and CT2. All solutions are based on the TMN architecture and other ITU and ANSI standards for fault, configuration, and performance management.

NetExpert is scalable, migratable, and adaptable. Small cellular providers can smoothly emerge as large, national providers through uncomplicated software and hardware additions to existing network infrastructure. NetExpert's robust architecture eliminates the need for extensive equipment and software upgrades or replacements. Regardless of individual technologies, NetExpert will withstand network expansion because it has been designed to adapt to the technologies of tomorrow.

14.1 DEVICE INTEROPERABILITY BY INTELLIGENT GATEWAYS

NetExpert's object-oriented architecture and user-defined rules and attributes define its interface and management behavior. NetExpert manages the network elements and the link interoperability of the wireless infrastructure. NetExpert currently manages MSC, BSC, and radio equipment from nearly all major manufacturers, including Motorola, Nokia, Northern Telecom, Siemens, Ericsson, Samsung, NEC, and Fujitsu. OSI also has developed interfaces to more than 100 different models of transmission equipment and ancillary network elements used within the cellular environment. This includes test equipment, Digital Access and Cross-connect Systems (DACs), microwave, Fiber Optic Terminating Systems (FOTS), SS7-based components, and environmental systems such as Dantel and UPS.

For example, with a NetExpert managed cell site, thresholds for the signal-to-noise ratio can be set and monitored. If a threshold is exceeded, a maintenance alert is immediately sent to operations and a technician is automatically dispatched to the cell site. Another example is the detection of a primary facility losing service because of the loss of a T1. NetExpert can dynamically reconfigure DACs or transmission equipment to route calls to a backup facility, and at the same time, open a trouble ticket and dispatch maintenance personnel to repair or restore service, if necessary. Only NetExpert provides instantaneous and simultaneous reporting, authorized notification, and automatic configuration based on events or failures in a customer-defined environment. This minimizes infrastructure and development costs by enabling systems growth as needed and what-if testing of new services without additional spending.

14.2 SCALABILITY OF THE NetExpert FRAMEWORK AND ITS COMPONENTS

Network management application components means end-to-end services can be managed adeptly, ensuring that maintenance and operations goals are met. Multiple logical and physical views of the network and service configurations enable seamless management of systems, services, switching, and delivery so that customers receive ubiquitous service, independent of network equipment. NetExpert easily addresses the need for inventive new services and unanticipated interface requirements to capture new markets and generate new revenue streams.

Figure 14.1 shows a typical network map indicating node's status of the wireless network.

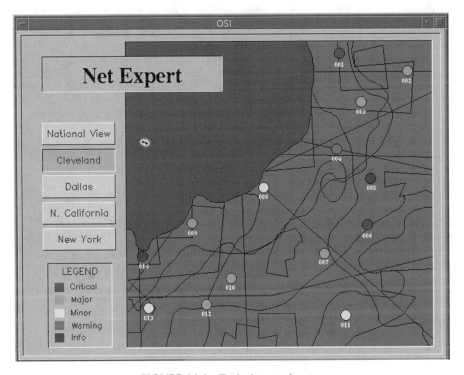

FIGURE 14.1   Typical network map.

Network experts can integrate existing elements into a framework that will gracefully accommodate growth as standards and market demands evolve. NetExpert's modular design, easy-to-use rule editors, and distributed architecture make possible almost limitless growth. Distribution is based on operator, gateway, and server software.

Network segments also can be displayed in detail as shown in Figure 14.2. Color-coded elements indicate segment's status on the map. If necessary, operators can drill even deeper and display device status to support fault detection and repair (Figure 14.3).

NetExpert uses distributed gateways to control traffic congestion caused by the convergence of data traffic from multiple remote networks. Gateways filter and selectively parse data from the raw data stream. These messages, recognized as events by NetExpert, then are sent to the server software for analysis. If an alert is required, the server software sends a message to the operator software, which displays the alert on the operator workstation.

Customers use rules and dialogs to build a system to desired specifications. All of NetExpert's software modules can operate on different UNIX-based platforms, allowing customers to add modules or distribute processing as necessary. There are no restrictions to the number and combination of devices and applications that NetExpert can manage. And, NetExpert easily manages large traffic volumes.

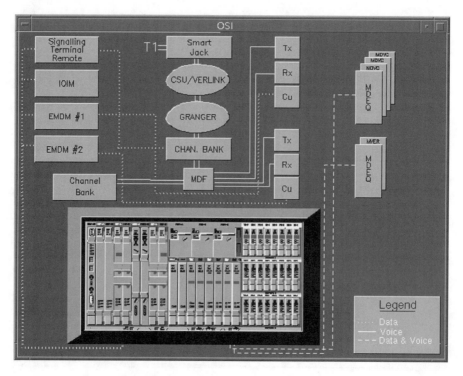

FIGURE 14.2   Display of network segments.

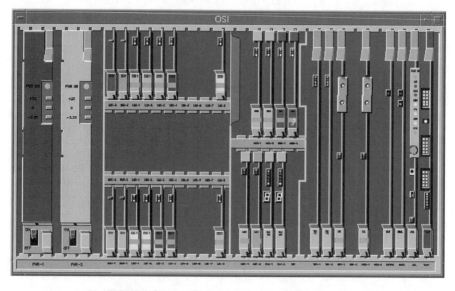

FIGURE 14.3   Display of detailed device status.

14.3   SERVICE MANAGEMENT WITH NetExpert

Service Management — Configuration, Provisioning, and Activation   NetExpert demonstrates superior benefits through rapid and economical service deployment, minimal human reaction to network events, increased scalability through distribution of managed functions, isolation and containment of network faults, and reduced network management traffic. NetExpert's object-oriented architecture supports the precise modeling of relationships between elements within the managed network. User-defined rules and dialogs check service requests from Service Management System (SMS) clients for completeness and consistency, assuring integrity of service data. Information is routed through normal event processing and is easily retrieved through easy-to-generate reports.

NetExpert's dynamic environment simplifies complex resource and service provisioning tasks. Through its object-oriented architecture, expert systems, and flexible gateways, NetExpert conquers the challenge of managing complex services spread across multiple networks, easily scaling to accommodate operations architectures and proprietary protocols. Service provisioning and activation are greatly enhanced through logical views of network and service information, which are user-definable, allowing operators to provision circuits or trunks between switches, base stations, and other network elements.

In the cellular world of customer-directed service request and activation, the functions of distributed management and service provisioning are interdependent. Effective service delivery requires simultaneous access to customer profile information, configuration information, and the service information itself. NetExpert integrates service provisioning and connection management with many of the same network components as the operations and data collections systems. Through NetExpert's mediation gateway and knowledge-based expert systems, customers get an integrated view of the network and can direct the requested services in real time, enabling on-demand service activation.

With NetExpert, service registration by subscribers is limited only by the capability of their devices. They may roam to other networks while using all their home features and have access to originating and terminating communications.

Performance Management   NetExpert operates in real time to report activities, events, and anomalies the moment they happen. With a comprehensive view of the network, NetExpert automates analysis and real-time alarm trending. Alerts, such as detection of degraded network service or fraud, can be reflected in text in hourly and daily registers and in graphic display. The expert system displays customized advice on possible solutions or, if desired, can automatically correct problems by imposing predetermined controls via rules and dialogs. Consequently, user-defined solutions are implemented in a fraction of the time required by other technologies. Collected switch traffic data and cell usage statistics can simultaneously be analyzed in real time, as with event thresholding, and sent to NetExpert's DataArchiver for processing and statistical analysis. This information can be provided in flexible summary reports and graphical histograms, allowing management to make informed decisions about the health of the network.

Real-Time Billing and Data Management   OSI's data collection products incorporate tools to rapidly address the integrated, multivendor environments of the wireless market. Centralized billing collection through DataArchiver allows real-time billing and quick capture of call record information for preprocessing, reformatting, and hot billing. Users can extract database information via NetExpert's ReportMaker or use any standard SQL interface software. Extracted data can be moved onto a fraud analysis system to review call patterns that may affect performance and to map calls to limit or profile use. With DataArchiver, a more robust and real-time recognition of possible fraudulent access and call attempts is possible, which allows faster action when fraud is detected. NetExpert's Data Message Handler (DMH) can be used in intercarrier communication for billing rating protocols. Incoming service order information can automatically be distributed to appropriate systems to activate service and modify customer billing records.

Alarm Reporting and Fault Management   Through logical system diagnostics, NetExpert displays real-time data analysis, automated escalation, alarm collection, event correlation, trending, and trouble ticketing. Wireless network managers can establish user-defined thresholds and alarm correlation for different event types, with specific notification display when thresholds are exceeded or root cause correlation is determined. Distributed operations are easily supported by NetExpert's powerful Peer-to-Peer Gateway, which provides ubiquitous management and centralized or distributed control and allows events to be forwarded. For instance, users may send events from the wireless portion of a network to a digital segment or to a public-switched telephone network connection. NetExpert resources in one market region, such as Dallas, can serve customers in Chicago, and share information about the health of the application, along with knowledge about each region's customers.

Operational and Environmental Security   Network security is always a concern. NetExpert's Authorizations Editor allows network managers to create authorization groups based on application, dialog, and event lists, and to set up partitions based on collections of attributes that are attached to network elements. Security can be established at all levels of the network, from an operator login point to a cell site. NetExpert validates every operator request by user name, location, and login, and then applies limitations that constrain the assigned authorizations. The Authorizations Editor reinforces security by allowing various levels of access for NetExpert users. Where appropriate, network administrators can restrict operators to specific classes of devices and cut-through access, establishing a "firewall" between users and unauthorized NetExpert capabilities.

NetExpert demonstrates an exceptional record of providing the fastest and most accurate information available on environmental network problems. Regardless of the source — natural disaster or human error — NetExpert activates and manages recovery solutions to maintain business processes and services to customers in real time, with or without operator intervention.

14.4 SUMMARY

NetExpert advantages for the wireless industry can be summarized as follows:

Speed of delivery NetExpert integrates required network applications, reducing the time and cost invested in service delivery to get wireless services to the market first.

Multivendor, multiprotocol network management Success in developing interoperability gateways means OSI can provide any device or protocol interface to NetExpert, standards-based or proprietary.

Object-oriented development Users can define and model classes of objects and see a comprehensive view of the exact relationships between network hardware, circuits, and subscribers.

Rules and dialogs NetExpert's consistent escalation of user-defined rules and dialogs provides an optimum combination of rapid deployment, scalability, and control.

Real-time data collection NetExpert provides a real-time reflection of network performance that incorporates tuning or modifications before revenue is lost.

Graphical alert display NetExpert's VisualAgent delivers the most efficient and consistent representation of network management through highly detailed, flexible, customized displays.

The next chapter addresses a domain level integrator application component, called mobileMASTER. In its implementation form, the average needs of wireless operators are addressed. Special needs can be met by customization which is part of professional services.

15 mobileMASTER

CONTENTS

FIGURE

The complexity and diversity of network technologies and services is increasing as a result of escalating industry competition. Service providers require the ability to integrate existing equipment and systems with new elements and applications, customize systems and applications to match business models, and scale management solutions to accommodate expanding networks and exploding traffic volumes. In response to these needs, OSI has developed tools and methodologies to minimize the risk and associated expense that normally accompanies the development and modernization of network management systems.

OSI's mobileMASTER provides wireless network operators with state-of-the-art tools and applications to support a variety of integrated functional requirements including fault, performance, configuration, accounting, and security management. Using the power of NetExpert, OSI's application-creation environment, mobileMASTER delivers wireless network expertise in the form of packaged rulesets for major wireless network elements. These rulesets can be plugged into the framework in a mix-and-match fashion to create custom-tailored, user-specific solutions. This approach allows optimization of OSI's rule-based, object-oriented, distributed systems and facilitates rapid solution deployment in any wireless application environment.

mobileMASTER is TMN compliant and supports most currently defined network management standards including SNMP, CMIP, and the GSM 12.xx specifications. mobileMASTER also excels at interfacing to and managing existing nonstandardized legacy network scenarios such as customer care and billing systems, as well as nonstandardized network elements (switches, digital cross connects, and transmission equipment). mobileMASTER offers significant existing application libraries for most currently available wireless network elements and support systems as well as many that are no longer in production. Current application feature sets are segmented along the TMN model.

15.1 CONFIGURATION MANAGEMENT

mobileMASTER includes a GUI or batch interface-based expert system for auto-mating network element configuration command sets and dialogs. This system can be connected via open application interfaces to existing legacy systems or operated with a Java-enabled GUI. Configuration management features include:

- Mobile Switching Center (MSC)/wireless switch trunk, routing, and trans-port provisioning, with complete trunk and dial-plan management
- Radio Frequency (RF) environment provisioning of cells, operations, and maintenance centers, radio base station controllers, and base transceiver stations, including complete network element parameter and topology/RF propagation management
- Subscriber provisioning of home location registers, voice mail systems, short message systems, roaming systems, and complete subscriber acti-vation transaction mediation
- Service activation for voice mail systems, short message systems, cellular digital packet data systems, and other wireless intelligent network ser-vices, such as complete mediation service activation
- Load management of network element software
- Network inventory management; shadow database; and tools for engineer-ing, planning, and asset management
- Network element backup and restoration, business-continuity planning
- Integration with business planning and engineering tools

Configuration management is structured to provide the wireless network operator with an expert system capable of interfacing to all network elements. This system provides a much simplified methodology and work process flow of network and subscriber configuration tasks and resource inventory. Configuration management can be applied across the entire network or specifically focused to individual func-tional areas such as switch trunk side and transport configuration, SS7 circuit con-figuration, RF environment (cell or bsc/bts) configuration, and subscriber configu-ration/service activation. The system provides automation of current configuration tasks with the flexibility to rapidly evolve to accommodate changes in business process, customer needs, or network elements. Configuration management provides a platform for the network operator to ensure all aspects of configuration are managed as efficiently as possible, supporting functionality in the following areas.

- Provisioning
- Status
- Installation
- Initiation
- Inventory
- Backup and restoration

An example of a typical segment of configuration management would be the addition of a new cell site to a network. This is a fairly complex set of network element dialogs that requires an expert network element operator to translate and engineer work requests into the proper command sets to build the cell site. In a conventional wireless network, these dialogs would be to the switch to actually configure the physical connection and attributes of the new cell. In a GSM/CDMA network, the dialog to the OMC-R mediates for the appropriate BSC. The new cell's neighbors also must be manipulated to facilitate appropriate handover functionality. The network transport facilities (DACs, MUX, Channel Banks) also must be manipulated to support the new cell connectivity.

15.2   FAULT AND TEST MANAGEMENT

mobileMASTER provides real-time network event collection, correlation, archiving, and desktop presentation. Fault management functions include the following:

- Alarm collection and archiving
- Alarm analysis, correlation, and presentation
- Trend analysis
- Fault location and isolation
- Integration with system test capabilities and with customer care, business planning, and engineering tools

Data feeds from network elements are streamed through terminal server (Xyplex, for example) and collected by the NetExpert Gateways. Most applications will opt to archive the streams, thus providing a data store for other processes. The data streams are also vectored through the rule-based event manager for identification and classification. This allows the customer to build a set of rules specific to the network elements that will perform the fault detection, isolation, notification, and reconfiguration required to maximize system uptime. As the data stream is an ASCII feed, identification and classification is based on pattern matching event stream subelements. Identified events are then processed using a rule-based expert system to detect critical, major, minor, warning, and information types of events. Notification occurs in two modalities, textual and graphical. The text-based alert displays all of the pertinent information for that alert. The graphics displays are active, real-time representations of the status of each component. Graphics give the operator a rapid assessment of the state of the system.

Once an alert has been issued, operators can take the needed actions or, if desired, the system can be configured to engage in a series of dialogs with the network elements to effect reconfiguration toward fault resolution.

The entire system from collection, through event ID and alert generation in text and graphics is extensible by the customer using the graphical, easy-to-use rule creation editors and interfaces provided. This gives the customer the power to create and modify the rules, the graphics, and the alerts as required. Due to the intrinsic flexibility of the NetExpert system, a prudent path for the user is to engage in a degree of analysis to establish standards for rulesets, naming conventions, network connectivity, and alert severity prior to development. This will give the user's system room to grow while minimizing the problems encountered when growth and integration of dissimilar systems eventually occurs.

15.3 PERFORMANCE MANAGEMENT

mobileMASTER features an open system with a standard relational database at its core. This facilitates integration into existing environments and leverages standard prepackaged components for data storage. mobileMASTER's performance reporting user interface is Java-enabled and can be run from most desktop platforms that support HTML browsers. Performance management functions include the following:

- Performance data collection and archiving
- Data filtering
- Trend analysis
- Network or business-specific report and analysis tools
- Functional group-specific report segmentation
- Ad hoc query tools
- Real-time analysis and threshold alarming
- Integration with customer care, business planning, and engineering tools

Wireless systems require a means for monitoring the performance of the system on many levels. Each system is a dynamic entity that requires monitoring and modification to manage the increases in customer base and use. A means to effectively provide a series of reports that are generated automatically based on the input streams of network element data is required. It is only through the use of the "window" into the behavior of the network that operations, engineering, and other departments will be able to tune the system to maximize customer satisfaction and cost effectiveness. A reporting system must take into account the multiple aspects of the wireless environment, data collection, transmission, processing, and utilization to be effective. The reports generated should be based on well-developed formulas derived in concert with the network element vendors. This is not a trivial task and can take extensive effort in some cases. Once formulas are derived, the implementation should allow for extensibility and ease of administration. There should be a broad array of reports, such as half-hour and hourly, busy hour, busy period, daily, weekly, monthly, and trending period statistics. The statistics for each one of these aspects of the system can and should include call completion/failure, blocking, dropped calls, cell/site level call, and hardware telemetry statistics. This is a minimum inclusion of the reports needed.

Specific report issues are:

Length of Report Data Retention   How long will raw data be stored?
User Access   Who will have access to the raw and processed data, and
    reports?

High Availability   Can we stand to have the reporting system offline in the event of hardware failure? For how long?

Training   Do we wish to have online training available?

Documentation   Online documentation and explanations of the formulas and processing

Standard Naming Conventions   Specific standards for network element names

Standard Formulas   All installations need to be using the same formulas (apples and apples)

Scope of the System   Regional, market, national, global? What is the scope?

Distribution of Finished Reports   HTTP/HTML?

Reporting Languages   Perl, C, Java

Data retention is an issue that will dictate the size of the system to be installed and how much disk space will be configured. The report suite and sizes will impact how much raw and processed report space is needed. Some industries are required by FCC regulation to retain raw call records for a certain period.

User access is important because users who can do the most good for the system need access to the reports, and in some cases, even to the raw data. The productive employee will find new and innovative ways of looking at the data, given the chance.

High Availability is an issue if the company comes to rely on the reporting information and makes high-level decisions based on this information on a daily basis. It should be decided what type of redundant systems can be implemented to ensure that reporting system downtime is minimized.

Training   The best of reporting systems is worthless if no one uses it. Usefulness also decreases if the users do not know the power of the system. Therefore, some form of training would be worthwhile whether it is the standard class/instructor and/or online hypertext training. General and administrator report creator training is advised.

A means for incorporating online documentation is advised to allow the users to rapidly find answers to their basic questions. This has been proven to be most effective in an HTML-based system using hyperlinks.

Standards are important to the growth and extensibility of the reporting system. Without them there is a one hundred percent certainty of major work in the integration and modification of the base reporting system.

Scope of the system will help define specific architectural issues. Essentially, the reporting system can be implemented at the market level, but if the scope is national or even global, the reporting system should be designed with that in mind. This will allow the overall performance of the wireless system to be reported on more easily.

Report Distribution   Since users of the system will be many and sit at multiple types of workstations, a simple presentation methodology for the reports is a great advantage. The current industry standard HTTP and HTML is the most effective means for distribution of statistical reports. This allows the customer to distribute reports to any authorized user through the use of an HTTP/HTML Web browser.

Source Code for Reporting   The choice for report coding has been C and Perl and now Java. Each of these languages offers distinct advantages and disadvantages. A combination of all three may be the solution with the primary systems coding being done in Java due to the use of an HTTP/HTML standard for the system.

The performance management system is an essential component for the successful management of a wireless telephony system. It is critical that the system be implemented as an integrated, extensible, reliable, and wholly automated entity. The system must manage itself with regard to data collection and storage. It must be robust, yet easily manageable by the local administrators.

15.4   SECURITY MANAGEMENT

mobileMASTER supports internal user security and enables access points for integration into customer security systems such as fraud management or roaming applications.

mobileMASTER was designed to facilitate rapid deployment of open, standards-based wireless network management functionality. The ruleset package contains most TMN element and network management layer features, including element-specific dialogs and graphics. Features that are specific to the service provider, such as network-level graphics, tailored correlation rules, and MIB integration, can be customized to specific needs.

Security of the systems is split into two distinct functional areas, telephony system security and LAN/WAN security. Both aspects must be addressed as revenue and productivity are at risk. Telephony security involves assessing, prevention, monitoring and detection of customer phone fraud. LAN/WAN security involves the assessment, prevention, monitoring and detection of unauthorized intrusion into the company data network.

In order to achieve a reasonable degree of security in both areas, assessments must be conducted to determine the degree of risk for each system. The assessment should include the following:

Access points   How can someone gain access to this system? (Physical, dial up, Internet, eavesdropping, theft, cloning, etc.)

Procedures   What procedures are in place with regard to system access and are they followed?

Loss of data   What is the impact to the revenue stream if data is lost?

Loss of access   What is the impact to the company if access is disrupted to reports, alarming, data?

Topology   What is the LAN/WAN topology and will it support firewalls while remaining accessible to users?

Corporate philosophy   Is it a closed shop or open system?

Users   Who are the actual users of the system and its data?

Administration   Who administers the system?

Prevention is extremely important for achieving a secure system.

In the area of prevention, the system must be assessed with and designed to establish challenged access and limited access from outside the enterprise. This must be accomplished while retaining remote access for enterprise users. Common behaviors of intruders should be planned for and designed to foil, such as IP spoofing, OS vulnerabilities, and user password compromises. Each system should have basic security measures such as file permissions, login challenges, and password encryption. Users of each system should be made aware of the need to exercise common sense with regard to company system passwords and account permission, and avoid leaving a system unlocked. The goal of prevention is to minimize the chance of intrusion through a combination of specific hardware, software, and wetware (people) configurations and behaviors, and to minimize any damage that may result from an intrusion.

15.5 ACCOUNTING MANAGEMENT

The ability to collect, process, and use billing information is essential to the revenue stream of all wireless carriers. What is often missed is the extensive utility of the billing data in other areas. Billing, or Accounting Management Application (AMA), data contains customer call information that can prove valuable to many departments inside an enterprise. Customer Service Representatives can quickly view not only customer billing information but also customer call behavior and status of customer calling history. Problem resolution can be enhanced if the representative can access call termination information commonly contained in call records. Proactive customer service can be implemented through analysis of call records to identify customer or system equipment problems that may be evident in the call termination codes. Fraudulent or cloning activities can be detected through analysis of call records for duplicated subscriber unit use in different geographical locations. RF Engineering can find use for calling patterns and loading behaviors throughout the daily cycle. Trending of usage patterns and growth rates can be useful for projected expansion plans.

Billing validation is another key issue for wireless companies. Often, subscriber billing is outsourced, and the cautious company will need a means for comparing the billing records at the switch and from the outsourced vendor. Errors are common and can be costly. Each of these scenarios can be managed through integration with NetExpert and the development of custom modules. AMA data collection and archival can be achieved with the NetExpert Gateways and DataArchiver, provided the AMA stream and parsing algorithms are available.

15.6 SUMMARY

NetExpert framework is a high-level application creation environment that eliminates traditional software programming. Figure 15.1 shows mobileMASTER for wireless networks. This figure includes the intelligent gateway and the management applications running in the NetExpert server. Both the intelligent gateway and the server are using prepackaged rulesets for a variety of components common to wireless networks.

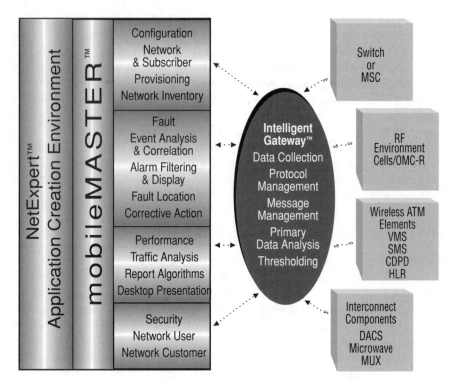

FIGURE 15.1   Architecture of mobileMASTER.

Part IV

Management of High-Capacity Transmission Systems

This part of the book shows first, how SDH/SONET can be managed using the NetExpert framework. A specific example shows root-cause fault correlation for the case of cutting a SONET cable. This chapter is followed by a domain level integration ruleset to manage high-capacity transmission systems. transportMAS-TER supports management functions for SDH/SONET transport networks. It can be used to provide management applications across multivendor fast packet data networks, including ATM, frame relay, SMDS, X.25 switches, PADs, FRADs, and IP routers.

Finally, the North Carolina Information Highway is shown as a full implementation example to manage ATM switches and high-capacity transmission systems. NetExpert is being embedded in the management systems of the principal operators of the North Carolina Information Highway.

The marriage of the popular Internet Protocol (IP) network protocol, Asynchronous Transfer Mode (ATM) switching technology, and Synchronous Optical NETwork (SONET) transmission technology promises to provide enormous benefits to both end users and service providers.

While few end users will be able to directly use the entire bandwidth of this technology, they will be able to acquire substantially more bandwidth than they currently have.

The technology's combination of speed and isochronous (nonvariable data rate) transmission will enable widespread deployment of applications that today exist primarily in laboratories. Such applications include:

- Desktop multimedia
- Video conferencing and collaborative development
- Bandwidth-intensive distributed computing

The technology presents service providers with a much more scalable and flexible environment than they have today. The ability to handle image, voice, and data within a single, scalable transmission facility greatly simplifies the planning and building of future networks. The technology provides a growth path for today's networks which are rapidly running out of capacity and capability.

However, providing this functionality is not "a piece of cake." These items are on the leading edge of technology and many things are undecided.

This lack of firm specifications is especially true for ATM. For example, the Internet Engineering Task Force (IETF), in conjunction with the ATM Forum is presently in the process of defining how the Internet's IP and Address Resolution Protocol (ARP) work over ATM. It is expected that ATM deployment into the Internet community will take several years, with ATM acting initially as a "wire replacement."

To accelerate general ATM development, the ATM Forum meets monthly to work on specifications for the following items:

- Signaling
- Network-to-Network Interface (NNI)
- Congestion control
- Traffic management

Also, for reasons of expediency, ATM specification activities have concentrated on Permanent Virtual Connections (PVCs) and have deferred work on Switched Virtual Connections (SVCs).

The area most in flux is the critical area of management of the newer ATM and SONET technologies. The management of IP devices is more well known. Not only is the work in progress, but also there are multiple management protocols to choose from. In general, the current SNMP solutions are restricted to monitoring activities, while the other management protocols can control as well as monitor. As SNMPv2 is deployed, and SNMPv2-specific Management Information Bases (MIBs) are developed, it is expected that more controlling functions will be added.

16 SONET/SDH Management

CONTENTS

FIGURES

The following sections will introduce SONET, its advantages, network elements, and topology. This information provides only an initial understanding of portions of SONET and is not intended as a complete introduction to all aspects of SONET. Concepts like the advantages of synchronous communications, SONET overhead, the distinctions between line, section, and path will need exploration via alternative information sources.

16.1 SONET/SDH AND NETWORK ELEMENTS

Industry-standard SONET was developed to provide fault-tolerant communications for service carriers whose livelihoods depended upon consistent network services. To maintain mission-critical networking, developers designed SONET, the optical transport standard, to survive fiber cuts and other network disasters. This ability to recognize fiber cuts and reroute traffic before service was interrupted or degraded, has earned SONET the reputation of a self-healing network.

SONET is the defining U.S. standard for the Synchronous Optical NETwork and is compatible with Synchronous Digital Hierarchy (SDH), the European standard for synchronous optical networks. SONET supports the Telecommunications Management Network (TMN) architecture's functionality specifications for the network element and network management layers.

A SONET is constructed using the following basic network elements. The transport portion of the network is comprised of add/drop multiplexers (ADMs), Regenerator/Repeater, and Optical Amplifiers. The access (or on ramp) portion of the transport network provides connectivity to the transport aspect of the SONET network and includes synchronous and asynchronous network elements like the broadband digital cross-connect (DCS), wideband digital cross-connect, and digital loop carrier (DLC).

Many, but not all of these, network elements are intelligent and provide a certain amount of accessible intelligence, making the elements themselves directly manageable. The intelligent network elements retain cross-connect and other information regarding their status and performance.

Intelligent network elements, such as the ADM, are built to support remote access by the network operators who can communicate with the intelligent network elements to send commands or retrieve the data which the elements have been designed to store and report. The information can then be used to remotely provision, maintain inventory, diagnose, and manage the network.

The world of telecommunications, and thus the SONET itself, are changing rapidly as technological advances are made, bandwidth needs increase, and consumer demands grow. As this book is being written, most ADMs run at rates ranging from OC3 to OC48. However, new network elements, higher data rates, and new functionality such as Wave Division Multiplexing (WDM), OC-192, and SONET over Microwave bands are being introduced. For the foreseeable future, SONET will continue to grow in capacity and complexity.

16.2   SONET/SDH NETWORKS

SONET's flexible architecture accommodates various configuration models so that SONET service carriers have the luxury of configuring networks in a number of efficient, fault-tolerant topologies, including emerging popular multipoint configurations. The SONET add/drop multiplex and hub configurations can provide efficient traffic management capabilities, both in the intermode and access environments. (See Figures 16.1 and 16.2.)

The most frequently found SONET topologies are Point-to-Point (or chain) and Rings. Variations, such as Drop and Continue Circuits, Meshing, Subtended Rings, and Optical Interconnects, allow the development of SONET networks to provide the advantages of SONET in a variety of configurations.

SONET rings are deployed in two-fiber and four-fiber configurations.

Drop and Continue   SONET's key network ring topology plays a big part in correcting failures. SONET configurations enable drop and continue — a key capability in both telephony and cable TV applications. With drop and continue, a signal terminates at one node, is duplicated, and is then sent to the next node and to subsequent nodes. In ring configurations, drop and continue provides alternate routing for traffic passing. If the connection cannot be made through one of the nodes, the signal continues along an alternate route to the destination node. For example, dual homing allows ring extensions (that is, customer hubs) to be fully protected against

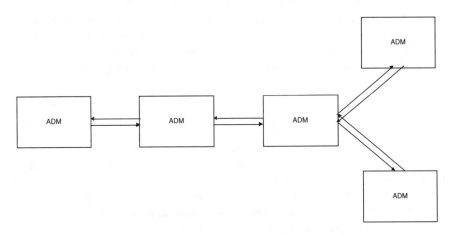

FIGURE 16.1 Point to point (chain) ending in a multidrop.

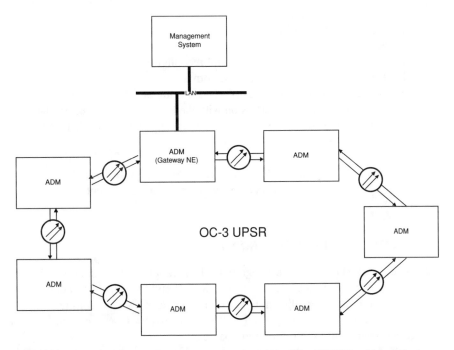

FIGURE 16.2 Two fiber OC3 unidirectional path switched ring (UPSR) with attached management system.

single node failures as well as fiber cuts, as if they were part of the same ring. If this were to occur on an asynchronous network, the traffic element would have had to return to the same point of origin, and if something happened to the element (that is, the node), then the data, and thus the communication would be lost.

Protection Switching   SONET rings use automatic protection switching to provide for automatic fault tolerance while supporting high volumes of network traffic. SONET employs two types of protection switching: line-protection switching and path-protection switching. Both provide for fault-tolerant backup service.

Deployed over two fibers   normally found in unidirectional path switch ring (UPSR)
Deployed over four fibers   found in bidirectional line switch ring (BLSR) topologies.

In each configuration, the SONET payload travels in both directions around the ring. As the payloads reach an end point, SONET selects the best signal based on the performance data in the payload overheads. Should one signal path drop, say from a fiber cut, the ring automatically switches over to the other signal path and network traffic continues uninterrupted. The transition is so smooth and seamless, that the end user is usually unaware of any network trouble.

SONET Advantages   Some of the major advantages of SONET include:

- Operations, administration, maintenance and provisioning (OAM&P) abilities are an integral part
- Efficient add/drop multiplexing
- Multivendor compatibility
- Remote, on-demand provisioning and reconfiguration
- Real-time network alarming and performance monitoring
- Fault-tolerant network services (survivable ring architectures)
- Nearly error-free data transmission with the lowest bit-error rate available
- Flexible architecture supports key technologies such as ATM, B-ISDN, and FDDI

Operations Support Systems (OSSs), such as NetExpert, provide the necessary tools to capitalize on SONET's advantages while permitting efficient service carrier operations.

16.3   SONET/SDH MANAGEMENT

A key benefit of SONET is the integration of operations, administration, maintenance, and provisioning (OAM&P) tasks within the network. This integration allows communication between intelligent nodes on the network, thus permitting administration, surveillance, provisioning, and control of a network from a central location without interfering with the information payload.

An ADM in a Central Office is connected back to the management system. This ADM is referred to as the Gateway Network Element (GNE). The GNE provides access between the management system and all of the nodes on a SONET ring. This allows remote management of a SONET ring from a centralized operations facility like a Network Operations Center (NOC).

Using remote management, an operator can access all of the information that the intelligent network elements store. This means that inventory information, such as hardware contents of an ADM and cross connections, can be gathered, commands such as cross connection setup can be transmitted, and fault analysis can be performed.

SONET ADMs provide many options for connectivity between the GNE and the management system. Most ADMs have RS-232 ports for Craft access as well as X.25 or 802.3 ports for remote management. Remote management systems can connect to any of these ports in order to manage the associated SONET network.

Most current SONET devices accept commands in Transaction Language 1 (TL1). However, since TL1 is a Bellcore standard, the flexibility of implementation with the constraints of the standard cause each manufacturer's network elements to speak a different "dialect" of TL1. Occasionally, devices from the same manufacturer speak different variants as well.

In order to support migration to TMN compliance and replace the TL1 babble, some devices interact using the Common Management Information Protocol (CMIP).

In order for management systems to interact with SONET network elements, the management system must be able to handle any set of the connectivity options (RS 232, X.25, 802.3) and the standard protocols (TL1 or CMIP). There are hardware and software companies that provide mediation products to convert among these options. Some management systems can support all of these options; some only support a limited subset and require many mediation devices to be used in order to cover this shortcoming.

The following two sections provide more detail on additional areas of connectivity complexity and issues that service providers and network management systems must conquer. Although dealing with a complex subject, these issues will be addressed in a simplified manner.

802.3 Connectivity   Most ADMs have a standard 802.3 port. However, Ethernet and ECP/IP are not supported on this port. In order to support the industry movement in a CMIP direction, a variety of standards based on the OSI stack have evolved. The three phase evolution started with TL1 over TCP with modified UDP on a 4 layer OSI stack. The industry is currently supporting TL1 on a 7 layer OSI stack with TARP (Target Address Resolution Protocol) at the 4th Layer. The final step is CMIP on a 7 layer OSI stack.

Thus, there are three possible types of 802.3 connectivity which the gateway ADM may use to connect to the management system.

X.25 Connectivity   Many ADMs have an X.25 port as well. The X.25 is usually a PVC channel. As many networks are Switched Virtual Circuit (SVC), the PVC must be mapped onto the SVC network. In an X.25 SVC network, the path for the X.25 packets occasionally switches. This leads to the connected management system thinking that network element connectivity has been lost. However, the network element's X.25 PVC circuit is still up and the network element does not think connectivity has been lost. This means that the network element does not reset its status. In order for the management system to reconnect, the network element must reset the X.25 PVC channel and this means, frequently, a second connection from the management system to the network element is required.

16.4 SONET/SDH MANAGEMENT IN PRACTICE

The following section describes the functionality which SONET network elements provide and that SONET management systems must provide. Where appropriate, examples will be provided of systems that have been constructed to support SONET management.

Fault Management   SONET networks are verbose. Because of the OAM&P overhead channel in SONET, the networks monitor themselves and announce, via messages, both problem and status information. Both ADMs and Regens can provide alarm messages. SONET alarm messages include:

- Alarms (TL1 REPT ALM)
- Events (TL1 REPT EVT)
- Housekeeping or environmental messages ((TL1 REPT HKA)
- Performance management data (TL1 REPT PM)
- System status information ((TL1 REPT BACKUP, COND, SYS)
- Network element database changes (TL1 REPT DBCHG)

The system reports on threshold crossing for performance as well as failures. ADMs report both failures that occur in themselves and on signals that have seen failures elsewhere in the network (sympathetic alarming).

Remote proactive maintenance is possible by setting the quality of service thresholds and monitoring the service performance. SONET relays to network operators what is broken and lets network operators perform remote alarm diagnostics.

Protection Controls   SONET rings allow remote control of protection switching. The operator can force a ring to protect mode or lock out protection switching. These controls are used for remote fault resolution and to support network modifications and testing.

The management system for SONET must make it easy for the operator to access these functions and to track and display the status of the network.

Root Cause Analysis   As one gateway network element can report all alarms reported by an ADM or Regen in the ring (or chain) that it is connected to, a great deal of messages can be reported. These are standards that begin to provide guidelines on how to eliminate unnecessary messages (for example, Bellcore SR-TSV-2671 subnetwork Root Cause Alarm Analysis).

A SONET Network Management System must view a subnetwork as a unit in order to avoid reporting excess alarms. Additionally, the OSS must use some type of containment correlation to avoid swamping an operator with excess messages. As an example, if a fiber cut occurs, all of the S\DS3s in the fiber will alarm. There is no reason to overwhelm an operator with all of the contained (or implied) messages.

In addition, root cause analysis suggests that an intelligent OSS can analyze a set of messages from the network and provide the operator with a small number of messages (ideally one) that can point directly to the underlying problem. NetExpert systems, using the combination of the filtering in a gateway, the object model in the MIB (that supports containment), and the intelligence of IDEAS, are ideally suited

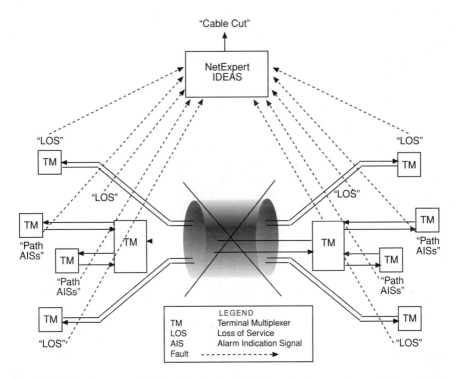

FIGURE 16.3 An example with cable cut and NetExpert as the central, coordinator of fault correlation and resolution.

to SONET management and have been used in this environment for years. (See Figure 16.3.)

Testing   As well as reporting any problems (local or remote), some SONET ADMs provide the operator with the ability to support testing. In addition to being able to establish and remove loopbacks, some network elements support diagnostic testing. This allows the operator to discover the existence of a problem and analyze the problem remotely.

Configuration Management   TMN configuration management provides the basis for all of the other functional areas. Without configuration information, containment correlation in fault is impossible. Without provisioning of circuits (a part of config-uration management), performance management is unnecessary.

Inventory   Technicians are no longer needed to physically check every piece of equipment to discover what is used, where it is used, and how it is being used. The SONET network elements store and report this information.

   A management system can perform autodiscovery of the contents of a network element by requesting this information and appropriately building and storing a database representation of the network element. Where supported by the network elements, a sufficiently intelligent management system, like NetExpert, actually can autodiscover the number of network elements on a ring, the ring topology (the order of network elements), and the contents of all of the network elements on the ring.

A production NetExpert transportMASTER customer simply tells the system the new ring number, a connection to the gateway network element, and some security information. Based on this information and conversations with the network, this system correctly builds the entire ring and displays the appropriate graphics down to the card/shelf level.

Provisioning   By permitting the operator to remotely provision circuits in shelves rather than dispatching a field technician to a remote site to turn up circuits, SONET facilitates activation of circuits. This ease of reconfiguration makes the system responsive to changes and allows circuits to be remotely enabled or disabled to carry or remove traffic. In addition, the circuits can be dynamically reconfigured to remedy trouble situations, traffic variations, or meet customer needs.

Using the records keeping and assignment capabilities embedded in the SONET intelligent network elements, the service carrier can provision complex ring configurations, such as dual-homed topologies, that enable bandwidth management to be extended beyond just a single ring architecture and provide an additional layer of fault tolerance.

Consequently, with a sufficiently intelligent management system, service carriers can provide bandwidth-on-demand performance to customers by taking full advantage of the bandwidth management capabilities found in SONET.

Activation consists of the turn up and removal of circuits (or cross connections) on SONET devices. Some management systems (especially EMSs) simply allow the user to enter (directly or via short cuts) all of the necessary commands to each device in the SONET ring. This functionality is frequenstly called provisioning. In fact, as the Bellcore definitions of EMS require subnetwork level activation, this functionality is only one small component of activation (the turn up of circuits).

NetExpert features rulesets for activation at the EMS level. From one user information entry panel, the rulesets provide the functionality to: travel the ring in the appropriate direction, issue commands to all affected network elements, and rollback (that is, delete any work done) upon encountering an exception condition. This functionality allows complete management of the activation process at the subnetwork level and facilitates end-to-end provisioning activities.

Rerouting consists of moving an existing circuit to free up some or all of the underlying network resources for reuse (or repair). Having the ability to activate at the subnetwork level allows the operator to reroute circuits at a subnetwork level. This rerouting ability, if accomplished without service interruption, permits the operator to manage bandwidth.

Bandwidth Management   Using SONET's flexible architecture, carriers can meet customer (and internal) traffic demands by mapping resources accordingly to fill up bandwidth, and can track and manage bandwidth in real time. By re-allocating time slot assignment abilities, carriers can achieve higher consistency in performance and move bandwidth from slow areas to high-traffic areas. SONET's remote provisioning ability lets network operators turn DS1s or DS3s on or off at certain network sites to increase the bandwidth at other sites.

Bandwidth management saves money by helping carriers cut the cost of hardware deployment. SONET uses about one third less hardware than comparable Async to carry these services. These savings can then be passed on to customers, potentially giving carriers a competitive edge through competitive pricing.

The management system for SONET must support bandwidth management by allowing easy-to-use and reliable activation as well as inventory and reporting abilities. In order to facilitate bandwidth management, the SONET management system must be able to track free bandwidth, show how bandwidth has been allocated in the network, and provide tools to move circuits (without traffic disruption).

Flow-through provisioning involves the electronic processing and distribution of customer service information to appropriate departments. Flow-through processing begins when the customer service representative enters the customer's requested dates, service types, and service attributes into the system. The support for the service is designed (the route for the circuit determined) and assigned (the exact system resource allocation is defined). Then the service is activated (the appropriate network facilities are initialized and service is turned on). Optimally, this flow-through can be accomplished in a matter of minutes with little human intervention. The customer request information is electronically forwarded to the appropriate departments so fewer employees are needed to handle the order, or manage the process. Flow-through processing promotes prompt customer service initiation, accurate billing records, and quicker problem resolution and dissolution of billing disputes.

Using NetExpert transportMASTER (see next chapter) EMS rulesets for activation, transportMASTER NMS and integrated service activation controller (iSAC) functionality can be integrated to provide most pieces of a flow-through provisioning system.

Performance Management   SONET is constantly checking the network with each point "talking" to each other to ensure the points are functioning properly and the payloads are being delivered — without affecting network traffic since this information is relayed separately. This error monitoring allows a full set of performance statistics to be generated, as well as the segmenting of alarms. These built-in, error-checking abilities allow speedy detection of failures before they degrade the system to a serious degree.

As service quality degrades and explicit, measurable thresholds, such as Bit Error Rate (BER), are crossed, SONET network elements issue alarms called Threshold Crossing Alarms (TCAs) that announce the problem and attract attention. The monitored thresholds are called Grade of Service (GOS) attributes or tables.

A SONET management system must allow the operator to set GOS values, see the current thresholds and network reported values, observe TCA alarms, and allow the collection and analysis of performance data. There are two aspects to performance data analysis. The real-time portion allows the comparison (at the management system) of reported values against more stringent levels of monitoring (management system level GOS monitoring). The real-time portion also should allow specific ports to be monitored more frequently to support higher quality service (for higher paying customers) or to watch for degradation in a suspected trouble area.

The analysis portion of performance data consists of the collection of performance data for trend analysis. In order to provide proactive maintenance following a failure, a service provider might examine historical performance data to determine the pattern of performance values that indicate impending trouble. Once these trends can be determined, the real-time portion of performance analysis should be modified to warn before problems occur.

If a future problem is predicted, the rerouting ability can be used to remove traffic off the failing portion of the network. Using remote testing, the potential problem can be examined and appropriate repair action initiated.

16.5 SUMMARY

The goals of a SONET management system are:

- Allow complete access to all of the abilities, information, and intelligence available from the SONET
- Facilitate the management of SONET to maximize the return on the benefits SONET offers
- Minimize the specialized training needed for operators to monitor and run SONET

The bottom line is that SONET management is necessary and worthwhile. In order to take advantage of and capitalize on the benefits of SONET, a service provider must actively manage the network.

By providing intelligence at the network element and the capacity for remote management, SONETs lend themselves to centralized management.

The complexity of managing SONET comes from the necessity of understanding the SONET environment, the service provider's needs, and the requirement of having a good network management system.

Companies with expertise in building high quality SONET management systems that facilitate (rather than hinder) the operator's job are rare. Products with the required abilities, including the ability to grow into the future, are also hard to find. The combination of expertise, experience, and ability allows OSI's NetExpert-based transportMASTER application ruleset to uniquely meet the needs of service providers and offer rapid solutions in real-world networks.

17 transportMASTER for the Management of High-Capacity Transmission Systems

CONTENTS

FIGURES

TABLE

The complexity and diversity of network technologies and services are increasing, driven by escalating industry competition. Service providers must be able to integrate existing equipment and systems with new elements and applications, customize systems and applications to match business models, and scale management solutions to accommodate expanding networks and exploding bandwidth demands. In response to these needs, OSI has developed tools and methodologies to minimize the risk and associated expense that normally accompany the development and modernization of network management systems.

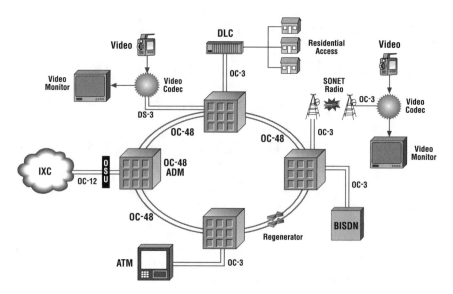

FIGURE 17.1   Architecture of a network with device support.

transportMASTER uses the power of NetExpert, OSI's application-creation environment, to provide management expertise for high-capacity transmission systems in the form of packaged common components for major transport network elements. This approach allows OSI customers to use the full potential of rule-based, object-oriented, distributed systems and to capitalize on OSI's transport environment operational expertise. Advantages include centralized, consistent provisioning of all equipment; flexible integration of new network elements and operations support systems (OSSs); the ability to deal with generic activation requests, rather than vendor-specific ones; cross-vendor, cross-function, and cross-domain correlation; shared maintenance and operation procedures; integrated problem resolution; and reduced software costs.

17.1   DOMAIN AND SUBNETWORK MANAGEMENT

Figure 17.1 shows the architecture of a network with hybrid transport. All typical devices of SONET and ATM are represented. Also multiring SONET networks may be managed by the NetExpert framework and its rulesets. (Figure 17.2)

To efficiently manage multiple element types in a transport network, service providers must be able to correlate messages across the entire network. Functioning at both the element and network management layers (EML and NML) of the TMN architecture, transportMASTER provides a strategy for integrated subnetwork domain management. It distributes intelligence closer to the element layer and frees transportMASTER's NML module for proactive tasks such as network wide testing and provisioning.

For example, transportMASTER provides domain-level management of network elements, with each domain controlled by its own EMS. Because domains can be

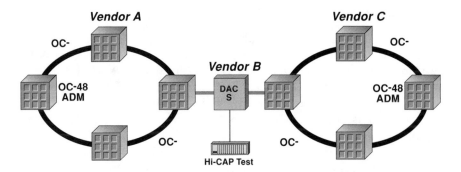

FIGURE 17.2 Support of multi-ring.

geographical, functional, or organizational, it is easy to map this network management system into anyone's specific business model. Using NetExpert's peer-to-peer technology, all domains report to a central transportMASTER NMS where operators have complete visibility of both the EML and NML paths, regardless of element type or vendor. Consequently, operators can easily perform test and provisioning across an entire network. transportMASTER goes beyond simple fault management, correlating performance degradation between subnetworks and across multiple domains to prevent faults and reduce service outages.

OSI's approach to managing transmission systems that provide high-capacity services is encapsulated in the transportMASTER application package. Utilizing the power of OSI's NetExpert framework, transportMASTER delivers transport network expertise in the form of packaged rulesets for all major transmission system hardware. transportMASTER allows a customer to capitalize on OSI's operational expertise in the transport environment.

OSI's transportMASTER application package is designed for rapid deployment of open, standards-based network management functionality. transportMASTER has been used by many customers (including a major United States Inter-eXchange Carrier (IXC) to manage SONET ring, chain, and point-to-point configurations as well as the associated digital cross connects. transportMASTER has been used in two fiber optic unidirectional and four fiber bidirectional SONET environments.

Most TMN, EML, and NML features are packaged, including network element specific dialogs and graphics. Features that are specific to the network operator, such as network level graphics, tailored correlation rules, and MIB integration, can be customized to fit the specific needs of the service provider and are quoted separately as part of professional services.

It is important to point out that the transportMASTER package maintains the flexibility of the NetExpert framework, empowering the network operator to quickly and easily add or change functions using expert system rule editors and graphics built with OSI's VisualAgent. transportMASTER solutions have been implemented in a variety of network architectures, ranging from "umbrella" management systems composed of individual EMSs to direct communication with network elements.

17.2   transportMASTER FEATURES

transportMASTER maintains the flexibility of the NetExpert framework, empowering the network operator to readily add or change functionality using expert-system rule editors and graphics built with OSI's VisualAgent. transportMASTER solutions have been implemented in a variety of network architectures, ranging from "umbrella" management systems composed of individual element management systems to network layer solutions with embedded element management layer (EML) functionality.

17.3   transportMASTER FUNCTIONS

Following TMN standards, OSI's transportMASTER abilities are divided into the functional areas detailed below. Graphical displays provide NML and EML (sub-network/domain level and network element level) views of the network. Logical and physical representations provide efficient management functions necessary to support these areas.

Configuration management is an inventory of all relevant managed objects that make up the transport network. This inventory is the basis for the Management Information Base (MIB), which represents not only the objects within the network, but relationships between objects. For intelligent network elements that support inventory reporting, transportMASTER incorporates an autodiscovery feature that populates the MIB with little or no operator intervention. On a periodic basis, autodiscovery can be used to update the status of network equipment.

Provisioning occurs at the subnetwork level through graphical tools that automatically issue provisioning commands to each network element or node on the network. Provisioning also can be initiated via an external data feed of design layout records or circuit layout records from an assignment-type system. Node provisioning quickly establishes cross connects for drop-and-insert and drop-and-continue circuit allocation. Cut-through access to the network elements is a basic NetExpert function.

Fault management collects, identifies, and parses fault events from all nodes on the network. Expert system correlation and root-cause analysis intelligently process these events so that network operators only have to deal with significant alerts that require intervention. These faults are visible at the network, subnetwork, and element level, with zooming capability to quickly traverse levels to identify the root of faults.

NetExpert functionality provides alarm information in text and hierarchical graphical displays. This alarm information can be integrated with a trouble ticketing system for managing customers' problems. Combined with test-head management, transportMASTER fault management can be used to further isolate and resolve network problems.

Alarm suppression is furnished for network testing, network additions, and whenever a "message storm" — resulting from a fiber cut or other significant fault — threatens operation of the management network.

Performance management functions coordinate setting parameters, as well as scheduling and collecting performance management data. Near real-time display of data is available at the NML and the EML. OSI's DataArchiver can store this data for historical reporting and trend analysis. Specific functions of the performance management functional area will depend on the abilities of the individual network elements. Suppression of data reporting and collection occurs on demand or through correlation rules based on network loading.

Test Operations   Because test operations are necessary in a variety of areas, including provisioning fault, and performance, testing is a function that requires special attention. Use of test management applications will vary, depending on the test equipment employed. However, automation of testing abilities is a powerful feature that can significantly improve the efficiency of the network operations staff. Because the transportMASTER MIB maintains circuit and test access-point information, this is a natural extension of OSI's application package. Typical operations include activating loop-backs, performing digital tests, and displaying or storing results, all from a graphical user interface.

Security Management   In addition to the security functions of the NetExpert framework, transportMASTER maintains user names and passwords for each individual node. NetExpert's authorization and partitioning functions restrict access to specific managed-object classes or instances, based on the user's defined security group.

17.4   transportMASTER IMPLEMENTATION

OSI understands the impact of disruption from the installation of any new system in the network. Our tools and methodologies help reduce the risk and associated expense of such changes. NetExpert's gateway processes and programming interfaces lets users replace only the systems that make financial sense for their business. Packaged common components and easy-to-use rule editors enable rapid deployment of a customized solution. Intuitive, user-friendly GUIs simplify training on and use of NetExpert.

17.5   transportMASTER CONNECTIVITY

Using NetExpert protocol agents and gateways, transportMASTER can interface directly or indirectly to network elements. Direct connectivity options include asynchronous (dial-up and dedicated), X.25, 7-layer OSI stack (TL1), CMIP/CMOT, and third-party TBOS conversion (into ASCII).

On SONET configurations, transportMASTER connects to one ADM and uses the SONET internal data communications channel to connect to and receive messages from all nodes on the connected SONET ring. It also handles chain or point-to-point communication.

17.6   LEGACY INTEGRATION

transportMASTER is built on and takes advantage of the benefits of the NetExpert framework. A standards-based, object-oriented package with open interfaces, transportMASTER easily integrates with other OSI application components, such as switchMASTER and trafficMASTER.

Designed for hybrid carriers, transportMASTER supports multiple equipment types from multiple vendors: add-drop multiplexers, digital cross connects, and test heads. As a result, customers gain extensive flexibility in implementing their needs.

In addition to element management layer (EML) and network management layer (NML) features, transportMASTER provides subnetwork or domain management. It can be integrated with legacy systems in an existing OSS or serve as the foundation for a new OSS.

17.7   CONFIGURATION OPTIONS

Flexible Configuration   Regardless of the configuration of the NetExpert servers, visibility and control can be achieved anywhere in the network. A transportMASTER implementation can take several forms. Carriers that maintain central control of the network and administration staff may backhaul all network management data to a central location, while others may need to distribute management functions. transportMASTER maintains control over any logical subset of the network, including provisioning activities, geographic domains, all functions within a regional boundary, and those based on element type or vendor.

Depending upon the architecture of the carrier's management network, as well as the distribution and responsibility of operators and network analysts, a transportMASTER implementation can take several forms. Below are two common types of configuration. However, because of the flexibility and distributability of the NetExpert framework, other options may also exist.

In either case, the concept of domain manager is introduced. A domain manager controls a logical subset of the whole network. For example, one domain may consist of only provisioning activities, while another may be geographic with the control spanning across all functions within a regional boundary. It is important to point out that, regardless of the configuration of the NetExpert servers, visibility and control can still be accomplished anywhere in the network. (See Table 17.1)

Centralized   For a carrier that maintains central control of the network and administration staff, it may be necessary to backhaul all of the network management data to a central location. However, depending upon the size of the network managed, the entire system may be located on a single server.

Distributed   Many carriers want to bring configuration management closer to principal network nodes. It offers more protection against single points of failures and offers more flexibility to regional management teams. This solution, however, requires a security solution of software distribution and version synchronization.

TABLE 17.1
Configuration Options

Options	Advantages	Disadvantages
Centralized	Central Adminstrative Staff	High Management Network bandwidth for backhauling
	Single location for software upgrades	Possible single point of failure
Distributed	Correlation/filtering closer to the network	System Administration less centralized
	Less likely to be the single point of failure	Multiple locations for software upgrades
	Provides abstract view to the central network manager	
	Management responsibility is maintained with a region.	Distributed Adminstrative Staff

17.8   transportMASTER ADVANTAGES

Reduced Business Costs   Minimize capital expenditures by deploying a fully integrated framework that includes intrinsic applications to handle alert and network displays, report generation, and command path management.

Improve operator efficiency, lower training expenses, and reduce staff requirements by automating routine tasks to free network operators for analytical and other value-added or high-value activities.

Take advantage of freedom to select hardware and database platforms, making it easier to negotiate with vendors.

Quality Assurance   Manage the network proactively, rather than reactively, with transportMASTER's performance management features.

Improve mean-time-to-repair by implementing fault correlation and automated testing.

Reduce human error and use personnel resources more efficiently with automated and scripted network element interfaces.

Service Competitiveness   Accelerate revenue recognition from new services offered and improve customer service with transportMASTER's automated provisioning abilities.

Get new services or service technologies to market ahead of the competition with transportMASTER's built-in extensibility.

Internal Process Improvement   Modify and expand existing applications to model your preferred business processes.

Results   Eliminate dependence on inflexible, hard-coded support systems and support the goal of a totally integrated network management system with the power of the NetExpert environment.

Manage hybrid, multivendor, synchronous, and asynchronous networks with a single integrated solution.

17.9   SUMMARY

transportMASTER is one of the domain level integration rulesets to support rapid service deployment, easy customization, and seamless management of a multitude of services and network technologies.

18 Management of ATM Networks — The North Carolina Information Highway

CONTENTS

FIGURES

TABLE

18.1 CHALLENGES AND STATUS OF MANAGING ATM NETWORKS

The management of ATM networks needs more advanced management techniques to ensure end-to-end visibility of a variety of physical and logical conditions. The volume of expected messages, events and alarms, and simple overloads forced management solutions. ATM uses one technology for transmitting diverse traffic types. ATM is the integrating technology for bringing WANs, LANs, and MANs together. SONET/SDH can serve as the underlying transport for ATM, which in turn can be used to support a variety of other services, such as frame relay, voice, SMDS, and switched digital.

ATM management requirements can be summarized as follows: (DESA95)

- **Fault Management** is defined as a consolidated management system that handles and correlates fault management data from all network devices, including ATM switches, frame relay equipment, routers, bridges, and hubs. This system also isolates faults and initiates local and remote recovery actions.
- **Topology Management** provides a graphical end-to-end view of the entire network. This capability provides the logical and physical views on one screen which can be customized to an individual user's needs.
- **Configuration Management** is the ability to configure, install, and distribute software remotely. This function also tracks switch configuration changes.
- **Accounting Management** is the capability to collect and parse accounting data in order to generate network usage reports.
- **Performance Management** offers a single source from which to monitor network performance and resource allocation.

Management is again an afterthought. ATM equipment is usually available much earlier than its management. Status of ATM management can be summarized as follows: (DESA95)

- **Fault Management** is probably the main concern of managers. The popular SNMP management platforms lack the performance and sophistication required to handle and correlate the volumes of ATM management traffic to accurately identify and react to fault conditions. SNMP is just not designed to handle the large amounts of traffic generated by an ATM network. The switch vendors have provided proprietary solutions that offer rudimentary management capabilities. These work well in a private, LAN-sized, homogeneous environment, but break down as soon as the user attempts to incorporate and manage multiple vendors' ATM switches. Additionally, a majority of these vendor offerings do not integrate adequately with SNMP platforms or with key third-party applications. Accordingly, handling faults, performing automated correlation and initiating recovery mechanisms all still remain unaddressed.

This has driven network managers to rely on patchy solutions that incorporate third-party applications with extensive in-house code. Such solutions, as managers are discovering, are generally expensive and hard to maintain. But certain vendors start to provide built-in capabilities, such as rerouting in the event of a network breakdown. The concept of seamless integration also is looked at closely by the SNMP platform providers. Service providers are offering their customers a capability to poll in real-time, as well as access non-real-time fault data via a combination of SNMP and FTP protocols. When such concepts are realized with matured products, network managers can expect to treat their ATM switch managers the same way they currently treat router managers to obtain a consolidated network management system.

- **Topology Management**   As customers implement ATM networks, the need to obtain an end-to-end topology has become a necessity. Currently, in most heterogeneous ATM networks, network managers must develop such a capability. Lack of interoperability between ATM switch management systems — coupled with loose integration at the topology and event database level with the SNMP-based management platform — has resulted in fragmented views of the network. The different topology files and databases are hardly synchronized with each other. Such solutions are unacceptable for managing large networks efficiently.
- **Configuration Management**   In switched, connection-oriented networks, maintaining and tracking network configuration is a complex task. It requires sophisticated tools to monitor changes in port and Permanent Virtual Circuit parameter settings. Information such as the type of connection made, Committed Information Rate and current state of network need to be remotely accessible and presented in a graphical manner. Switch manufacturers usually offer some automatic way of configuring the devices. But these solutions are proprietary. Integration is not yet visible. Service provisioning is one of the most critical areas. Without the proper tools, ATM network configuration management is very difficult to handle. There are currently four major classes of service to choose from:
  — Available Bit Rate (ABR)
  — Constant Bit Rate (CBR)
  — Unspecified Bit Rate (UBR)
  — Variable Bit Rate (VBR)

Guaranteeing the desired service level involves configuring optimal connection parameters on each switch in the network. In order to do so, traffic priorities, selection of routes, network status, trunk availability, and other prioritiess must be properly set.

- **Accounting Management** The huge volume, coupled with different types of traffic generated in an ATM network, place a great stress on accounting and billing. Private organizations, as well as carriers, need to determine and map network traffic to individual users. A billing system based on actual usage is required to generate accurate cost accounts. Vendors are becoming aware of this need and are beginning to address it in their management solutions. Monitored data can flow into a standards Bellcore Automatic Message Accounting system or into customized billing records. It is a great progress, but still represents a proprietary solution. Network managers require this type of application from a third-party-vendor in order to integrate with their ATM switch management system. Unfortunately, this does not exist. Custom development is extensive and expensive.
- **Performance Management** In an ATM network, there are an increased number of connections to be dealt with. In order to manage and to evaluate performance, massive amounts of statistics must be collected. The inadequacies of the existing SNMP management platforms and the SNMP protocol also are applicable here. There is too much network traffic and too many statistics to be collected and presented in real time. Although several switch vendors claim that their management software can collect and process volumes of ATM network data in real time, they still need to present it in a manner that makes sense for operators and analysts. Accessing statistics from the switches using the fast and efficient TFTP or other techniques is just the beginning. Making this data available for network managers to easily interpret, store, and integrate with other management data is the more difficult job.
- **Operations Management** This is one area where network managers need to think carefully in order to identify their operations concepts, processes, and procedures. Hardware and software vendors provide management capabilities, but it is the operators, technicians, and support staff who really manage the network. Accordingly, each organization needs to define and address issues such as staffing and equipment sparing policy, trouble-ticket correlation and escalation process, internal roles and responsibilities, and a detailed operator's certification program on an individual basis.
- **Resource Planning** ATM networks require intelligent design and capacity planning tools capable of network modeling. If design applications are optimized to leverage management data provided by the specific types of switches, the carriers and end-user organizations deploying these switches will benefit. These applications can run "what if" scenarios based on such parameters as line cost, bandwidth availability, and service outages — allowing customers to prototype new services and get a handle on how growth will affect the presently installed assets. Careful planning is the only way to avoid management problems later on.

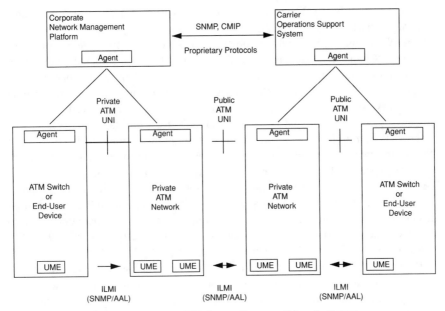

FIGURE 18.1   Standardization of ATM management.

18.2   EVOLVING ATM MANAGEMENT STANDARDS

ATM management standards are in progress. Both the ATM Forum and IETF Atom MIB Working Group are providing specifications for management (ALEX95).

The ATM Forum intends to give at least some opportunities for management. The result is ILMI (Interim Local Management Interface). It is a temporary solution that does not answer every need, but that could help those who are implementing ATM network now. ILMI is using a prespecified ATM virtual connection on the ATM UNI (User Network Interface) to communicate with a management application using SNMP messages. Every ATM switch or endsystem, and every private or public network that deploys UNI must have a UNI Management Entity (UME), which supports an ILMI MIB. The UME resides between the switch and the network or between a private and a public network. It is responsible for maintaining management data and responding to SNMP commands received over the ATM UNI through the ILMI (Figure 18.1).

As in SNMP-based applications, these messages are sent over the network via TCP/IP user diagram protocol packets. These packets are segmented into ATM PDUs, which are then incorporated into ATM cells using ATM Adaption Layer (AAL) 3/4 or AAL 5. The ILMI MIB collects information concerning the physical-layer group, the ATM-layer group, the ATM-layer statistics group, the virtual path connection (VPC) group, the virtual connection (VC) group, the network prefix group, and the address groups. The physical layer group includes such information as the port type on which the ILMI interface resides, the media type (single-mode fiber, multimode fiber, or Category 5 unshielded twisted pair), and the operational status of the port.

The ATM layer group includes objects that specify the maximum number of VPCs and VCs that can be defined for the interface. It also specifies the number of VPCs and VCs that have been configured, and details information concerning the configurations. The ATM-layer statistics group includes information on the total number of ATM cells transmitted and received on the interface, as well as the total number of incoming cells that have been dropped at the interface.

The VPC group includes information concerning the operational status of each VPC on the interface. It also contains data on the traffic descriptors and service parameters specified by both the transmitting and receiving ends of the virtual path. The VC group includes the same information for all VCs using the managed interface. The network prefix and ATM address groups make up the address registration portion of the ILMI MIB. Address registration occurs when the network is set up and when a new end-station is added. The network prefix group specifies the ATM address prefix for the specific network.

ILMI is able to collect statistics on a per-connection basis. Each ATM switch interface card gathers the ILMI MIB statistics and forwards them to the SNMP management console. Once collected, statistics are parsed and inserted into an SQL database or other formats.

The Internet Atom Working Group defines and develops the necessary management objects using standard SNMP protocol for managing ATM devices. The original goal was to concentrate on ATM PVCs that provide management of the ATM interface as well as an end-to-end view of the ATM service. The group has based its efforts on the SONET MIB, the DS1/E1 MIBs, and the ILMI MIBs. Since ATM technology can be implemented both in LANs and WANs, the group needed to understand and define managed objects for multiple views, including a local view of ATM-based communication over the local interface, a network view of the local interface, private and public ATM switch management, private and public ATM-based service information, and an end-to-end view of the communication path.

The current version of the Atom MIB defines object groups that can help to construct some of the views. It specifies several SNMP managed objects to manage ATM interfaces, ATM virtual links, ATM cross-connects, AAL 5 entities, and AAL 5 connections supported by ATM hosts, ATM switches, and ATM networks. The MIB uses SNMP structure of management information to describe these objects. Objects can be accessed by SNMP management applications. This specification is limited in comparison to the ILMI MIB; it does not manage SVCs, signaling functions, and ATM services.

The Network Management Working Group of the ATM Forum defines and specifies the end-to-end management model that includes both private and public network services and that lays out standards for interworking between the two. The model also defines gateways between SNMP and CMIP systems, and between standards-based and proprietary systems. Five key management interfaces are defined (ALEX95).

Figure 18.2 shows these interfaces, labeled M1 through M5. M1 and M2 define the interface between the network management system at the customer site and an ATM endstation or network switch. Because SNMP is so widely used by endusers, M1 and M2 embrace SNMP-based specifications. These include MIB II and relevant

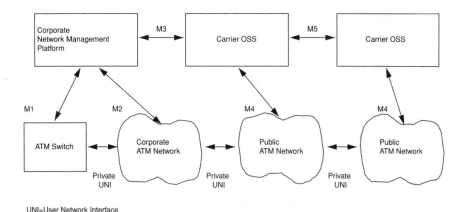

FIGURE 18.2   Management interfaces.

standard MIBs for DS1, DS3, and SONET/SDH connections. Also included is the Atom MIB.

Also encompassing these MIBs is M3, the Customer Network Management (CNM) interface. M3 describes the interface between the customer and carrier management systems that gives the customer a view into the carrier's network. The merger of public and private networking technologies begins at M4. M4 is the management interface enabling Network Management Level (NML) views and Element Management Level (EML) views to the carrier's network management system and the public ATM network. Consequently, this is where the differing approaches to management must converge, since both the private network manager and the carrier want to be able to monitor the service. The manager of the private enterprise wishes to take advantage of public services yet retain control over them to guarantee quality of service. The carrier, on the other hand, wants oversight of customers' networks, which would then give it the ability to offer network management as a value-added service.

M5 is the management interface between a carrier's own network management systems. M5 is the most complicated of all interfaces. Finalizing it will take a considerable amount of time on behalf of standardization groups.

Furthermore, the group specifies OAM cells targeting fault management including alarm indication signal (AIS) cells and far-end reporting failure (FERF) cells which communicate failure information throughout the network. Also OAM loopback capability cells have been specified to verify connectivity and diagnostic problems that AIS or FERF cells cannot. These cells, and many more, provide the information basis for standardized or proprietary management solutions.

TMN layers can be applied to ATM management. For managing ATM resources, only the bottom layers are important (Figure 18.3) (FOWL95).

Network Management Layer (NML)   The NML provides a management view of the ATM network that is under one administrative domain. Through the EML, it can manage subnetworks or network elements. It can provide detailed views of the portions, or segments, of the ATM connections within its domain.

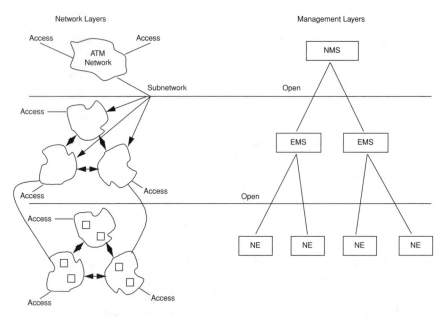

FIGURE 18.3   Open and proprietary interfaces for ATM.

Element Management Layer (EML)   Network elements of similar type are man-aged at this layer. These network elements may be managed individually, or may form a subnetwork. Aggregating nodes into subnetworks for ATM may be advanta-geous in order to manage equipment from the same manufacturer. The EML may filter message traffic going to the NML, such as network element alarm correlation).

Also the element layer contributes to the management by performing basic functions for ATM equipment, such as detecting faults and counting errors.

18.3   BASIC CHARACTERISTICS OF NORTH CAROLINA INFORMATION HIGHWAY

The North Carolina Information Highway (NCIH) is the first wide-scale public deployment of ATM technology. In its initial architecture, users of the NCIH are connected via 155 Mbps fiber to 12 ATM switching systems. It started with approx-imately 30 universities, community colleges, schools, hospitals, prisons, and gov-ernment facilities. It is expected that more than 3,000 state government sites will be connected over the next decade.

18.3.1   NCIH SERVICES

The services provided by the NCIH are designed to be able to support numerous emerging applications that are expected to provide significant benefits in the areas of

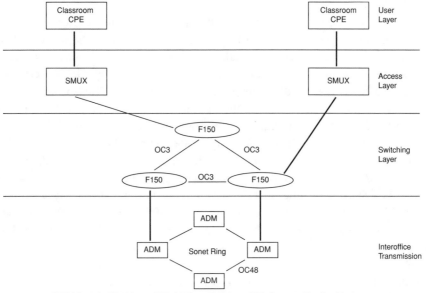

SMUX=Statistical Multiplexor ADM=Add/Drop-Multiplexor CPE=Customer Premises Equipment

FIGURE 18.4   NCIH architecture.

education, health care, crime control, economic development, and government. The initial services include ATM cell relay, SMDS, and circuit emulation. These services require multipoint video bridging, scheduling, and advanced network management.

18.3.2 NCIH ARCHITECTURE

The services described above are provided by a flexible architecture, shown in Figure 18.4. This architecture is expected to support the initial as well as future applications. The layout of the network is shown in Figure 18.5.

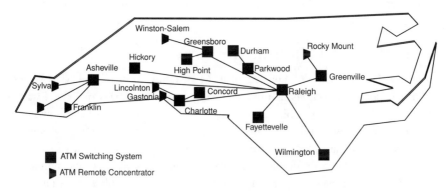

FIGURE 18.5   Location of ATM switches and concentrators.

Single mode fiber is deployed to each site to provide a high degree of environmental isolation, extremely high bandwidth capacity, excellent bit-error-rate performance, and high reliability. SONET transmission is used on both the access and the interoffice fibers. This provides a standardized interface for connecting optically to customer equipment, to other local exchange carriers in the same local access and transport area (LATA), and to interexchange carriers. SONET also provides the capability to monitor the status of the optical links continuously and to provide standardized operations and maintenance capabilities. Initially, the 155 Mbps OC-3 rate is used between each customer site and the central office, with the capability to grow to 622 Mbps, 2.4 Gbps, or even higher access rates as needed in the future. The rates used between LATAs will be dependent on the availability and cost of inter-LATA SONET facilities that the state will obtain from interexchange carriers.

ATM cells are the fundamental currency for information exchange in the network. At the customer premises, existing interfaces such as DS1s, DS3s and SMDS are converted to ATM cells by the ATM multiplexer. The ATM service multiplexer is customer-owned equipment with interface cards provided based on the customer's specific needs. Initially, two types of interface cards are in use: one for DS3 circuit emulation using AAL Type 1 and one for SMDS using AAL Type 3/4. The DS3 circuit emulation card is used for DS3 video connections, and the SMDS card is used for connectionless data communications. Other interface cards such as DS1 circuit emulation, FDDI, Ethernet, Token Ring, and ATM Cell Relay for connection to premises-based ATM switches are expected to follow.

The cells from the ATM service multiplexer are carried through one or more ATM switches to the destination, where the original interface is regenerated by another ATM service multiplexer. In some cases, the ATM connections are terminated in the network for processing. For example, cells carrying SMDS information are carried to an SMDS message handler that routes the SMDS packet, based on the destination address contained in the header of the SMDS packet. Also, for distance learning service, an analog baseband video signal is generated from the DS3 digital video information at the Multipoint Control Unit to enable functions such as quad split video bridging.

18.3.3   NCIH ATM EQUIPMENT

The NCIH ATM network consists of ATM switching systems and ATM remote subscriber concentrators that are provided by Fujitsu's Fetex-150 Broadband Switching Systems. All main locations are equipped with these switches. These switches are interconnected by SONET transport facilities. The Fetex-150 Broadband Remote Line Concentrators are located separately from the Fetex-150 switches to support the connection and concentration of remotely located subscribers to the switches.

An ATM Service Multiplexer is provided by Fujitsu's SMX-6000 at the customer premise for the data link switching (DLS) application. The SMX-6000 converts the subscriber's services into ATM cells before transporting the traffic to the ATM

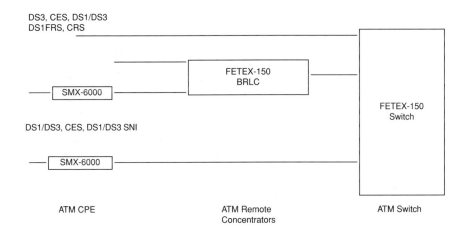

DS3, CES, DS1/DS3
DS1FRS, CRS

FETEX-150
BRLC

SMX-6000

FETEX-150
Switch

DS1/DS3, CES, DS1/DS3 SNI

SMX-6000

ATM CPE

ATM Remote
Concentrators

ATM Switch

BRLC=Broadband Remote Line Concentrator
CPE= Customer Premises Equipment

FIGURE 18.6   NCIH ATM nodes and user interfaces.

network. NCIH user interfaces reside remotely on the SMX-6000 and directly on the Fetex-150. DS3 Circuit Emulation Service can be provided directly on the Fetex-150 switch, the Fetex-150 line concentrator, or the SMX-6000. Figure 18.6 shows the basic topology for the ATM nodes in the NCIH and the user interfaces.

The Fetex-150 switch includes a highly reliable OAM&P platform with full redundancy at critical component levels throughout the system. When a duplicated system component failure is detected, the switch will automatically place a standby unit or slave unit online, initiate appropriate alarms, and generate a fault isolation message. The software design uses hierarchically structured memory to provide control and protection against loss of data. The Fetex switch supports provisioning, fault diagnosis, performance monitoring, site growth planning, and traffic management for ATM PVS service, SMDS, and frame relay. It also supports functions for automatic testing, diagnosis, reconfiguration, and recovery. There are three interfaces:

1. Craft interface provides access to the switch for command input and message output for various operation and maintenance activities. It supports both local and remote operations, using either a simple command and terminal display or by going through a graphical user interface. The command line craft interface uses TL1 format commands developed by Fujitsu for broadband network operation. This interface facilitates configuration, fault, performance, security management, and data collection capabilities.
2. Switch control center systems interface helps remote monitoring and control of the Fetex switch. All sub-interfaces, such as those used for

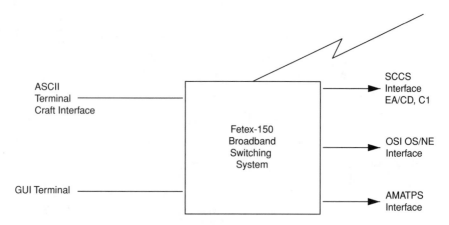

SCCS=Switching CenterControl Systems
EA/CD=Emergency Action/Control Display Channel
CI=Critical Indicator Channel
OSI=Open System Interconnection
OS/NE=Operations System/Network Element
AMATS=Automatic Message Accounting Teleprocessing System

FIGURE 18.7   Fetex-150 operations interfaces.

emergency actions (EA), critical action indicator (CI), and maintenance
channel, are supported by the Fetex switch. It also supports command/port
assignment for both autonomous messages and input commands.

3. OSI-based OS/Network Element Interface is supported as well. This interface has been defined by Bellcore. Its requirements have been refined and
simplified to facilitate rapid implementation.

Figure 18.7 shows the operations interfaces of the Fetex-150 switch.

18.4   MANAGING THE INFORMATION HIGHWAY

Management of the network is segmented due because of switch ownership. Bell-
South, Carolina Telephone, and GTE are the owners of the Fetex-150 switches. Since
broadband signaling standards are not yet complete, connections are managed using
a CNM-interface into the broadband operations systems of BellSouth and GTE. Both
management solutions, based on NetExpert of Objective Systems Integrators, will
be described in some depth.

18.4.1   MANAGEMENT COMPONENTS AND FUNCTIONS

The information highway is a complex environment consisting of multiplecarriers
and vendors. Management consists of the following principal components:

• Management of users
• Management of accesses

- Management of connections
- Management of switching and multiplexing devices
- Management of signaling
- Management of the infrastructures
- Management of services

18.4.2   BELLSOUTH SOLUTIONS (DAMO94)

BellSouth is using a combination of existing and new products for OAM&P of the information highway.

The NCIH operations support strategy involves the following support functions:

- **Physical Transport and Switch Support Functions**   The provisioning and maintenance of SONET physical transport network components are handled by existing regional and North Carolina state centers that traditionally support these functions. The majority of broadband switch maintenance functions are supported by the newly created Regional Bell Operating Company (RBOC) center.
- **Service Support Functions**   The service support includes operations support for logical resources associated with the broadband switch including user-network interfaces, virtual-channel connections, virtual-path connections, virtual-channel links, and virtual-path links. The majority of these operations functions will be supported by the new RBOC center.
- **Application Support Functions**   This application support is an end-to-end support layer starting at the customer premises equipment (CPE) on the customer premises, extending through the network, and terminating at the far-end CPE. The functions include end-to-end trouble analysis, end-to-end service order issuance and coordination, trouble receipt, trouble isolation, and help-desk functions. This layer is supported by the Data Customer Support Center (DCSC).

BellSouth uses a simple billing strategy for NCIH. The NCIH customer bill consists of the fixed, flat-rate component per 155 Mb/2 ATM access interface and a usage component per giga cell of usage. For initial implementations, the cell usage is determined based on reservation of DS3 circuits as metered by the Broadband Operations Support System (BOSS).

Existing OSSs were used for OAM&P functions associated with the SONET physical transport network. Table 18.1 summarizes the use of the major existing OSs.

BellSouth has developed BOSS to perform OAM&P functions for the information highway. A high level view is shown in Figure 18.8. The BOSS consists of two distinct components: the Scheduler and the Operations System.

The concept of the BOSSScheduler evolved due to the lack of industry standards for ATM signaling and routing, and the vendor capabilities for supporting switched ATM virtual connections. Since only ATM PVC capability was available, BellSouth developed the concept of the scheduler to simulate switched ATM virtual channel capability. The distance learning application demands on the time required to set up

TABLE 18.1
Use of Major Existing OSSs by BellSouth

Existing OS	Brief description of use of existing OSs in NCIH
Service order control system and service order analysis and and control (SOCS/SOAC)	Used for processing orders for physical circuits.
Trunk integrated record keeping system (TIRKS)	Used for both loop and trunk physical circuit (SONET) design, physical circuit inventory and assignment provisioning of physical circuit, and inventory/assignment of ATM switch ports.
Plug-in inventory control system (PICS)	Stores and maintains plug- in inventory and supports provisioning process.
Network and service data base (NSDB)	Stores and maintains customer service records describing physical circuit facilities and transport equipment.
Network monitoring and analysis (NMA/Facility)	Used for physical circuit facilities monitoring.
Work force administration (WFA) family of OSs	Supports consolidated work force administration functions for service activation and service assurance.

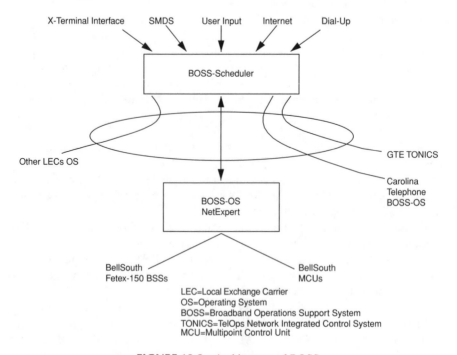

LEC=Local Exchange Carrier
OS=Operating System
BOSS=Broadband Operations Support System
TONICS=TelOps Network Integrated Control System
MCU=Multipoint Control Unit

FIGURE 18.8   Architecture of BOSS.

ATM connections could only be met by the scheduler. This arrangement serves as an interim step for the migration to switched ATM connections. However, the use of the scheduler for setting up full-service classroom scheduling still will be required, even with the availability of switched ATM virtual-connection capability.

The BOSS Scheduler supports a customer interface designed specifically for the distance learning application. The BOSSScheduler is operated by BellSouth, and its use is shared by all participating carriers. The end-user-interface is menudriven and supported on a simple, low-cost TTY-terminal interface. It can be accessed by the end-user via dial-up line over modem, Internet, SMDS, or via direct terminal interface.

It performs scheduling, routing, and resource leveling, and generates usage information for billing. The routing of virtual channels across the multicarrier ATM network and location of the video bridging by the multipoint control unit (MCU) is determined by the scheduler. It computes the minimum cost routing across the network. Also usage-based measurements are available at the scheduler. The BOSSScheduler generates the billing information electronically to be downloaded to BellSouth and other carriers' mainframe-based billing systems. It interfaces to the other carriers' OSSs via an agreed upon Informix database schema, which is accessed using standard SQL language over an FTP or a TCP/IP interface. The same interface is used by the scheduler to communicate with BOSSOS.

Its platform consists of SunSparc hardware, Solaris operating system, Informix database engine, and Informix 4GL language.

The BOSSOS configures and maintains Fetex-150 BSS, and the MCU. This network management application will configure the network for the distance learning sessions. It communicates with the MCU controller at each central office using a BellSouth proprietary protocol, called Simple Connection Management Protocol (SCMP). The BOSS OS conveys the information on required video sessions to the MCU controller using SCMP, and the MCU controller in turn establishes the required DS3 cross-connections in the MCU. The BOSSOS also communicates with other carriers' operations systems for end-to-end multinetwork distance learning service configuration and management.

BOSSOS uses the point-and-click, object-oriented system building tool, NetExpert. The NetExpert tool set includes SNMP, CMIP, and generic gateways. Generic gateways allow the application expert to build user interfaces and applications to bridge between modern standards based management systems, existing OSSs, and vendor-specific proprietary systems. This tool set allowed rapid development of the BOSSOS.

The overlay operations system supports both service activation and service assurance processes. Service activation refers to activities associated with the establishment and/or modification of customer service. From the NCIH customer's perspective, the service activation process begins and ends at the data customer support (DCSC) center. It accepts customer requests for service and ensures that it is provided as desired. In order to complete its task, the DCSC communicates with other operations centers and organizations including the RBOC, responsible for controlling broadband switch management activities. Service activation consists of three phases: service inquiry process, service request form process, and service provisioning process.

Service assurance refers to a set of activities required to maintain the network and associated services on an ongoing basis. The NCIH service assurance process was targeted for troubleshooting customer problems and maintaining service. From the NCIH customer perspective, the service maintenance process begins and ends at the DCSC, which is responsible for ensuring that each customer problem is properly completely corrected and/or repaired. The DCSC will coordinate/trigger maintenance activities and has control over closure of the customer's trouble. In addition to existing centers, the RBOC assumes responsibility for maintaining the ATM switches. ATM remote network elements, MCUs, service multiplexers, and the associated service level aspects. The service assurance has two aspects:

1. Reactive maintenance process
2. Proactive maintenance process

In summary, both BOSS management systems manage not only network components, but also networking services.

18.4.3   GTE Solutions (PATH94)

To support the NCIH and its services, GTE took an integrated approach and developed new capabilities for integrated management of broadband networks. These include:

- **Functionally Integrated Network Management** enabling GTE to monitor and control ATM, SONET, and other related networks in real time; these capabilities are provided by the TelOps Network Integrated Control System (TONICS).
- **Automated Service Management** enabling GTE to provide service-level configuration and fault management for new broadband services; these capabilities are provided by the Broadband Service Management System (BSMS).

Both provide a complete and integrated solution for managing broadband networks such as the information highway. The following goals have been pursued during the development and deployment phase of the management systems:

- Integrated network view should be provided for all contributing network elements and services
- Integration of network and service management
- Integration with the existing operations infrastructure
- Integration on the basis of a management framework
- High availability of the management systems
- Flexible and modular design to support new systems

Successful operation of the services depends on the ability to make services available to customers rapidly and provide customers with flexibility for customi-

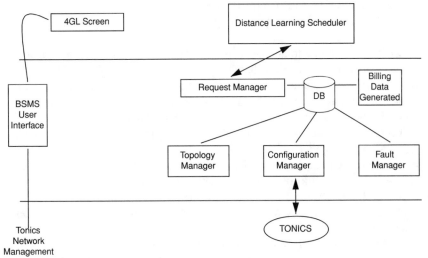

BSMS=Broadband Service Management Systems
TONICS=TelOps Network Integrated Control System
DB=Database

FIGURE 18.9   Architecture of BSMS.

zation. To achieve these goals, BSMS has been developed. It is designed to manage the operations of GTE's nationwide broadband service offerings. In the NCIH, it provides the management of the distant learning service. For this service to be successful, it must be provided ubiquitously across the multiple telephone companies (Telcos) participating in the NCIH network. Consequently, the operations must be provided cooperatively. Figure 18.9 illustrates a high-level architectural view of the BSMS. To provide administrators with the same "look-and-feel," BSMS is also accessed through the TONICS user interface. From this interface, the user can administer the BSMS and monitor ongoing activities in the system.

The *Request Manager* handles the interactions between the BSMS and the Scheduler. Multiple, redundant Request Manager processes can be run simultaneously for additional robustness. Scheduler requests are read into shared memory and immediately written into the BSMS database. Any of the Request Managers can pick up work from this database. The Request Manager dispatches configuration-related work, such as PVC setup and tear down, to the Configuration Manager, which directs TONICS to effect the configuration. The messages from the Scheduler include a high level of specification of the end-to-end virtual connections, indicating the classrooms, serving switches, and the MCU.

The *Configuration Manager* maps these data to the next lower level of abstraction in terms of virtual channel links (VCLs) — virtual connections from the SMUX to the switch, MCU to the switch, and between the switches. TONICS, in turn, maps the VCLs to switch and MCU-level entities, creates appropriate commands, and sends them to the switch or MCU. The Configuration Manager is independent of the manner in which TONICS accomplishes its tasks. TONICS has implemented two alternatives to configure the Fetex-150 switches:

- Using an OSI CMIP interface on TCP/IP
- Using a TL1-like interface through a serial maintenance channel

Switching from one to another is transparent to the Configuration Manager.

The *Fault Manager* handles changes in operational status of the PVCs. TONICS notifies the BSMS of status changes in VCLs and their cross-connections. The Fault Manager determines which PVCs and their distance learning sessions are affected, and notifies both GTE personnel as well as the Scheduler.

The *Topology Manager* is responsible for notifying the Scheduler, through the Request Manager, about changes of GTE topology, such as the addition and removal of classrooms, MCUs, switches, and trunks. This is essential, as the distance learning scheduler needs network topology information to decide PVC routes and resource assignment. Internally, the BSMS topology database is synchronized with that used by TONICS. The Topology Manager is activated only when changes are made to the topology database.

The *BSMS Billing Data Generator* runs daily to extract distance learning session information from the database and to store it in a repository, so that billing records then can be created by the GTE Billing Center. Since the switch is unable to provide usage measurement, the burden of collecting usage measurement for billing purposes resides with the operations support system.

The BSMS is administered and monitored from a Motif-based user interface. Typical administrative tasks include starting, shutting down, and checking the health of the system. Online monitoring can be done for all the activities performed by the BSMS, such as the interaction with the Scheduler and TONICS, and the setup/tear down of PVCs.

TONICS enables GTE to offer ATM-based services in a reliable, cost-effective, and expeditious way. Because TONICS was developed as an integrator, the cost of adding support for the information highway network was minimal. TONICS is based on a network model that descibes all network objects pertinent to network management, their attributes, and their status and dynamic relationships. In this model, the network is partitioned into functionally-like network components. TONICS also includes a data model of the network operations. The data model is needed to identify names and addresses of managed objects, group them into monitoring units, track subnetwork controllers and related OSS, and administer the identities and privileges of TONICS users. Data in the network and operations models form the TONICS MIB, which comprises individual element MIBs as well as higher-level network and operations objects. Some of the data require persistent storage in the TONICS database, whereas other data can be retrieved from remote database servers.

TONICS is a distributed client/server-based system as shown in Figure 18.10. Servers support communications with network elements and subnetwork controllers, as well as basic application-specific data storage and manipulation. They also provide basic fault, performance, and configuration capabilities. TONICS clients supply a graphical user interface for individual login sessions. Clients transparently log into graphically distributed servers. Depending on the types of windows opened by the

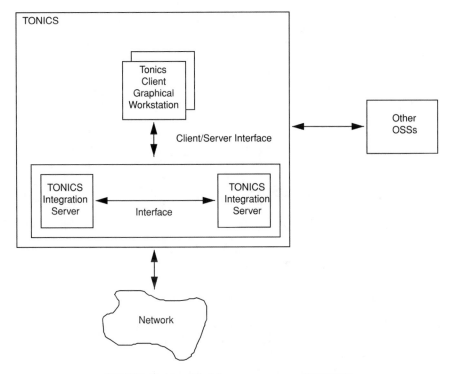

FIGURE 18.10   Client/server structure of TONICS.

user, the client software queries for historical data, establishes real-time data flows, or requests other network management services.

The TONICS server is a multiprocess system that provides communications, storage, retrieval, routing, and correlation functions for network management.

Network interfaces have been developed to communicate with intermediate systems such as subnetwork controllers or mediation devices, and with network elements directly. Supported protocols include X.25, TCP/IP, UDP/IP, SNMP, and CMIP. To support the information highway, it was necessary to develop two new network element interfaces:

- Fetex-150 interface based on CMIP
- Maintenance channel

Figure 18.11 shows some of the interfaces of TONICS.

The *Object Manager* decodes raw messages from network elements and associates them with managed objects in the TONICS MIB. It reads and parses incoming messages and identifies the reporting object. For newer generation equipment, this can be done using the vendor-supplied MIB. For legacy equipment, the TONICS MIB must be derived from an analysis of the network element. The Object Manager makes it possible to offer more detailed information to clients. For example, once

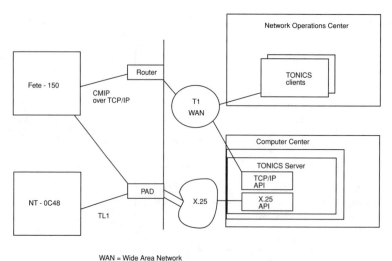

WAN = Wide Area Network
TL1 = Transaction Language One
API = Application Programming Interface

FIGURE 18.11   TONICS interfaces for NCIH.

a specific managed object has been identified as the source of a critical alarm, the client can graphically display a logical or physical configuration diagram, coloring the faulty component.

The *Database Layer* supports inserts, queries, updates, and deletes to and from the persistent TONICS data store. The database layer handles inputs both from the network and from clients. Also garbage collection is built in. The separate database layer provides advantages such as insulating other software from changes in the schema, providing optimized memory-based access to frequently used objects, and hiding cross-server database distribution from clients.

Network Management Applications provide automation and decision support for the key network management functional areas. For example, in fault management, there is support for automatic clearing, thresholding, escalation, holding, and other special processing for alarms that match given regular expressions. Other facilities include user notification of network elements that have fallen silent, and automatic or manual escalation of unresolved network conditions.

The *Message Router* routes incoming real-time network data to subscribing clients and other server processes. Subscriptions can be for fault, performance, or configuration data from specific network elements. For example, a client may subscribe to a stream of critical alarms from all ATM switches, or a client may get specific performance measurements and exceptions from SONET ADMs. Users work with windows and filters; the mechanics of subscription request, cancellation, and hand-off in case of server failure, are transparent to the user.

The *System Manager* initializes the other server processes. It is responsible for creating and managing the basic interprocess communication resources used by communicating processes. After startup, it monitors the health of server processes and the interprocess communications subsystem, providing recovery if necessary. It also handshakes with other servers and performs backup and restore operations as necessary.

The server also includes system administration tools for managing the runtime state of the server, administering the database, and generating reports on system performance. Different data flows through the server are triggered by user operations, network and scheduled events. Display objects are selectable with operations tied to them through context-sensitive pop-up menus.

18.5 SUMMARY

The carriers are expected to work more closely together. Next releases of the four principal management systems, such as BOSS-Scheduler, BOSS-OS, TONICS and BSMS, will incorporate the latest extentions of the NetExpert offerings:

- More applications for the framework
- More distribution of functions, including gateways
- Better peer-to-peer capabilities using standard interfaces for information exchange

Altogether, scalability will play a key role because the targeted number of end-users requires additional ATM switches, more SONET rings, and more interaction with signaling networks.

Part V

Management of Broadband Services

The broadband demand is constantly growing because of the deployment of new applications such as video networking, high-speed Internet access, and telephony over cable. Huge and complex networks are built that bring broadband into residential areas. Today, customers demand easy network access, entertainment, and interactive services.

Everything about broadband is new — consumer expectations, transport protocols, business processes, and vendor alliances. These factors make it one of the most difficult telecommunications and networking challenges today. Network managers are not simply faced with integrating various hardware components, but must create an entire industry from scratch. With NetExpert's flexible rule-based dialogs and expert system technology, service providers can perform well in the areas of service creation, activation and assurance.

The first part addresses Telco mass-market broadband solutions. Telco mass-market broadband networks can be characterized by rapid hardware and software changes, phased deployment, a number of pilot implementations, evolving business processes, uncertainties within the regulatory environment, and new competition from the CATV networks.

Chapter 20 concentrates on voice and data over cable. Additional competition is made possible through the anticipation of continued deregulation combined with recent technological advances. Technological advances have created an environment in which alternative service providers such as competitive access providers (CAPs), cable operators, and wireless operators can deliver telephony service to the same home for potentially less than half the cost of the local exchange carrier (LEC).

However, alternative providers still face many challenges before they can effectively compete in the telephony market. The key challenge is of deploying operations support systems (OSSs) that are low in cost, easy to implement, and open and flexible enough to allow changes as businesses grow and change.

Chapter 21 introduces loopMASTER. It is a prepackaged rule-based application component that provides configuration, performance, and fault management with root-cause correlation of alarms, for access network elements including HTDs, NIUs, video headend equipment, cable data modems, and HFC transport elements at the subnetwork level. Multiple loopMASTER modules can be implemented across regional domains and can build a national or enterprise OSS environment providing a single view of the network. It integrates well with other rulebased application components, as well as as iSAC, and with trouble and workforce management systems.

19 Telco Mass-Market Broadband Solutions

CONTENTS

FIGURE

As the rush to provide integrated voice, video, and data continues, many service providers are finding the term "digital convergence" to be something of a misnomer. Rather than leading to a merging of existing cable and telephone technologies, the new paradigm may instead call for a total replacement of network infrastructure, service architecture, and business processes. Operations and engineering groups have discovered that existing management and operational support systems simply cannot handle the wide range of new hardware, software, and services.

At the same time, consumer expectations have been heightened by the prospect of having the much-touted information superhighway pass through one's own living room. Reliability, ease of use, speed of repair — all will be closely watched — both by customers often skeptical of new services and by governmental agencies weary of service providers spending rate payer funds on unproven technologies. And with the eventual provision of lifeline services such as voice telephony and business data across a common conduit with television, system uptime must equal or exceed that of the current telephone networks.

Regulatory changes promise to bring digital convergence and with it a new age of competition and ingenuity to the market. Regulators also promise to force service providers to shorten product-development time and to accelerate the service creation cycle.

19.1 CHALLENGES IN MANAGING MASS-MARKET BROADBAND

The challenge, then, is to provide exceptional service from day one — exceptional service over a network composed of largely untested components, often far from their final forms, supported by business organizations and processes still under

323

development. Such an environment demands support systems capable of immediate change and reconfiguration, offering full mission-critical duty cycles, and able to grow from managing a small pilot implementation to running a full metropolitan system with a minimum of rework.

To meet these needs, Objective Systems Integrators has developed a suite of mass-market broadband solutions based upon its NetExpert framework. OSI's solution set provides full network management and OSS functionality for telco multimedia broadband networks.

Characteristics of telco mass-market broadband networks are:

- Hardware and software change rapidly. Since much of the technology is at present unproven in the field, functionality is often modified almost weekly. Management interfaces in particular remain in flux long after the hardware is delivered; rarely does the initial release of a product fully support an industry-standard management interface such as SNMP or CMIP.
- Implementations are phased and mutable. No service provider has proposed to deliver a complete next-generation network on initial turn-up. Neither technology nor business projections are there to support such an effort. Instead, most networks offer one or two aspects of a full-service network — analog and digital video, for instance, or perhaps high-speed Internet access — and plan to provide additional services as the business matures.
- Initial implementations are pilot studies. Often, first offerings are in areas covering merely 5,000 to 10,000 homes passed, with anticipated subscriptions of 20% or less.
- Business processes must be created. Lately, the emphasis within the telecommunications industry has been upon business process reenginering. Mass-market broadband networks, on the other hand, call for process engineering. There are few existing benchmarks to emulate in developing a support infrastructure, and early proponents are finding that neither cable nor telephony models are sufficient.
- Multimedia networks will give way to integrated broadband networks. Today, the force driving mass-market broadband is the promise of consumer-oriented multimedia services: Internet access, video-on-demand, interactive gaming, home-shopping. At the same time, the sheer power of the networks being installed make it inevitable that they will eventually be used to carry voice and video telephony, business data, telemetry, and other services.

Objective Systems Integrators has assembled solutios suites that address the needs of the major mass-market broadband configurations today. The suites include the NetExpert platform, one or more gateways and ruleset modules that can be combined to address a number of networking needs, and the necessary support required to tailor the implementation to a particular set of components and applications.

NetExpert may be used as an integrator for end-to-end management. Figure 19.1 shows an example for end-to-end management, showing the full connectivity between selected entities.

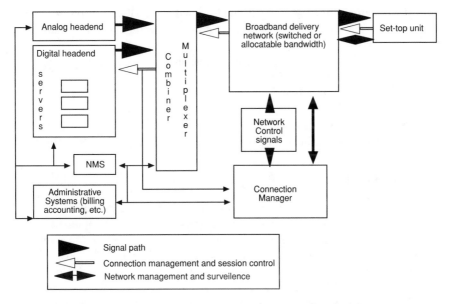

FIGURE 19.1 Mass-market broadband network, regulated environment

19.2 MANAGEMENT IN AN UNREGULATED ENVIRONMENT

The components of OSI's Solution Suite for full-service networks include the following modules:

- **Network Management** Allows for monitoring and control of analog and digital head-end and delivery network components, event analysis and correlations, automatic testing, process automation, and trouble tracking.
- **Connection Management** Provides bandwidth allocation on demand through control of ATM switches or other delivery devices.
- **OSS Integration** Allows network management and business support systems to share data and messages.
- **Data Collection** Supports the aggregation and normalization of fault, performance, and billing data from a number of sources.

19.3 MANAGEMENT IN A REGULATED ENVIRONMENT

For multimedia broadband networks, OSI offers the following:

- **Network Management** Allows for monitoring and control of analog and digital head-end components, control and service delivery networks, event analysis and correlations, automatic testing, process automation, and trouble tracking.

- **Connection Management**   Provides bandwidth allocation on demand through control of ATM switches, cross-connects, and other delivery devices.
- **OSS Integration**   Allows network management and business support systems to share data and messages.
- **Data Collection**   Supports the aggregation and normalization of fault, performance, and billing data from a number of sources.
- **Spectrum Management**   Controls the allocation of spectra within the delivery network, including the placement and sizing of return paths.

19.4 SUMMARY

In an environment where hardware and software is upgraded, enhanced, or retired nearly every day, service providers need a powerful and flexible solution to effectively beat the competition. NetExpert's robust architecture provides that solution. Current investments in nonswitching network twisted pair, and CATV coax systems are migrating toward hybrid fiber-coax. Future implementations will include ATM, SONET/SDH, and FTTC. NetExpert is experienced with those technologies. NetExpert's open suite of network management applications effectively manages hardware like video servers and MPEG-compliant mux/mods still in the prototyping stages. NetExpert gets the edge on handling the difficult network terrain required to support the emerging broadband industry.

The last chapter of this segment — will introduce a product called loopMASTER that is a domain-oriented prepackaged ruleset on top of the NetExpert framework.

20 Voice and Data Over Cable

CONTENTS

FIGURES

Fueled by competition and deregulation, information service providers are rapidly designing and deploying network infrastructures to bring the promise of the "Information Superhighway" into reality. With billions of dollars in future revenue at stake, the winners in this race will be those organizations that can get to market fast with the right services at the level of quality the customer has become accustomed to with the traditional telephony network.

Competition made possible through continued deregulation combined with recent technological advances has thrown information providers, such as cable providers and the local telephone companies, into a race for market share. Technological advances have created an environment in which alternative providers such as competitive access providers (CAPs), cable operators, and wireless operators can all deliver telephony service to the home for potentially as little as half the cost of the local exchange carriers (LECs).

20.1 CHALLENGES TO CABLE COMPANIES

Cable services providers can enter the competition for voice and data services. Depending on the country, there are millions of households and businesses with cable television connections. In the majority of cases, cable television is a distribution channel supporting one-way communication only. Using cable modems, channels could be provided for two-way communication, allowing consumers to send back data or use the cable for phone conversations.

Figure 20.1 shows the structure of such an arrangement. The structure is simple and requires just a few additional networking devices. The components are:

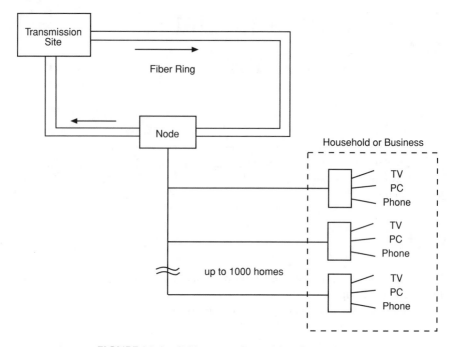

FIGURE 20.1   Cable access for multimedia services.

- The equipment at the transmission site generates television signals and houses switches that route phone calls and monitor the network.
- Parallel fiber optic lines spread out over the area. If a line is out, traffic moves almost instantly to another route.
- Electronic nodes convert signals for transmission via coax cables. One node can serve between 500 and 1,000 homes.
- Coaxial cable runs into a box on side of the home. Electronics in the box split signals, sending phone traffic over regular phone lines, and television signal over cable lines to the television set.
- Phone traffic would use on the existing wiring. Television could be interactive, allowing signals to flow back up the cable line. A personal computer also could be connected through the cable-modem, which transmits information up to 1,000 times faster than phone lines.

Turning cable television systems into local phone networks is not easy nor cheap. Plans are ambitious to compete with local service providers. A basic cable system starts out as copper coax cable that carries signals in a line from the head-end — where TV signals are generated — to each consumer. Signals only go one way; in case of cable cuts or other damages, all the consumers from that point forward are out of service.

A system designed to handle phone calls and interactive TV looks much different. The main trunks of the network are set up in interconnecting rings of fiber-optic lines, which can carry thousands of times more information than copper cable. If a line is cut, traffic moves to another ring in a microsecond and practically no one loses service. Because phone calls are two-way, a phone-cable system must be two-way and include the sophisticated switches that send calls to the right places.

About 40% of today's cable TV systems are still copper. The other 60% have some fiber in their trunk lines, but most of them need an upgrade. Just a few are set up in rings and have switches.

Power could cause another problem. Telephone lines have just enough electricity running through them to power the phone and the network, independent of the power grid. If a storm knocks out the main electrical grid, typically the phone will still work. Cable lines don't carry power, and if electricity goes out, so does cable service. In this scenario, a phone attached to an unpowered dable network would not work. Cable lines, however, are capable of carring power. Adding power from a node over the coaxial cable also could add noise. There is a debate today in the industry about whether power should be provided on the cable-phone network, or another solution used, such as attaching backup batteries to the side of the households.

Once the pieces of the network come together, cable companies will face other issues. One is number portability — the ability to let consumers keep their phone numbers if they change phone service from the local phone company to the cable company. Cable-phone systems also will have to prove to a skeptical public that they can be as reliable as existing phone systems, which almost never fail. Cable systems have to overcome a reputation for failing frequently.

From a technology standpoint, competition in the high-speed data services market will initially be between a dedicated architecture and a shared architecture. ISDN or other dedicated solutions, like xDSLs and cable-LAN services require infrastructure upgrades, will require significant investments on the part of cable service providers. And while most performance comparisons focus on peak bandwidth, other aspects of network usage, such as customer density and average session time, can also affect cost and quality of service.

The type of upgrade required to deliver high-speed data over cable networks should be fairly obvious: cable television transmission is typically one way, but a data and phone network must permit two way traffic. Limited forms of data service are possible using the telephone line as a return route, but this is not a long-term solution. Some cable networks already have migrated to a hybrid fiber-coax infrastructure, but many operators are still immersed in the upgrading process. Once the networks are tailored for two-way traffic, broadband LAN technology will be incorporated to allow digital data to be transmitted over a separate channel. Most of the vendors that supply technology to the cable operators will use an Ethernet-like approach, where the consumer's computer will be fitted with a network interface card and a cable modem for accessing the cable LAN. Access speed will depend upon the peak rate of the cable modem and the volume of traffic on the cable LAN. Subscribers probably will experience connection speeds that vary according to factors such as usage.

Cable LANs will operate full duplex with two channels, each sending data in a different direction. For customers using Internet or Intranet applications, one of these channels would connect to an Internet or Intranet Point of Presence (POP) router, which would then forward data packets from all users on the neighborhood LAN to and from all other systems on the Internet or intranet. The other channel would receive data from the Internet or Intranet service. Cable LANs are unlike ISDN-based services in that they furnish full-time connections, rather than switched or dial-up links. The obvious advantage is the need for dedicated transmission and switching resources is eliminated. Users can access the cable LAN when needed, and only the cable network makes use of the Internet or Intranet point of presence.

A shared cable LAN requires only a single connection to the Internet or Intranet provider. With ISDN, the lines of many individual subscribers have to be multiplexed and concentrated, and the number of subscribers online at any given time is limited by the number of connections between the provider and the telecommunication network.

In terms of geographical coverage, cable LANs can be quite large. Initially, they are intended to be implemented for residential broadband services. The future target is, however, corporate internetworking.

Currently, there are no standards governing the transmission of data over cable LANs, but this has not slowed down equipment suppliers, which bring proprietary products to the market as rapidly as possible.

The cable modem, which connects the consumer's PC with a coax drop linked to the two-way, broadband cable LAN, is specifically designed for high-speed data communications. When the modem is in operation, cable television programming is still available. However, the return path used by the cable modem is limited and must be shared among other digital services, including interactive television, video on demand, telephony, videoconferencing, and data services. One problem with the cable modem is the immaturity of existing devices. The majority of manufacturers are targeting peak bandwidth between 10 Mbps and 30 Mbps.

Competing with cable technology are ISDN and emerging technologies, such as ADSL, RADSL, HDSL, SDSL and VDSL. These technologies all use existing wires to corporations and to residential customers.

Figure 20.2 shows competing infrastructures.

Because ISDN delivers both voice and data, it makes use of circuit-switching and packet-switching technology. For high-speed data services, the local telephone plant must be upgraded so that two-way digital transmission — over the existing copper pairs that ISDN also relies upon — is possible. In addition, ISDN-capable digital switches must be installed at the telecommunications central office. ISDN may offer faster data transmission than analog modems, but the dial-up access model is similar. From the data transmission perspective, this is an inefficient approach

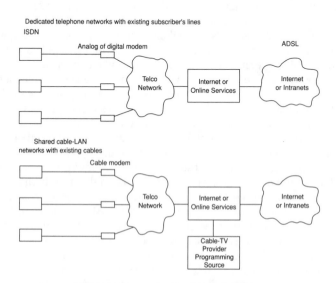

FIGURE 20.2   Competing infrastructures.

because circuit switching requires the service provider to dedicate resources to a customer at all times. In other words, it is not consistent with the bursty nature of data services. The result is wasted bandwidth.

Implementing an ISDN network could end up being more costly than deploying cable LANs. In telecommunications networks, e.g., an assessment of the incremental costs of providing ISDN Internet or Intranet access include the cost of initializing a subscriber's ISDN circuit at the central office; the cost of multiplexers, concentrators, and terminal servers; and possibly the cost of T1 lines from the telco to the Internet or Intranet provider.

ADSL technology sidesteps some of the inefficiencies associated with ISDN. Modern ADSL systems transmit voice and data over the same copper pair; however, the voice is transmitted on an analog signal in its native 0 to 4 KHz band, and the data is transmitted at frequencies above 20 KHz. At the central office, the two signals are split off. The voice signal goes to the PSTN and the data signal terminates at either a router or a multiplexer for transmission to ISPs.

ADSL, after many years of hype but no deployment, appears likely to be deployed in volume by telcos starting in 1998. ADSL was originally conceived as a video delivery technology in support of video-on-demand, but was "too little, too soon" (that is, not enough bandwidth to support a service with no established market demand). As a result, very little ADSL deployment had taken place until now.

A variety of market, technology and regulatory factors have made ADSL viable now.

- DSP technology has progressed greatly since the late 1980s; ADSL is now viable at speeds of 6MBps downstream and up to 600 Kbps upstream at projected costs far below previous estimates.)
- Internet access has exploded, along with the variety and wealth of material on the World Wide Web — as has, but to a lesser extent, working at home. Analog modem speeds are becoming unacceptable because web content has moved from text-based to image- and multimedia-based. This has created a market demand for high-speed access technologies far greater than was realized when targeting ADSL towards video-on-demand.
- Demand for Internet access has wreaked havoc in telco planning organizations, both on a loop "pair exhaust" basis — many carriers are in danger of running out of assignable loop pairs due to unanticipated demand for second, third, and even fourth lines to the home — as well as on a traffic engineering basis — local switching and trunk design were not engineered for dial-up Internet access' free local calls with very long holding times. This has created a demand for a technology which will allow voice and data to be supported on the same physical loop, but keeps the data off the voice network.
- ADSL is a rapid telco response to the competitive threat posed by cable modem deployment
- In North America, the regulatory climate has now become more favorably inclined toward large, incumbent Telcos offering ADSL-based services

Operationally, ADSL is the first time plain old telephone service (POTS) operations have been in juxtaposition with a true data service. Usually, business data services have operated completely outside of POTS, or mass-market operations environments. ADSL is positioned as a mass-market data service. The implications of this are that any operations processes for ADSL must be performed in close proximity to POTS operations. This means that legacy systems will be, at least peripherally, involved in support of ADSL-based networks and services.

The key carrier operations needs for ADSL are:

- Speed/time-to-market, even while operations requirements change (near term)
- "Cheat" the legacy systems (interface with them via surround methods) to automate what would otherwise be manual processes without full-scale legacy system integration (near term)
- Support high-volume deployment in concert with POTS/core operations processes (long term)

ADSL typically offers speeds of 2 to 6 Mbps in the downstream direction (that is, ISP to end customer), and 32 to 640 Kbps in the upstream direction. However, while the complexity of an ADSL modem is roughly on a par with that of a cable modem, it has a potential cost disadvantage in that a dedicated pair of modems are required for each sector. That is, while "n" customers require roughly "n+1" cable modems (one per customer plus one at the headend), the same "n" customers would require "2xn" ADSL modems.

Cable-LAN access might prove to be less expensive for data and online services. It is assumed that cable television providers are facing less financial risks than telecommunications providers when deploying data services. Lower cost translates into consumer savings. The cost per bit of peak bandwidth (ROGE95) in providing Internet or Intranet access is significantly lower for hybrid fiber/coax networks than it is for ISDN — about 60 cents for a 4 Mbps residential service as opposed to 16 cents for a 128 Kbps service. Shared fiber/coax networks also compete well against dedicated ISDN on average bandwidth and peak bandwidth. A 4 Mbps residential cable service, e.g., can provide the same average bandwidth and about 32 times the peak bandwidth of a 128 Kbps ISDN service for 40% less money.

While deployment costs for both technologies continue to decline, ISDN deployment still runs several hundred dollars more per subscriber than cable-LAN deployment. This difference, in addition to the higher performance of cable LANs, will be important factors as Internet, Intranet and online service access become a commodity product. But, the cable-LAN approach and the cable services industry have their shortcomings, too. Cable-LAN modems and access products are still proprietary, so once an operator has selected a vendor, it is likely to remain locked in. Also, shared networks function properly only when subscribers' usage habits are well known. The more subscribers deviate from an assumed usage profile, the more performance is likely to deteriorate. Big changes in usage could impact the cost of providing acceptable service levels over neighborhood cable-LANs.

Finally, the cable industry in many countries is greatly fragmented. This inhibits coordinated efforts, which are vital to the rapid deployment of services. Although telecommunications providers are in competition against each other, they will work better as a group if they act quickly in deploying the resources needed to make ISDN universally available.

In order to avoid these bottlenecks and other fault and performance related problems, cable services providers need a powerful management solution. A complete rethinking of presently used solutions also is required. The management solution should address high volumes of events, alarms, and messages, failsafe operations, traffic control and management, quality assurance of services, and collection of accounting data.

20.2 DESIGNING AND DEPLOYING THE NEW OSS USING NetExpert

The cable industry has one key advantage, the coax cable to the home. Over the last 2-3 years cable operators had continued to invest in upgrading their cable plants from the original "tree-and-branch" architectures consisting of coax feeders, amplifiers, line extenders, and taps to a "telephony capable" Hybrid Fiber/Coax (HFC) network. The use of HFC networks has improved the reliability of the network and lowered the cost to the home.

However, cable operators still face many challenges before they can effectively compete in the local telephony market. Converting from a subscriber based service delivered as one-way analog transmission to a transaction based service delivering two-way digital transmissions represents both a cultural change as well as presenting numerous technical challenges.

One of the key challenges cable operators face in improving the reliability of their service is the implementation of operations support systems (OSSs) that will enable them to deliver telephony services to a public that demands 99.9% reliability. OSSs in this new competitive environment must be low in cost, easy to implement, open and flexible enough to allow change as the business grows over time. To remain at a low price point, cable operators cannot afford the large, expensive, and inflexible OSSs that drive the existing telco operations today. Yet, they must match or better the quality of service provided by the local telco enabled by those seasoned OSSs.

The new OSS, like the networks they support, must be flexible enough to interoperate with a number of network elements, other systems or applications and must be distributable to users with an easy-to-use, common user interface. Individual systems with proprietary databases and protocols cannot fulfill the need to operate in this new operational environment. Just as the new network architectures have been developed in layers with each layer being dedicated for a specific function and communicating through standard interfaces with the layers above and below, the structure of the OSSs must conform to a similar strategy.

Working with some of the key telephony providers in the communications industry, OSI has built a suite of off-the-shelf telephony solutions based on the OSI NetExpert platform. This solutions suite furnishes an information provider with many of the critical OSS functions required to deliver full service telephony.

Packaged as a suite of NetExpert rulesets, these modules allow the information provider to quickly plug in a production ready solution for the targeted OSS. Additionally, because these application modules are built on the NetExpert technology platform, they easily can be molded and modified over time as the management and operations environment changes.

The Telephony-Over-Cable Solutions Suite consists of several modules and interfaces to provide the following:

- Line Side Provisioning
  - AT&T, Northern Telecom, Ericsson, Siemens, Fujitsu Switches, POTS, ISDN, PIC/CARE ...
  - Includes flow-through provisioning of RDTs & HDTs (TR08 & TR303)
- Switch Fault Management
  - Switch "Domain" Manager monitors all switch events and correlates to generate critical alarms. Alarms are reported up to the central switched network console.

- Port Management
  - Manages multiple applications access to limited ports on the switch.
  - Features Time Queueing and prioritization of requests
  - Maximizes switch access, reduces the need for "dedicated" ports.
- Configuration Management
  - Includes Load balancing, Thresholding, Monitoring of Load Parameters.
- Integrated Fault Management
  - Includes collection of alarms from switches, HDTs, and outside plant monitoring systems for rapid fault correlation, sectionalization, and root-cause analysis. Also includes integration with trouble management and workforce management systems.

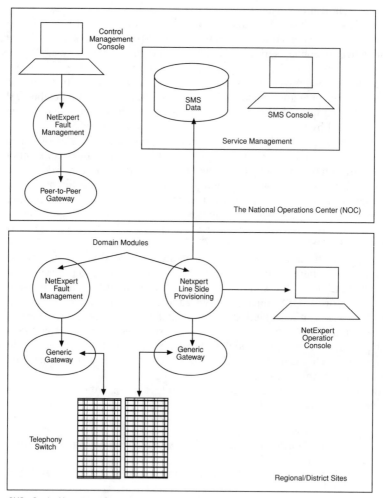

SMS = Service Management System

FIGURE 20.3   Management of telephony over cable using NetExpert.

The Cable/Telephony Solutions Suite are all developed using the editors provided with the NetExpert technology platform. With their easy-to-use interface, network knowledge experts can enter rules that define the interface and management behavior for each of the devices or systems.

Once installed, each of the solution modules provides an immediate, functional and productive environment, and using the editors, changes can be made rapidly, tested in an off-line environment, and redeployed online without any loss of network visibility.

Figure 20.3 shows the NetExpert installation in central and regional offices. This arrangement supports state-of-the-art service management.

20.3 SUMMARY

Telephony over cable is not yet a mature technology. It is an opportunity for further accessing and serving business and residential areas. Reengineering the networks can be done, but it takes some time and requires a lot of investment. The outcome of quality and price battles cannot yet be predicted. But, in any case, without flexible and service-oriented operations support systems, alternative providers cannot compete for a reasonable market share. NetExpert technology may help to shorten the time to market services, to supervise service and network components, and to rapidly prototype in a dynamically changing environment.

21 loopMASTER

CONTENTS

FIGURES

Nowhere is competition in telecommunications being felt more keenly than in the new local loop. The key element for service providers in delivering voice, video, and data services to consumers is local access to the customer. Success will be measured by quality of service at the best possible price.

However, the new local loop is complicated, with many unknowns. Who can say which technology will win the quality vs. value contest. Hybrid fiber/coax (HFC)? Fiber to the curb (FTTC)? Hybrid fiber/wireless (HFW)? Or DSL? Which services will ultimately capture the consumer? Video-on-demand? High-speed data access? Or Interactive services? One thing is certain: the complexity of delivering these services over any of the new infrastructures has escalated dramatically.

21.1 loopMASTER IN THE DOMAIN MANAGEMENT ROLE

Meeting the technological and marketing challenges of the new local loop requires a domain management approach with a loopMASTER solution. loopMASTER easily integrates a broad range of element types and equipment, regardless of interface protocol or supplier.

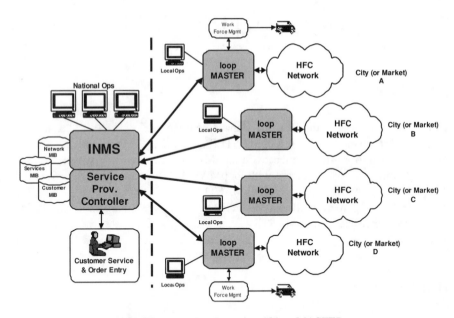

FIGURE 21.1   Configuration of loopMASTER

An ITU-T, TMN-compliant application component, loopMASTER advances one step further, using a sophisticated subnetwork management strategy that distributes intelligence closer to the element level, freeing network management systems for higher, service-related tasks. Figure 21.1 shows multiple loopMASTER modules logically implemented across regional domains.

The loopMASTER application component is an integrated network and element management layer (NML/EML) product which provides comprehensive configuration, fault, performance, and security management of the domain of the "new" local loop. OSI has developed the loopMASTER product, recognizing that the new local loop may consist not only of network elements from multivendors but also of multiple network architectures, such as HFC, FTTC, and HWC.

The application component was built using OSI's NetExpert framework. loopMASTER uses a common graphical user interface (GUI) to combine the functionality and information from multiple element management systems (EMSs) and/or network elements into one integrated application component, which consists of rulesets, dialogs, graphics, and supporting material required for "plug and play" delivery of functionality to the customer.

loopMASTER provides a single interface point to higher-level, operations support systems (OSSs), that is, to other network management layer (NML) or and/or service management layer (SML) OSSs, such as OSI's integrated Service Activation Controller (iSAC) and to other components such as work force management, trouble ticketing, and spectrum management. This single interface point is one of the key value-added benefits of the loopMASTER product, especially for equipment configuration and service provisioning.

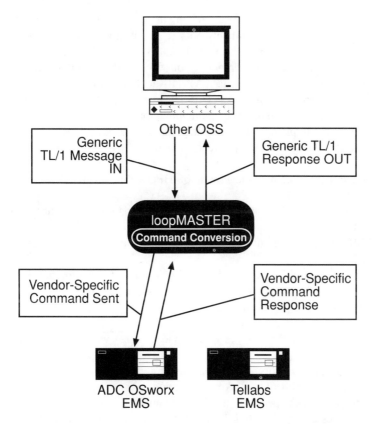

FIGURE 21.2   Command conversion in loopMASTER.

The system defines and utilizes a standard communication protocol interface, common to all equipment EMSs under the loopMASTER umbrella to perform centralized, multivendor, "flow-through" provisioning tasks. The system will take input from other applications via this common protocol interface and manage the process of converting it into the vendor-specific commands required to provide the desired functions (that is, flow-through provisioning of new services). This process is accomplished by coordinating the messages and commands with the respective EMS using standard communication protocols, such as CMIP, SNMP, and TL/1. For example, loopMASTER allows provisioning of new cable telephony services through a standard provisioning protocol across multivendor EMSs for telephony equipment (Host Digital Terminals, Network Interface Units, etc.). This significantly simplifies the internal processes required to support telephony services over the HFC cable plant. Figure 21.2 depicts this concept.

loopMASTER provides root cause correlation of alarms and faults generated by any device or set of devices in the network. This product also allows the automation of corrective procedures as defined by the customer network operations personnel. loopMASTER will receive alarm and event information from various EMSs via standard communications interfaces. Then, it performs intelligent correlation of alarms across all equipment subnetworks to isolate the root cause problem and generate the appropriate network level alarms. These alarms are immediately forwarded to various systems for proactive follow-up and resolution. Alarms are displayed to the network operator via a VisualAgent GUI in near-real time. Alarms may also be forwarded to higher level OSSs, work force/fleet management systems, trouble ticketing, spectrum management application, technician paging systems, etc., depending on the implementation requirements of loopMASTER.

The generic features of the product include:

- Integrates domain management of HFC cable plant, FTTC, or HWC networks
- Provides fault, configuration, performance, and security management for HFC networks
- Offers a single point of management for multiple cable/phone vendor equipment.
- Integrates multiple status, monitoring vendors for comprehensive management of cable plant transport elements, such as amplifiers, end-of-line monitors, and power supplies
- Integrates cable modem management
- Supports multiple services (voice, video, data) via a single OSS infrastructure
- Integrates with workforce management applications
- Integrates with trouble ticket applications
- Integrates with spectrum management applications
- Provides a common GUI display for complete network infrastructure

Integrated domain management is shown in Figure 21.3.

The loopMASTER product embraces the layered approach represented by the TMN model adopted by the international standards bodies to provide integrated, interoperable network management solutions.

loopMASTER functionality is divided into the following modules:

- Fault management
- Configuration management
- Performance management
- Security management

Figure 21.4 is a high-level diagram that illustrates the processes supported by OSI's loopMASTER.

21.2   FAULT MANAGEMENT WITH loopMASTER

loopMASTER provides root-cause correlation of alarms for network elements including HDTs, NIUs, video servers, cable data modems, HFC transport elements

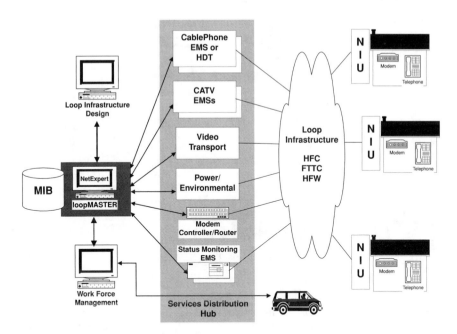

FIGURE 21.3   Integrated domain management with loopMASTER.

(amplifiers, power supplies, optical nodes, etc.) at the subnetwork level. At the network management level, loopMASTER presents related details to the network operations staff, maximizing effectiveness and minimizing network down time.

loopMASTER receives events from all of the network elements in the HFC loop through standard (CMIP, SNMP) and nonstandard (TL/1, proprietary) interfaces. loopMASTER receives events by either interfacing directly with network elements or through an existing EMS. For example, alarms from HFC transport elements, such as amplifiers, end-of-line monitors, and power supplies, are forwarded onto loopMASTER by head-end/status monitoring systems such as LanGuard from AM Communications.

Alerts generated in the network are correlated across multivendor equipment or subnetworks by the loopMASTER fault management module and, depending on the alarm type and network element involved, may also take steps to further analyze the problem, such as automatically launching a test routine to gather additional information about the element and/or root cause of the alert.

Once loopMASTER has determined the root cause of the problem, one or more options for notification may be invoked. A color-coded alert may be sent to the network operators by updating their graphical display of the network model; the alert also will open an alphanumeric, alert-display window, and an automated page can be sent to one or more network operations personnel.

loopMASTER supports the forwarding of alerts to one or more other OSSs based on the type and severity of alert, or the type of managed object. This enables automated routing of alerts to different systems to speed the problem resolution process.

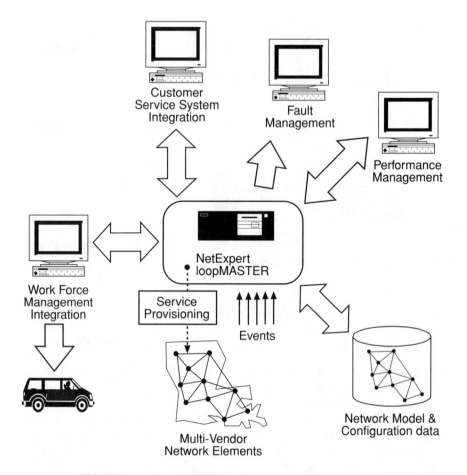

FIGURE 21.4   loopMASTER in the role of service integration.

loopMASTER interfaces to other, higher-level OSSs via multiple communications protocols, including TL/1 messaging over TCP/IP or through OSI's Peer-to-Peer Server.

Fault management features include:

- Provides alarm root cause correlation across multiple network elements, including telephone equipment (HDT/NIU), HFC transport elements (amplifiers, power supplies, end-of-line monitors) and cable data modems and modem controllers.
- User-definable alarm correlation — enables alarm correlation to be extended to meet specific business processes
- User-definable alert forwarding to other OSSs
- Alarm correlation with existing and logged trouble tickets
- Graphic and nongraphic display of alarms, color-coded according to severity
- Customizable alert displays for severity, time of alert, network element type, subnetwork, etc.

21.3   CONFIGURATION MANAGEMENT WITH loopMASTER

Automating end-to-end service management is one of the major challenges to be solved when implementing an OSS infrastructure for a multivendor, multi-technology, broadband network. loopMASTER is a key component of OSI's advanced operations environment which provides a layered, open architecture designed to enable flow-through provisioning of service across the domains of the local loop, switch, and transport networks.

loopMASTER's configuration management module enables vendor independent provisioning of network elements. loopMASTER interfaces either directly to the network element or the vendor-specific EMS to direct the provisioning action.

loopMASTER directs the configuration of network elements through either a Motif-based GUI or by receiving service requests from another OSS and acting on these incoming messages.

The GUI consists of pop-up and pull-down menus, pop-up forms, and a rich graphical display with both logical and physical views of the network and the network elements. Using the full capacity of OSI's VisualAgent GUI technology, the loopMASTER user can provision DS0s, cross-connects, and other parameters by pointing and clicking on an object, which then triggers a pop-up form where the user enters the pertinent information. loopMASTER then takes this information, converts it to a vendor-specific messaging format, and sends the configuration command to the network element via a NetExpert dialog. Confirmation (or rejection) is then displayed on the graphical display of the device.

Configuration requests coming from other, higher-level OSSs, via either TL/1 or OSI's Peer-to-Peer Service, are handled by loopMASTER as follows:

- The incoming service order is evaluated for validity
  — If the order is not valid or the information incomplete, an error message is generated, sent back to the originating OSS, and an alert displayed at the loopMASTER console.
  — If the service order is valid and the information complete, a service order managed object (MO) is created, assigned a priority, placed in a queue, then scheduled for processing. Service orders are processed in order based on priority.
- When the service order is next for processing, it is converted into a vendor-specific command and a dialog is sent to the appropriate network element.
- Once the service order has been processed, a positive (or negative) confirmation message is returned to the originating OSS.

The configuration management module maintains a complete history log of all configuration actions taken in the network. This information is stored in archival form and may be reviewed or browsed at any time.

Configuration management features include:

- Offers a common protocol interface to support multivendor HDTs/NIUs
- Provides a common GUI for provisioning telephony service for multiple HDT/NIU combinations

- Gives an automated population of inventory and configuration database (as supported by the equipment)
- Allows setting of parameters within multivendor network elements and/or passing information to existing EMSs
- Has the ability to initialize and close down network elements
- Collects information on demand on network element configuration and status
- Modifies network element configuration automatically based on pre-defined conditions of the network
- Collects and maintains the equipage and configuration of network elements
- Maintains audit trail on historical changes to the network
- Manages an assignable inventory (hardware and software) for specific network elements and performs the design and assign function, including setting port options based on definition of service designs supplied by network management personnel
- Discovers physical connections between network elements automatically within vendor subnetworks and updates the topology database (as supported by vendor hardware)
- Provides a report facility for managing network element equipped and spare capacity, including information such as CLEI, version, part number, location, spare and unequipped capacity
- Provides network management personnel with ability to locate an unused plug-in, by plug-in type, on demand
- Enables the definition of capacity or fill thresholds for all consumable subnetwork resources (that is, assignable plug-ins and facilities)
- Allows the operator to set common value options to one or more network elements using a single operation

Additionally, the configuration management module maintains a complete backup of the network element's configuration and provides the ability to download a complete restore of the network element's current configuration should an element go down. loopMASTER also enables network element software revisions to be automatically downloaded on a vendor-by-vendor basis, depending on the capability of the equipment itself.

The configuration management module enables autodiscovery of network element equipment and its configuration (abilities may also vary based on the vendor's equipment). This information is maintained in the network model with the loop-MASTER management information base (MIB) and can be formatted and sent to the central corporate inventory management system in either a relational data model or a flat file form.

21.4 PERFORMANCE MANAGEMENT WITH loopMASTER

loopMASTER collects comprehensive, statistical data and enables the setting of the various performance thresholds supported by the different equipment vendors into one common environment. Network operators can define which monitoring criteria they wish to modify, including parameters such as start/stop times, polling frequencies, alarm thresholds, which events and/or objects will be monitored, and so on.

These thresholds are monitored in real time and performance alerts are generated if any of the thresholds are exceeded. Additionally, loopMASTER automatically adjusts the operation of the system, based on exceeding certain performance-related thresholds.

The setting of threshold levels, and the actions taken when these levels are exceeded, whether automatic or manually controlled by a network operator, are defined by the user.

loopMASTER's GUI, based on VisualAgent (VA) technology, will also display performance criteria defined by the user via a variety of real-time graphical displays. In addition to the real-time updating of the hierarchical display of the network model itself, a number of standard bar and line graphs are supported with the standard loopMASTER product that provide a rich array of graphical feedback to the network operator of performance-related information.

loopMASTER maintains a historical log of all network events and performance criteria that provides a complete historical view of the state of the network.

Performance management features include:

- Gathers comprehensive statistical information
- Maintains a log of system state history
- Allows real-time graphical display of user-defined performance data
- Automatically adjusts system operation based on performance criteria
- Allows user to define monitoring criteria, such as start/stop times, polling frequencies, alarm thresholds, events/objects to be monitored
- Supports performance monitoring of parameters such as the following DS1 items, depending on specific vendor abilities:
  — Path coding violations
  — Path errored second violations
  — Controlled slips
  — Path severely errored seconds
  — Controlled slip second
  — Severely errored framing second
  — Line errored seconds
  — Provides a database of all performance management events for specified ports and paths within its subnetwork

21.5 SECURITY MANAGEMENT WITH loopMASTER

loopMASTER supports multiple levels of security. Security measures begin with a proper "logon" authorization onto the UNIX workstation. In addition, the user must also "login" to loopMASTER with a valid user ID/password.

Based on the user login identity, access can be partitioned as follows:

- By available commands
- By network element type
- By subnetwork

loopMASTER supports four levels of functional authorization:

1. Read only
2. Read and provision
3. Read and fault management (manages alarms, performance, and test)
4. Full access (system administration)

The system administrator maintains control of user authorizations and passwords. The authorization editor, within the NetExpert framework, is used to administer all system security authorizations.

Security management features are as follows:

- Provides management of multiple levels of user types based on system login
- Allows partitioning of access by function, subnetwork, managed object, or type of network element
- Enables dynamic collection of security events
- Provides automated corrective action recommendations based on security alerts

21.6   TEST MANAGEMENT WITH loopMASTER

loopMASTER incorporates all of the testing abilities of each of the manufacturers supported by the product into the easy-to-use GUI of loopMASTER. Test procedures, or events, can also be incorporated into key management tasks, such as provisioning new services to automating the process of verifying system availability.

As part of the standard loopMASTER implementation, test procedures are accessed via a pop-up menu containing the valid tests based on a specific network element or subnetwork. By selecting one of the tests to be performed, the user may be prompted to fill in any outstanding information required to complete the test. Complete results from the test are captured and an alert is generated indicating the test's completion with a summary of the results indicated, such as, "NIU ID XYZ loop back test completed ... results successful."

loopMASTER also supports the scheduling of preventive maintenance tests: which tests should be run and the parameters that surround these tests, such as the scheduled times and intervals. However, the customer defines these parameters and this is determined at the implementation phase.

21.7 USER INTERFACE OF loopMASTER

The user interface to loopMASTER is built using Visual Agent GUI technology. It features the ability to display information both graphically and textually through a wide variety of data forms. The interface is designed to present information to the user in an organized manner, but without a rigid structure. The TMN model of fault, configuration, accounting, performance, and security is used as the basis for organizing the menu structure of the initial user interface.

loopMASTER's interface can be easily customized on a customer-by-customer basis using the GUI editing tools available with VisualAgent. The hierarchy of the network model graphic contents of the pop-up command menus can be modified by the customer via simple GUI editing files.

The graphic displays of a customer's network are easily configured using predefined model templates that are included with the basic loopMASTER product. Additionally, graphics may be imported from a variety of popular graphics and/or CAD applications such as MapInfo and AutoCad. Once these graphics have been imported, additional modifications and updates can be performed using the SL-DRAW1 graphics editor.

Figure 21.5 shows a detailed graphical area with the layout of the wires. Also, customer-defined graphics can be generated to view the network (Figure 21.6) In order to support repair, maintenance, and change management, detailed physical equipment views can be generated as well.

21.8 SUMMARY

Seamless integration with OSI's application components, transportMASTER, switchMASTER, iSAC, and other prepackaged rulesets, allows efficient management and control of the service activation process affording the service provider significant operations and maintenance savings. loopMASTER completely automates an OSS environment by easily integrating other leading vendors' applications, such as those for trouble management and workforce management.

loopMASTER integrates other critical engineering applications, such as outside plant design and layout tools, to maximize the investment in the reuse of data. Plant design data is the basis for the network model within loopMASTER.

loopMASTER provides a unified service assurance process by integrating with trouble management systems and workforce management applications. Network problems can be tracked through their life cycle from the network alert to the generation of the automated dispatch and trouble ticket.

loopMASTER offers the following benefits:

- Investment protection from a flexible domain manager that can be tailored to current demands and reconfigured as the network evolves
- Rapid implementation of a customer defined, easy to implement, ruleset package means it starts working today without internal development

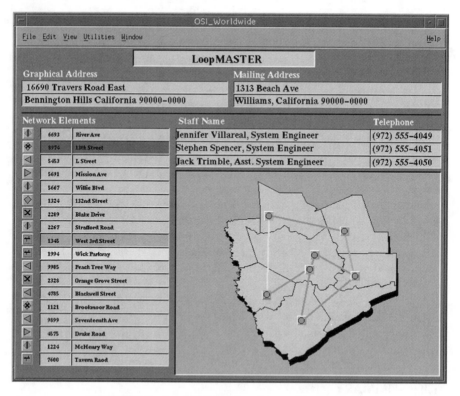

FIGURE 21.5 loopMASTER interface — "domain" view.

- Efficient use of computing resources because only the needed functions are implemented
- Complete visibility and control through root-cause correlation and problem identification across all elements in the local loop network
- Ease of use and minimum operator training with only one interface point between the local loop infrastructure and all other OSSs
- Competitive advantage by getting new services to customers first, with flow-through provisioning that is independent of vendor hardware
- Reduced complexity and faster response time for operations personnel because trouble management is integrated throughout the OSS infrastructure
- Integration of elements from within one application, despite vendor EMS or supplier

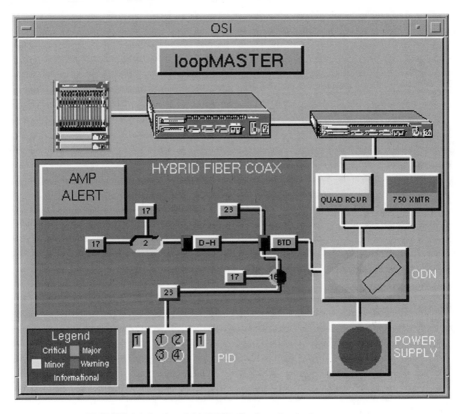

FIGURE 21.6   loopMASTER display physical components.

Part VI

Management of Access Networks and Their Services

The cooperation between telecommunications services providers and their customers is deepening. Providers are offering several gateways and APIs to their customers to pursue business electronically. It may be implemented by connecting the OSSs to the NMSs of the customers.

End-to-end management requires that each piece of a network and its equipment are managed in coordination. In several cases, access networks are not getting enough attention from suppliers and vendors.

This part of the book first addresses customer network management. Driving forces, challenges and flexible solutions are discussed first, followed by the product accessCNM provided by Objective Systems Integrators. Access networks, the internet and intranets require significant cultural changes within the corporation. Making the necessary changes for many obvious benefits is only the first step. In the second step, seamless management of these new kinds of networks must be guaranteed to customers.

Existing solutions are ususally SNMP-based, requesting that any system in the integrator or umbrella management role should be able to connect to SNMP managers or directly to SNMP agents residing in devices. OSI maintains a special SNMP gateway; this gateway will be addressed in some depth explaining how the conversion of SNMP commands into NetExpert internal commands and vice versa is solved.

Internet Service Providers (ISPs) need strong management solutions to compete successfully. ISPs usually build their offer on router-based-networks. They need extreme flexibility, rapid implementation of configuration changes and real-time fault and configuration management. The NetExpert framework is shown in the role of managing a large number of different kinds of Cisco routers utilizing both public and private entries of the Cisco-MIB. Pre-packaged rulesets address configuration, fault and performance management.

This part is concluded by a case study provided by Verio, the nations's first ISP to combine the power of a high-speed national network with local, home-town providers. Verio has selected NetExpert to handle all the disparate functions of the network from one basic framework.

22 Customer Network Management

CONTENTS

FIGURES

TABLE

Customer Network Management lets corporate users of communication services view and alter their segments of a provider's network. Once such a standardized and open interface is in use, both the service provider and corporate users benefit.

The advantages to the service providers are:

- Keep network loads to a minimum, despite the inexact nature of traffic prediction.
- Provide customers safe access to pertinent OSS and network data, from port assignments to billing and account details.
- Isolate individual customer domains without revealing details of the carrier network configuration.

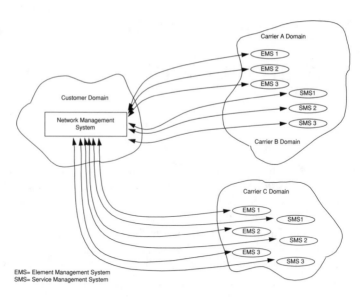

FIGURE 22.1   Interfacing multiple services offered by multiple providers.

- Accomplish even the most complex mapping by gathering values from across the network or among OSSs.
- Establish customer network domains with full assurance that customers can make only authorized changes.

The advantages to corporate users are:

- Alter data network configurations without the delays of paperwork or telephone calls.
- Produce any level of report, from performance on a single switch to comprehensive management overviews of account histories.
- Manage faults dynamically, reducing the need for carrier intervention.
- Streamline troubleshooting with easily generated reports and automatic fixes, even in multi-carrier environments.
- Integrate to the carrier network whether or not current end-user management systems are robust.

The way to a standardized and open interface is long. Today, there are many interfaces and manual information exchange is typical. The customer has to support multiple interfaces that are usually different for each of the providers. Figure 22.1 shows this typical case.

This solution can be characterized as follows:

- Support of many proprietary element management systems; most of them are legacy-type-systems. They address PBX management, multiplexer management, modem management, management of packet switching nodes, frame relay management, ATM management, wireless management, etc.

- Suppport of many proprietary service management systems that are evolving without any core management functions. They address service provisioning, bandwidth management, service assurance, etc.
- Lack of well understood management protocols
- No easy way of exchanging management information because database and MIB structures are very different.

22.1   CONCERNS OF CUSTOMERS

Customer Network Management has been a long time coming for several reasons. *First*, it is difficult to measure how much customer network management benefits the bottom line. CNM is generally on the cost-saving side which is very hard to sell to management. Selling is easier when the CNM user is actually a value-added provider, which makes CNM a critical componnet of the value-added-service.

Second, when it comes to CNM, many network managers simply do not know what services to require from the supplier. *Third*, some network managers have serious security concerns about letting an outsider get a detailed look at mission critical data. Others are afraid that CNM is an attempt by the carrier to lock the customer into a long-term relationship. And some others worry that CNM is the first step towards outsourcing.

Fourth, CNM is very complex to implement. Enabling customers to perform both read and write operations on the internal operations support systems of the telecommunications providers places a considerable stress on those OSSs. Most OSSs are not designed for extra transaction handling and security imposed by CNM. Further, integrating a CNM interface with the network management system of the customer is a difficult task.

Prior to the decision making about implementing CNM-functions and features, corporate users should complete a diligency phase consisting of the following tasks: (HOLL95)

- Which services come with CNM? Are different services integrated in some way?
- What software, hardware, and management platforms do CNM applications run on (Unix, Windows, Sun, HP, IBM, etc.)? Can they be easily ported to the company's current network management platform?
- What facilities are furnished to help integrate CNM functions into the existing corporate management infrastructure (CPE, management applications, accounting systems, databases, documentation systems, workflow solutions, etc.)?
- What is the end-user interface to the CNM applications (Windows, Openlook, Motif, etc.)? Have provisions been made for training users and technical staff on the CNM system?
- How is corporate data protected against unauthorized access and use?
- What is the cost of the CNM system on a component-by-component basis, including access charges, transport of CNM data to customers, initial installation, integration and ongoing support?

- How is the CNM system supported? What services are offered to help integrate CNM with other management systems?
- What are the procedures in case of significant changes of the OSS? What are the impacts on the CNM interfaces and gateways?

After completing this diligency phase, the corporate network manager is well informed about what management functions, databases and applications can be integrated into the corporate network management systems.

22.2   BASIC STRUCTURES AND CORE COMPONENTS

Corporate networks must be able to perform various tasks. Chapter 1 gives an overview in some depth. In a more compressed version, the following tasks should be supported: (HOLL95)

- Fault management, including fault detection, analysis and reporting, tracking and resolution
- Performance and quality of service management
- Configuration management including inventory management, service control, service ordering, and tracking
- Security management including the protection of the network and its management from both outside and within
- Accounting management including invoicing, maintaining user and usage profiles, scenario analysis, trend reporting, and exception reporting

Most of these tasks must be duplicated for equipment and services of the providers. This redundancy can lead to serious inconsistencies between the provider, corporation, and reality due to the lack of synchronization between inventory files and databases. Moreover, without near real time information about the provider's network, it is difficult to establish and maintain a coherent, end-to-end view of the network, its services, and its performance.

Corporations that buy services from multiple providers find that their problems multiply as the number of interfaces to the service provider rises: operational, fault reporting, inventory, service modification, accounting, and so on. However, even when these interfaces were unified into a one-stop-shopping-concept at the provider end, integration with the corporation's internal management systems remained a problem. There are a number of issues to be resolved:

- **Accounting Management**   If a customer wants to receive billing information from the the proivider in near real time (end of shift or end of day) to update an accounting system, some form of electronic interface between customer and provider is needed. Alternatives like e-mail or sending a tape via courier are not the best solutions.
- **Bandwidth Management**   Without integrated CNM and enterprise network management, customers that want to change the bandwidth of a service or add more channels to voice and data, have to contact the

TABLE 22.1
Core Components of Customer Network Management

Fault	Configuration	Accounting	Performance	Security
Reporting, tracking and resolution	View inventory of telco-provided CPE and services	Expenditure tracking on services in near real time	Monitoring of quality of service (throughput, delay and availability)	Access authentication and authorization
Interface to customer trouble-ticket or work flow system	Order new services	Interface to customer accounting system	Ability to generate reports and verify against service contract	Separation of customer data
Fault domain identification	Reconfigure services and network	Extract of histories and usage profiles by customer cost center;cost comparison of rival telco services (ISDN, leased lines, etc.)	Performance comparison of rival telco services	Separation of telco and customer data

provider through its interface. After confirmation which may take long, the customer can start to reconfigure their routers and other network devices. Ideally, using a CNM system, it would require a single application that would accept the request for additional bandwidth. A component of this application would wait for notification that the change has been made and then initiate reconfiguration of the customer network.

- **Quality of Service**   Many customers use their own network management systems to verify that the provider is meeting contracted quality of service commitments. Doing this properly involves a significant amount of resources. The provider on the other end is probably collecting the same data for the same purpose. It would be best if both parties were working from the same view of the service.
- **Fault Management**   Customers will likely perform initial detection and diagnosis of fault using their own network management and monitoring systems. Without CNM, they then must relay this data via phone or fax to the provider and track the progress of fault rectification using the same medium. Assuming the high level of sophistication on both ends, this is not the most efficient way to solve problems.

Table 22.1 (HOLL95) summarizes the core CNM components and high priority tasks.

In order to avoid redundancy and inconsistencies, state-of-the-art CNM solutions request a very tight connection between the management architectures and products of the provider and the corporation.

There are a number of ways in which a CNM system could integrate or fail to integrate with customer systems. The first alternative is no integration at all. The provider's CNM system could continue as an independent stand-alone system that provides a convenient point of access to services such as PBX management. Beyond

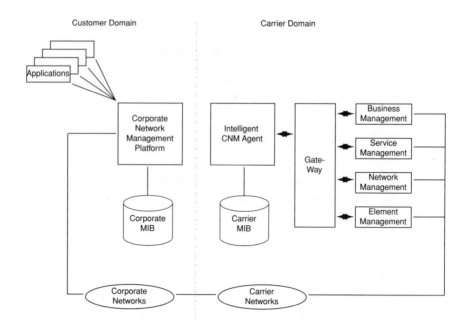

FIGURE 22.2 Customer network management.

that, the provider could supply customers with a standard interface that encapsulates a particular combination of protocols, information models, and behaviors such as a CNM agent and MIB. This will be the integration point for the management applications at the customer premises. But this still will cause problems if different providers define different interfaces with different information/object models for similar services. At the next level, integration could be achieved via a common graphical interface at the user interface level. The provider would provide a Windows or Motif CNM application that runs alongside the customer's management application on its network management platform. In some cases, the provider would furnish applications as part of the CNM system that use a private provider CNM agent MIB on the provider side. This is really an extension of the previous approach: the provider offers more of the application functionality to the customer.

In order to enhance efficiency and simplification at the same time, the network management platform of the corporation should be connected to a very intelligent "agent" on behalf of the provider. This agent unifies and coordinates the work of multiple managers that are responsible for business applications, service, network and element management. It also is responsible for synchronizing data files, databases and MIBs.

Figure 22.2 shows an integrated structure. There are two connections between the systems: one at the physical level and one at the network management level.

This high level of integration is expected to be reached in multiple phases. Telecommunications providers are on the move to select, customize and deploy powerful management frameworks that will play the role of the intelligent agent.

22.3   accessCNM FROM OSI

CNM service is a service which allows a customer to obtain information regarding their segments of a public network. Since the public network is shared by many customers and services, customer access must be controlled to avoid security and network integrity violations.

In support of this service, the CNM Server provides a segmented and secured view of the network elements to the Customer and to protect the network elements against improper and unauthorized use. It provides three basic functions:

1. Flow Control (to avoid overloading of network elements)
2. View Mapping (to translate between customer views and the network view).
3. User authentication (IP/Community String based)

The user authentication is not complete since part of the user authentication must be performed by the access network and the underlying access protocol (see Sec. 22.3.1). A simple operation model is presented by the following algorithm.

Get/GetNext/Set Operation

1. Receive customer SNMP Request PDU
2. Apply Flow-control
3. Invoke the MapFactor
4. Invoke any Handler functions requested by the Handler
5. Formulate a response to the Customer
6. Transmit a response to the customer NMS

Trap

1. Receive Trap/Alarm from Network Element (NE)
2. Invoke the NE Handler Function
3. Invoke the Trap MapFactor (if registered)
4. Form a new trap for the customer
5. Send the trap

The MapFactor and Handler functions are the open APIs which govern the behavior of the objects. They are described further in the following sections.

Note that the operation is to be done with as much parallelism as possible and the server does not wait for Network Element/Database reply when processing new packets. The handler routines are designed to manage the communication with the data repository; the repositories may be SNMP Agents in the network elements, local databases, Operations Systems, Element Management systems, etc.

22.3.1   SYSTEM ARCHITECTURE

The overall view of the system is depicted in the following diagram. The following sections describe each subsystem depicted in the diagram (Figure 22.3).

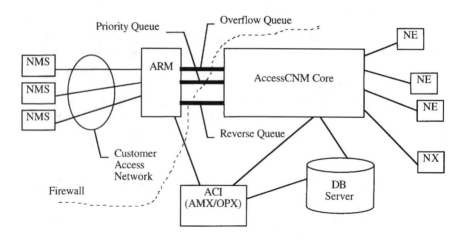

FIGURE 22.3   accessCNM architecture.

Customer Access Network (CAN)   While the primary purpose of this network is to provide transport for customer data from their location to the CNM server, it also serves as the main authentication vehicle and a rudimentary flow-control entity. The authentication is performed using the underlying network service's capabilities and features. Details of the authentication process are specific to the installation and underlying services and will be provided on a case by case basis. However, the CAN must authenticate the IP address and it must provide a hard limit on the amount of traffic submitted to avoid the overloading of the ARM (see Sec. 3.1) host hardware and operating system. An example is limiting the customer access to a 56kb/s Frame Relay channel. OSI can provide guidelines and support information for the specific type of underlying access network. In general, a Frame Relay or SMDS link should be provided with a static ARP entry in the router (to avoid reassignment of the Frame Relay or SMDS channel). Access over open Internet is less secure and use of the encryption option should be considered. PPP or SLIP access can also be provided, but either a dedicated leased line or a modem with dial-back capability is recommended. Other means of securing dial-ins, such as SecureCard™ may also be used.

Access Regulator Module (ARM)   The ARM is the gateway into the CNM server. Its main function is to regulate the load on the accessCNM Core and serve as a security firewall. In doing so it provides the following functions:

- Apply Customer-Specific Flow Control
- Discard any malformed or error packets
- Discard packets from non-registered customers
- Discard non-SNMP (or other registered protocol) packets
- Provide a firewall by hiding the accessCNM Core Host/Socket from the customer

In general, the ARM is an IP relay engine. It receives SNMP packets from the customer, performs its various checks, and transmits the packet to the next module of the CNM server, the accessCNM Core. Initially, only SNMP will be supported; the ARM will only relay UDP/IP traffic destined to port 161 (SNMP) from registered customers towards the server. A registered customer is one whose IP address is entered to the ARM database as a valid CNM user. Traffic control in the reverse direction is just as important even though such traffic is generated in the server. This especially is needed in the case where an intruder can disguise itself as another entity and discovers a mechanism to transmit data from the network to its original network. ARM provides a session mechanism to ensure only valid replies are being sent in the reverse direction. After the response is received (or a time-out reached), the association is closed and no more packets will be relayed to the customer. To support SNMP traps, traffic to UDP port 162 will be allowed towards all registered customers. Future versions of SNMP (e.g., SNMPv2) will provide for the support of TCP and other protocols for SNMP use; if use of such protocols becomes popular, then the method of association establishment and packet relay will be modified to accommodate the new protocols; alternatively, a separate instance of an ARM may operate to service each transport protocol.

In addition to packet parameter verification, the ARM also preserves the functional integrity of the CNM server by keeping the traffic levels below the serviceable limit. Considering the capabilities of typical network management platforms (e.g., HP Open-view), applications could be triggered to flood the server with requests. While the access network (CAN) will provide rudimentary flow-control, a more context sensitive customer flow-control and prioritization must be performed at this level. The ARM implements a priority queue mechanism with special handling to regulate the traffic to a known level. This flow control, considering the non-blocking nature of the SXF (or accessCNM Core) is quite complicated and involves reverse buffer load information from SXF. The ARM processes this information to modify its traffic acceptance criteria to ensure fair operation even when the number of outstanding packets is large.

The decision will be based on monitoring each customer's traffic separately using a Jumping window algorithm (an approximation to a Leaky Bucket Algorithm, but much faster and simpler to implement). If the customer is transmitting within its limits, his packets are handled with priority; if they exceed their limit (within the "window" time interval), their packets will be marked as overflow and will only be processed if no priority packets remain.

The overflow queue may be overflown under normal conditions. This is not considered an alarm and will simply result in rejection of requests which were originally considered above the customer quality of service parameters. However, due to the relatively short time-out interval for SNMP packets (compared to service times), the overflow queue will be serviced periodically to remove "old" packets. To keep the accessCNM Core clear of these issues, the ARM must also implement this queue grooming feature.

ARM, through its Usage measurement option provides statistics on Accepted, Rejected, and Overflown packets via a GUI and SNMP interface. This interface is only accessible from the SXF (accessCNM Core) for security reasons.

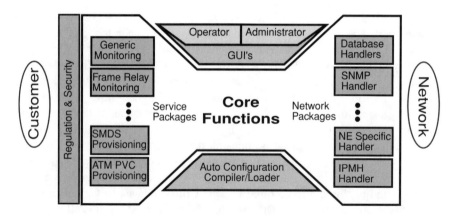

FIGURE 22.4 Core of accessCNM.

SNMP Translator/Filter (SXF or accessCNM Core) The accessCNM Core (Figure 22.4) is the core request processing engine for the server. As mentioned earlier, two queues (one priority and one overflow) connect the ARM to the access-CNM Core. The accessCNM Core will service the priority queue until it is empty and then begins to serve the overflow queue. For each SNMP packet popped from the queues, a virtual session is established; a session is necessary to properly group the various requests in a packet and also to correlate packets with their responses. Each customer request within an SNMP packet is mapped to one or more internal requests. The mapping is performed in three stages: first, a MapFactor function translates the customer view to the network view (e.g., interprets customer specific mapping information) for the requested object; second, the Dispatcher identifies a Switch Type for the underlying customer interface. Based on the switch type, a Handler is identified and, as the third stage, the handler is invoked to perform device specific operation. The handlers may be specific to a device or specific to an object, or both. Some handlers may use SNMP to contact the NE, some may use SQL to obtain database information, others may use IPMH to invoke requests (Dialogs or events) in NetExpert. The capabilities of the accessCNM Core may be expanded by addition of MapFactors and handler functions. These additions will be grouped as packages which are discussed below.

The MapFactor is considered the main steward for the SNMP Object. The handlers are invoked under the discretion of the MapFactor and, in cases where multiple queries are required to obtain an object value, multiple handlers may be invoked by the MapFactor. This assists in delineating between the Customer View of SNMP objects (e.g., CNM MIBs) and the Network Element details (e.g., NE Enterprise specific MIBs).

If the information is stored internal to the server, the accessCNM Core can obtain the data from internal databases by invoking a special SQL handler for local data.

In the reverse direction, the results are first processed by the handlers to remove any NE specific aspects. Then they are forwarded to the MapFactor. Note that the specific instance of the MapFactor will be on a "sleep queue" while waiting for the handler response. However, the accessCNM core will not be blocked and will continue to process other requests. Therefore, any slow external device (NE, OS, database, etc.) will not cause slow operation of the CNM server. The non-blocking nature has shown an incredible improvement in performance in the early tests.

The reverse communication channel (for responses) between the accessCNM Core and the ARM is established using only one queue since congestion is not expected to occur at the egress. The accessCNM Core will be designed in a fashion so that the remaining modules need not interpret SNMP messages and only be concerned with generic database functions and flow-control.

Database Server   The Database Server is where the logic and information for managing customer information resides. It provides the basis for the MIB traversals, MapFactor and Handler assignments, etc. In addition, it holds the tables of internal network topology parameters used for Dispatch. In doing so, it associates the real network view to the segmented and limited views of the customers and plays a major role in the security of the system; in fact, information will not flow from the server to the customer unless it has a corresponding mapping entry in the ARM. The Database server also includes a set of routines which will be compiled with the accessCNM Core and will be used to cache the database information; this is to avoid excessive DB lookups when processing packets. The Database server is a non-blocking server which will hide any database blocking, locking, and delays from the accessCNM core. The requests are asynchronous, allowing the accessCNM Core to process other requests which the DB functions are being performed.

22.3.2   ADMINISTRATIVE AND CONTROL INTERFACE (ACI)

The ACI provides the operator and maintenance interfaces for the system. ACI consists of two processes: the OPX (operator interface) and the AMX (Administrator or Maintenance interface). The operator interfaces are used for routine modifications (e.g., provisioning) of the system while the maintenance interface provides privileged commands used for maintenance, monitoring, initialization, and reports. The design of these interfaces is expected to evolve significantly to match the skill level and organizational structure of the operating entity. Initially, two simple interfaces will be provided for each function. Currently, the plan is to transition the monitoring function to NetExpert, thus providing for integrated management of the server.

The operator interface provides the capability to assign an interface to a customer, to modify flow control parameters, and to generate a report of the current configuration parameters. The addition/deletion of customer assignments are performed without a system restart. This information is relayed to the ARM and Database Server for proper updates in the operating parameters. Further, since the CAN may also be affected, the module may be expanded to provide a report for the necessary modifications to the access network.

The ACI is highly customizable and additional GUI routines may be added to support MapFactor and handler initialization and operation.

Autoconfiguration modules can be attached to the ACI to automatically update the database and avoid a re-entry of configuration information. This is useful in the case where information is already input into another OSS or where it might be available from the NE itself. The API includes hooks for attaching such functions; further, the DB Server may be accessed externally to update such information by applications invoking simple SQL statements.

22.3.3　CNM Packages

The features provided to the customer are grouped into Packages referred to as CNM Feature Packages (CFPs). Each feature package will represent a set of objects along with their view translation and handler functions. Addition of a CFP does not necessarily involve installation or development of a new MapFactor or a handler, since some of the existing MapFactors and handlers may suffice.

The simplest possible package includes one SNMP object and a MapFactor returning an internal value. An example of such a hypothetical package is a package providing the "sysDescr" MIB-II Object. The MapFactor simply returns a string such as "OSI accessCNM Server." Of course, this is an overly simplified case. In general, packages include several objects (e.g., MIB-II systems and Interfaces group), a few MapFactors (e.g., one for constant values and one for interface-specific values), and one or more handlers (e.g., one for the specific NE used). As support of more objects are necessary, objects and MapFactors may be added; and as support of more network elements are necessary, handlers will be added. Depending on the service and marketing plans, a set of packages may be developed such as: Basic Monitoring, Trouble Ticket Submission, FrameRelay Performance Measurement, SMDS Provisioning, etc.

Addition of a package may require simple point-and-click procedures (for example to add a simple object mapping feature) or may require some development (e.g., to access the database of an integrated OS).

22.3.4　Implementation Notes

For security purposes, the accessCNM server implementation requires two work-stations. One will run the ARM module and the other will run the remaining modules (SXF/accessCNM Core, AMX, OPX, DB Server, etc.). The following diagram (Figure 22.5) provides a simple view:

The ARM is considered to be the firewall; therefore, the network to the left of the ARM is considered to be unsecured (customer side) and the one on the right side is considered to be secure. This diagram shows an ARM with two network (e.g., Ethernet) interfaces. This may also be achieved with an ARM machine with only one interface; but the latter requires implementation of secure filters in the routers to ensure the firewall functionality is preserved (OSI can provide support in configuring most common routers, such as Cisco and Bay Networks). The SXF may be directly connected to Network elements or it may connect to an OSS. This depends on the types of handlers used. The configuration may be changed between a direct connection and an indirect connection (i.e., through an OSS) without impacting customer views. OSI can provide more detailed support in recommending a deployment architecture on a case by case basis.

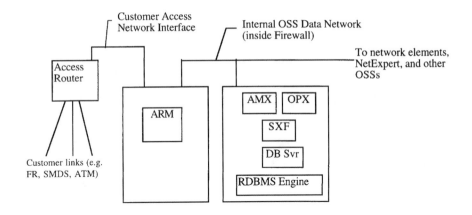

FIGURE 22.5   Structure of accessCNM workstations.

22.4   SUMMARY

The rising dependence of commercial businesses on internetworking and data communications presents an excellent chance for service providers to gain the loyalty and partnership of the largest spenders in their markets. Even in markets where competition has not yet occurred, providers are finding that customer relations and the efficient implementation of requested services directly impacts profit margins and the provider's general image. If competition is expected soon, the edge may be retained by providing CNM solutions today and expanding the features offered as market opens.

Offering comprehensive CNM services complements the most basic reason for being in business: getting and keeping customers so that revenue generation is assured. End users, demanding the power to view and alter their segments of the carrier network, gladly pay for the added value and, increasingly, are comparing CNM functions when deciding on service providers. The right CNM technology gives providers confidence that their networks and OSSs are secure and that customers have access only to their own domains. Risks are minimal and profit goes up with accessCNM because its functions can be enhanced on demand.

Empowering the customer by implementing a comprehensive CNM solution brings a hidden bonus to providers: end-user self-government decreases operator intervention and lessens trouble calls, thereby reducing the cost of doing business. The characteristics of accessCNM — standards-based, open, and object-oriented — are the very features required of operations support systems as an entity. They also are the traits of systems most likely to return investment quickly because such systems are ultimately less costly. accessCNM is flexible enough to integrate with existing OSSs and is ready for anticipated expansion.

23 NetExpert SNMP Solution

CONTENTS

FIGURES

The most important industry standards have been summarized in Chapter 4. Basic introductory remarks have been made about SNMP and RMON versions 1 and 2. Chapters 1 and 5 have addressed the NetExpert framework in some depth. The framework includes several generic and specific gateways. One of the special gateways is for SNMP. Although SNMP is not the typical protocol of telecommunication equipment, facilities and services, it is still a very important standard to manage certain interconnecting components that are widely used in commercial networks. This chapter gives further details on the protocol and on the protocol translation. Major emphasis will be on the "get" and "set" commands.

23.1 OVERVIEW OF SNMP

The Simple Network Management Protocol (SNMP) is a management protocol designed for Internet Protocol (IP)-based networks. A basic SNMP-managed network is comprised of managed nodes containing agents and devices, the IP protocol suite, the SNMP protocol, and a manager node. The User Datagram Protocol (UDP)/IP protocol is typically used to transport SNMP messages between the manager and agents. UDP/IP is a connectionless protocol that does not guarantee message delivery to the receiving (target) node.

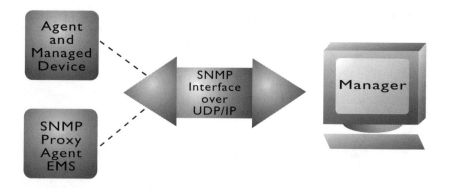

FIGURE 23.1   Simple SNMP-managed network.

In the SNMP approach, agents are kept simple and small. More complex tasks are performed by the manager or management applications. Thus, the agent normally performs minimal manipulation and the manager analyzes, manipulates, combines, or applies some algorithm to the values received from the agent. In the SNMP management of LAN or WAN devices, which include routers, muxes, bridges, workstations, etc., the SNMP Agent resides on the actual managed device.

Some device vendors are taking a different approach to SNMP management. In some cases, the SNMP agent does not reside on the managed device. The SNMP software actually resides on an Element Management System (EMS) which then manages the vendor-specific devices over a proprietary protocol. The vendor provides the EMS as a product to manage their specific equipment. (See Figure 23.1.)

These vendor EMSs usually provide vendor-specific reporting as well as performance, fault, and configuration management functions. Some vendors also provide SNMP functions on the EMS, thus eliminating the need to develop SNMP agents for each device. Proprietary events are mapped to SNMP generic or enterprise-specific traps. Fault, configuration, and performance attributes are mapped to the SNMP Managed Information Base (MIB).

OSI's goal is to provide a simple means of managing these SNMP devices. The NetExpert SNMP Interface is made up of two separate processes: the SNMP Gateway and the SNMP Trap Daemon. These two processes together allow NetExpert to initiate SNMP Gets and Sets and to "listen" for SNMP Traps.

The SNMP Community is an administrative relationship between SNMP entities. The community string is usually configured for each SNMP Agent and/or SNMP Proxy Agent. It is used to group agents of a particular region and to screen the receipt of traps from a certain region.

Within NetExpert, each SNMP Gateway is started for a particular SNMP Community. The Community must be defined within NetExpert as an instance of the SNMP_COMMUNITY class. IDEAS communicates with any SNMP Agent that is managed by the Community instance through this Gateway.

The SNMP Trap Daemon is able to receive traps from any SNMP Community string, as long as the SNMP Community string is an instance of the SNMP_COMMUNITY Manager Class.

The SNMP Gateway can communicate Set and Get primitives for only one SNMP_COMMUNITY Manager Class instance. Thus, one SNMP Gateway is required for every SNMP Community while one SNMP Trap Daemon is able to receive traps from any number of SNMP communities.

23.2   THE SNMP MANAGED INFORMATION BASE

Typically, SNMP managers are supplied with a MIB. This MIB usually is available from the vendor in an Abstract Syntax Notation One (ASN.1) format.

A MIB is like an object database in that there exists a definition of the data by type and structure that may be stored in the MIB. Different network management standards have strict rules regarding how data can be defined for the MIB. NetExpert has a MIB with rules that support loading the data in a way that is compatible with all the different network management standards, including SNMP.

The SNMP MIB II definitions are preloaded into the NetExpert MIB. SNMP Objects are treated like NetExpert Attributes defined for NetExpert's preloaded SNMP Node Class (addressed in NetExpert as SNMP_NODE).

Network element manufacturers may provide an SNMP enterprise MIB which describes information that is specific to their network elements. The MIB associated with the network element needs to be loaded to enable NetExpert to access and use this management data.

The NetExpert SNMP MIB Loader provides the flexibility to customize the MIB loading process. Some SNMP enterprise MIBs are very large (over 5000 lines) and the user may choose not to load the entire MIB.

An Object Identifier (OID) is a data type denoting an authoritatively named object. It provides a means of identifying an object type within a MIB tree. The tree consists of a root connected to a number of labeled nodes via edges. Each label consists of a nonnegative integer value and, possibly, a brief textual description. The OID contains a sequence of these integer values that traverse the tree.

Enterprise MIBs are loaded into NetExpert under the "enterprise" OID. These MIBs are loaded using the SNMP MIB Loader.

23.3   SNMP GATEWAY

The SNMP Gateway process provides bidirectional communication between the NetExpert system and the SNMP Agents in the managed network. This bidirectional communication consists of Get/Set requests and Get/Set responses. One SNMP Gateway process is run for each SNMP Community that is to be managed.

In a typical SNMP network, management information is obtained when the manager "polls" the agents by periodically sending them requests and receiving their responses. However, the agents can also send "traps" (unsolicited events) to the manager which may trigger the user to poll the agent to investigate the problem.

Each agent maintains the management information for the device for which it is responsible. These single pieces of management information are called "variables" in SNMP terminology. This list of variables makes up the agent's MIB. To monitor or control a device, the manager accesses the device's variables by issuing either a Get (view) or Set (change) command. The Set command can also be used to initiate an action by the agent.

SNMP Gateway Protocol Translation   SNMP Get and Set functionality are accomplished with the SNMP Gateway. These basic SNMP primitives are generated and the responses are received via the SNMP Gateway process.

First, the SNMP Gateway process translates the request sent to it from the OSI CMIP (Common Management Information Protocol) format into the SNMP Protocol Data Unit (PDU) format that is necessary for communication with the SNMP Agents. Next, it looks up the Managed Object name in the database and retrieves an Internet address to use in addressing the PDU to the appropriate SNMP Agent.

After the translation process is complete, the SNMP Gateway sends the request to the SNMP Agent and the expectation of a response from the agent is held in a list. Once the agent has responded, the SNMP Gateway matches the response to the request and returns the PDU information to the original requester. If the agent does not respond within the specified response time, the SNMP Gateway returns an error response to the requester.

Another feature of the Gateway is that it can generate a "trap" if it receives, in a response from an agent, an attribute that has not been loaded into the NetExpert database through the MIB Loader. This trap is translated into the SNMP_unknownObjectID event by a Trap Daemon if there is one currently running on the Gateway's host.

SNMP Trap Daemon   The SNMP Trap Daemon is the process that accepts unsolicited events called "traps" from any SNMP Agent. One Trap Daemon is run for each host computer to which SNMP Agents have been programmed to report their traps.

The SNMP Trap Daemon "decodes" the SNMP trap PDU and maps the trap to a corresponding event within the NetExpert IDEAS. Rules within the event are then written by the user. The default rules seeded within NetExpert will generate a default alert. However, the user has the option of enhancing rules to get additional information or correlate alerts.

SNMP Trap Daemon Protocol Translation   SNMP Traps are in PDU format, a standard nonprintable, byte-oriented unit. The SNMP PDU is decoded by the SNMP Trap Daemon. It is the SNMP Trap Daemon's function to take key elements of the SNMP Trap and store them into NetExpert attributes.

Another function of the SNMP Trap Daemon is to map a given generic-trap type to a specific event within NetExpert's IDEAS. These events correspond to each Generic Trap Type, which is standard to SNMP. All enterprise-specific traps are mapped to one single event, which provides the flexibility of generating other enterprise-specific events, which can generate enterprise-specific alerts or Get or Set commands to be sent to the SNMP Agent.

SNMP Trap Protocol Data Unit   A SNMP Trap Daemon translates the Trap PDU received from a SNMP Agent into an EVENT REQUEST structure. The Trap Daemon generates an EVENT REQUEST from the following list:

SNMP_coldStart-The agent is turned on.

SNMP_warmStart-The agent is reset.

SNMP_linkDown-The agent has an attached interface that has gone from an "up" state to a "down" state.

SNMP_linkUp-The agent has an attached interface that has gone from a "down" state to an "up" state.

SNMP_authenticationFailure-The agent has received a PDU with the community set to a community in which the agent is not a member.

SNMP_egpNeighborLoss-The agent has an EGP peer that has gone into a "down" state.

SNMP_enterpriseSpecific-The agent had an event occur that is associated with a manufacturer's proprietary, enterprise-specific MIB. Refer to the agent documentation for the possible values that can be associated with this type of event.

SNMP_unknownObjectID-The agent referenced an attribute that is not recognized by NetExpert. This implies a mismatch between the MIB for the SNMP Agent and the SNMP Agent itself.

Application of the SNMP interface between network elements and NetExpert is shown in Figure 23.2.

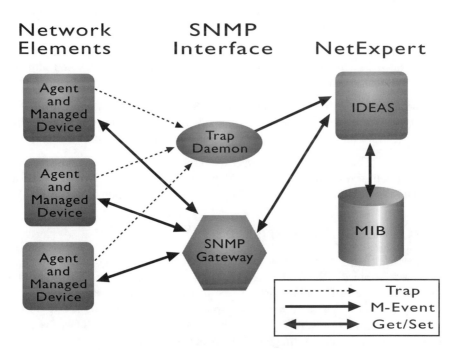

FIGURE 23.2   The SNMP interface and NetExpert system.

Managed Object Name for Table Instances   When an enterprise-specific MIB is loaded, the MIBII tables, such as ifTable and atTable, have a ConnectDown relationship to an "Entry" Class, such as ifEntry and atEntry. When a Get Event is configured properly, the SNMP Gateway automatically creates an instance of the xxTable class under the xxEntry class. To populate a NetExpert MIB with instances of desired MIB Tables, IDEAS must be configured to create managed objects on the fly.

23.4   IDEAS FOR SNMP

The SNMP agent sends a trap or PDU generated by the managed node to alert the network management system of an event or threshold. When the SNMP Trap Daemon receives a trap from one of its managed nodes, it creates an M-Event. This M-Event is directly related to the standard MIB II traps (i.e., Link Up, Link Down, Cold Start, Warm Start, etc.). These SNMP trap events are "seeded" into NetExpert under the SNMP Community manager class.

IDEAS loads information requests, gateway responses, and SNMP traps for the SNMP Trap Daemon just as it loads non-SNMP data into a generic gateway. For requests and response pairs, the SNMP Gateway interfaces all associated SNMP nodes from one runtime process. This one gateway loads into memory all the translated MIB variables, both standard and vendor specific. The gateway can perform multiple asynchronous SNMP requests.

The configuration information loaded by IDEAS uses the information that identifies an agent (or element manager) as an SNMP node.

Traps   When an SNMP Trap Daemon receives a trap, it creates an M-Event. The data is then passed to IDEAS for further processing. Associations and attribute definitions that are acceptable to IDEAS can be defined in the NetExpert Rule Editor.

Requests   IDEAS executes either a Get or Set operator on an attribute (SNMP object) that belongs to an SNMP managed node. The command is sent to the SNMP Gateway, which translates it into SNMP format and transmits it to the managed node.

Responses   The SNMP Gateway expects a response after a request. The request is repeated until a response is received. When the response is received, the gateway translates the response to the OSI Local CMIP format and sends it back to IDEAS.

23.5   SNMP GETS AND SETS

In proactive SNMP Management, NetExpert emulates a human operator who may Get predetermined SNMP MIB objects from an SNMP Agent and review this information. The MIB information can be tied to a scalar attribute or to a table. NetExpert can be configured to Get one specific attribute from one SNMP host or to Get an attribute from all hosts within an SNMP Community. This Get Event can be triggered on demand or triggered by scheduling the event.

Gets and Sets can also be used in configuring the SNMP Agent. The user, however, may discover that most SNMP Agents are limited to read only (i.e., Gets

FIGURE 23.3 SNMP interface with NetExpert IDEAS and MIB.

only) or limited to minimal read-write access (i.e., Set capability). This depends on the device and the maturity of the SNMP Agent.

Determining the Object to Get or Set When using NetExpert to Get or Set an object, the NetExpert rules should specify the Manager (SNMP Community String) and/or Affected Managed Object (SNMP Host). To specify the correct Manager and/or Affected Managed Object (AMO) the NetExpert intrinsics @AMGRCLASS, @AMANAGER, @ACLASS, and @AMO should be set appropriately. When executing an SNMP Set, the AMO should be set to the host name. (See Figure 23.3.)

Instance Identification The MIB objects are only template definitions of the actual objects within an SNMP Agent. It is the instances of the objects which are manipulated by the management protocol. This is analogous to the NetExpert Class Definitions vs. the actual Managed Object (MO) instances under each class. For SNMP, in addition to specifying the object name (Object ID or Class Definition), the instance-identifier (or Managed Object instance) must also be specified.

Within NetExpert, the instance-identifier is specified through the SNMP_IID event attribute. This SNMP_IID attribute is concatenated to the Object Identifier (OID). If the object type is a table, then specifying the SNMP_IID will specify the instance of the table which the user wants to Get or Set.

The SNMP_IID's format also depends on the device to be managed. Some vendors have incorporated the IP address as part of the SNMP_IID. The IP address(es) specified in the SNMP_IID will concatenate to the OID to uniquely identify an object instance within the SNMP_Agent, such as the interface on a router.

Besides an integer value and IP address, the vendor may choose to use a fixed or variable-length string for the SNMP_IID.

Get and Set Operator Syntax　The Set operator normally expects an OPERAND2 specification. However, for SNMP Get operations, the OPERAND2 column is left blank. Furthermore, OPERAND1 is always a Managed Object Attribute (#Attr_Name) for both Gets and Sets.

IDEAS executes the following logic for an SNMP Get:

> If OPERAND2 is blank, IDEAS determines the OID for #Attr_Name and sends a Get Request via the SNMP Gateway to the SNMP Agent (specifically designated as the Affected Managed Object). If OPERAND2 is not blank, then IDEAS treats the request as a non-SNMP Get operation.

The Get Request gets the object (attribute definition) from the SNMP Agent and the Get Response contains the object. If the Object is a scalar, the value is set to #Attr_Name. If the Object is a table, then the table is instanced under the appropriate table class.

IDEAS executes the following logic for an SNMP Set:

> If the Affected Managed Object (AMO) is an SNMP object, then IDEAS determines the OID for the #Attr_Name and sends a Set Request via the SNMP Gateway to the SNMP Agent. The Set Request sets the object (attribute definition) to the SNMP Agent's MIB object designated by the #Attr_Name OID.

Managed Object Attribute Values　In SNMP Set and Get operations, the default behavior of IDEAS is always to update the MO's attribute values. The default behavior of Set is always to do an external set operation.

If the OSI_SNMP_UPDATE variable exists in the environment and the rule is a Get, then the retrieved values are not placed into the MO attributes automatically. This provides finer control over the ruleset and improves performance when attributes are retrieved to make decisions or perform calculations.

When the OSI_SNMP_UPDATE variable exists and the Rule is a Set where the attribute used in both Operand1 and Operand2 is the same attribute, the Set does not perform the external operation.

Retrieval Of SNMP Rows　SNMP Get requests can be used to retrieve a scalar value, a column from a particular row, an entire table, or a single row in a table. The rule 'Get #<xxTable> SINGLE_ROW' tells IDEAS to use the SNMP_IID that was set to retrieve only the columns that are associated with that table for that row.

Command and Response Tool for Gets and Sets　Get or Set events can be triggered using the Command and Response (CARS) Tool in the Client Manager window or by selecting Device Management in the Alert Display window. This function allows users to Get or Set attributes on demand.

Scheduling Tools for Gets and Sets　Get and Set events can be scheduled to execute at specific times by using the polling feature in the Administration Editor. This process is transparent to the operator unless he or she chooses to monitor the status of the Gets and Sets by starting IDEAS.TRACE or Dialog Watch for the SNMP Gateway and Event Watch for the SNMP Trap Daemon.

23.6 NetExpert SNMP ASSISTANT

The NetExpert SNMP Assistant package provides a set of graphical user interfaces for performing SNMP operations and configuring and running SNMP data visualization applications. SNMP Assistant provides an application building framework that augments NetExpert functionality by offering an alternative SNMP view of a managed network. The SNMP Assistant tools satisfy basic customer needs for easy, on-the-fly access to SNMP network information for fault management and problem diagnosis.

The NetExpert SNMP Assistant package consists of a MIB Browser and four separate data visualization tools. The graphical user interfaces were developed for use in an X-Window/Motif environment and are available for use with NetExpert 3.5 on the SUN/Solaris 2.5 platform. Future ports for the IBM/AIX 4.1.4 and HP/HPUX 10.20 platforms will be available as well. SNMP Assistant consists of the following tools:

NetExpert SNMP MIB Browser   Traverse an SNMP MIB and perform SNMP operations (Get, Set, or Walk) and/or execute a Gateway Analysis Rule, using selected MIB variables.

Figure 23.4 contains the main window of the NetExpert SNMP MIB Browser. The operations menu contains selections for performing an SNMP operation (Get, Set, Walk) using the values entered in the Host, Community, and SNMP Variables fields.

Each of the data entry fields contain a browse option for accessing lists of related values. The following illustration contains the SNMP Variables: Browse dialog, which is invoked when the user selects the Browse button next to the SNMP Variables field.

The upper left section contains a list of MIB variables. The user may toggle between a list view of the variables, or a graph (tree) view. The user may select a variable from the left section and press the arrow to initiate an SNMP walk. The instances for the selected variable, and their corresponding values, are populated in the Instances:Values section. The arrows may be used to populate the Selected Variables section (bottom) with items selected in either the Variable List/Graph or Instances:Values sections. The user may select Apply or OK to add the selected variables to the MIB Browser's main window.

NetExpert SNMP List   Retrieve and display/present MIB information in a tabular list format. A list application may be configured to be updated on request, or by way of a poll (Figure 23.4).

NetExpert SNMP 2D Graph   Retrieve and plot/display scalar MIB information as a two dimensional graph, where the X-Axis is time, and the Y-Axis is either the actual MIB variable value, or differential (difference between current poll value and previous poll value). A 2D Graph application may be configured to be updated on request, or by way of a poll (Figure 23.5).

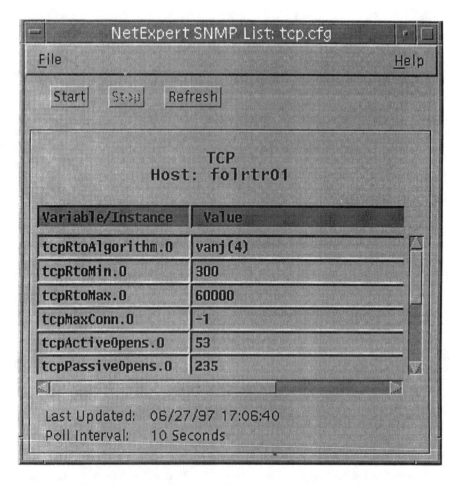

FIGURE 23.4 Retrieval of SNMP information in tabular format.

NetExpert SNMP Table   Retrieve and display/present tabular MIB information in a row/column format. A table application may be configured to retrieve specific columns from an SNMP table. The table data visualization tool will discover all instances for each column and display the individual row/column values in row/instance order. A Table application may be configured to be updated on request, or by way of a poll (Figure 23.6).

NetExpert SNMP Threshold   Retrieve and display/present numerical MIB information in the form of meters. The meters provide indicators that associate selected scalar MIB variables with configurable threshold values. Color is used to indicate when a threshold is crossed (value is greater than or equal to threshold), or reset (value is below threshold). Each threshold meter may be configured to pop up a message dialog whenever the specified threshold value is crossed. A threshold application may be configured to be updated on request, or by way of a poll.

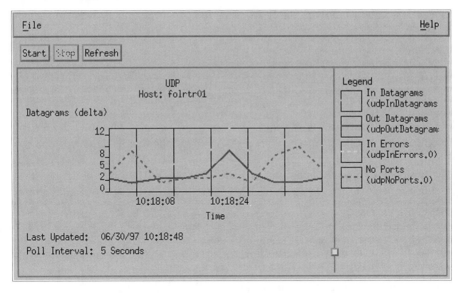

FIGURE 23.5   2D graphics to display MIB PDUs.

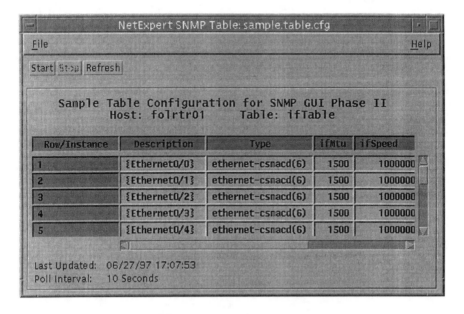

FIGURE 23.6   Instances for SNMP variables.

Operation   The NetExpert SNMP Assistant tools may be launched via the following mechanisms:

- From the Command Line
- From Visual Agent Pop-up Menus
- From the NetExpert MIB Browser's configurable Tools Menu

The SNMP Assistant tools require a client connection with either a NetExpert SNMP Server or SNMP Gateway process. All SNMP operations initiated from the SNMP Assistant tools are routed to the specified SNMP Agent Host via the SNMP Server or SNMP Gateway. The SNMP Server or Gateway will return either a successful response from the SNMP Agent, or an error condition to the SNMP Assistant client.

In addition to SNMP operations, SNMP Assistant may be used to select and forward a request to an SNMP Gateway to execute a Gateway Analysis Rule. The Gateway Analysis Rule may be either part of an SNMP operation (Get, Set, Walk) request, or by itself as a Gateway Analysis request. If the name of a Gateway Analysis Rule is selected to be included in an SNMP Operation, the SNMP Gateway will execute the Gateway Analysis Rule upon successful completion of the SNMP operation.

23.7   SUMMARY

By protocol conversion from and onto SNMP, NetExpert can easily manage SNMP, and non-SNMP-devices together. The result is that NetExpert may be considered as an integrator or umbrella manager for many commercial networks. But, NetExpert cannot change the "polling"-philosophy of SNMP. NetExpert can, however, take advantage of traps or events that are unsolicitedly sent to the manager, and concentrate on the economical use of "gets" and "sets."

24 Managing Router Networks

CONTENTS

FIGURES

TABLES

Intranets are going to be built and implemented on routers. All major router manufacturers offer solutions for intranets and for accessing the Internet. Practically, all ISPs (Internet Service Providers) are confronted with the problem of managing a large volume of homogeneous and heterogeneous hardware and software. Usually, router manufacturers provide management functions, e.g., configuration management or element management on the basis of SNMP.

Introduction   This NetExpert application provides fault, configuration, and performance management of Cisco equipment running the IOS operating system for network communication between hardware elements. The package includes an SNMP MIB, a NetExpert ruleset, and associated graphics. Modeling the rules on the IOS language makes it possible to interface to many different types of Cisco hardware with a single ruleset.

This NetExpert application is composed of a subset of NetExpert Framework modules including the SNMP Gateway and Generic Gateway with Shell Protocol Agent, IDEAS to provide object management, scheduling, authorization, process administration, and event correlation, Operator Workstation, Visual Agent and Remote Access to allow operators to retrieve management information across the Internet.

24.1 MANAGED OBJECTS

In order to understand the NetExpert ruleset provided for Cisco products, it helps to have a basic understanding of the Cisco products themselves. An overview is presented below of the Cisco hardware and software products to which the rules apply.

24.1.1 HARDWARE COMPONENTS

Cisco produces a variety of hardware products for Internet Service Providers. Products for which this ruleset is intended include the following:

- 2500 Router series
- 4500 Router series
- 5200 Access Server
- 7000 Router series

Cisco 2500 Router Series   The 2500 Router is a dual LAN/WAN router that combines two local-area network (LAN) routers with two synchronous serial wide-area network (WAN) ports in a single system. It supports a range of WAN services and routing protocols configurable through Cisco IOS software. It also offers IBM tunneling and conversion such as data link switching, bandwidth management and optimization, and security features, such as data compression, IPX/SPX spoofing, and packet filters. Software upgrades can be accomplished via PCMCIA Flash memory card.

Cisco 4500 Router Series   The CiscoPro CPA4500 routers deliver modular WAN solutions typically required by large branch offices, medium-sized businesses, and central sites for network access. In branch office applications, CPA4500 routers provide scalable WAN access for multiple LANs. In a central-site application, a CiscoPro CPA4500 can aggregate local-office access for maximized utilization of WAN links.

The configuration of the CPA4500 models can accommodate differences in WAN services in North America and internationally. Base modules support dual Ethernet LANs and synchronous serial, Integrated Services Digital Network (ISDN) Basic Rate Interface (BRI), and ISDN Primary Rate Interface (PRI) wide-area interfaces. A choice of add-on WAN modules for all CPA4500 products simplify network growth by supplying additional WAN ports for synchronous serial, LANs, ISDN BRI connections, and ISDN PRI channelized T1 or E1 circuits.

All CPA4500 models incorporate the Cisco Internetwork Operating System (IOS) software, which supports multiprotocol routing while maximizing bandwidth through compression, priority queuing, and Dial-on-Demand Routing (DDR).

Cisco 5200 Access Server   The Cisco 5200 Access Server provides a single solution capable of handling a mix of asynchronous analog users and ISDN users, in addition to high-speed local area networking. Multichassis, multilink configurations are available, allowing scalability and CPU load spreading. Features include Level Two Forwarding protocol for tunneling and virtual dial-up, modem management for realtime diagnostics and end-to-end troubleshooting, and IOS software control.

Cisco 7000 Router Series   The Cisco 7000 series of multiprotocol routers provide advanced levels of reliability, serviceability, and performance for core and distribution point internetwork routing. All of the 7000 series routers work with the Cisco IOS software. The Cisco 7000 and 7500 series support a broad set of media types:

- One- or two-port Fast Ethernet
- Two-, four-, six-, eight-, 10- or 16-port Ethernet
- Two-, four-, or eight-port Token Ring
- One-port FDDI
- Four- or eight-port serial
- ATM (single port: DS3, T3, TAXI, and OC-3c single mode or multimode)
- Single- or dual-port channelized T1 and E1
- Fractionalized E1 (G.703/G.704)
- Single- or dual-port ISDN PRI
- Single- or dual-port channel interface for IBM mainframes

Network interfaces reside on modular interface processors, which provide a direct connection between the highspeed bus — or buses, in the case of Cisco 7507 and Cisco 7513 — and the external network. Flash memory is standard.

The Cisco 7000 and 7500 series routers feature a range of processor slots: from five in the Cisco 7010 and Cisco 7505 to thirteen in the Cisco 7513.

The Cisco 7200 series also supports a broad set of media types:

- One-port Fast Ethernet
- Four-, five-, or eight-port Ethernet
- Four-port Token Ring
- One-port FDDI
- Four- or eight-port serial
- HSSI (dual-port)
- Dual-port ISDN PRI

24.1.2 INTERNETWORKING OPERATING SYSTEM (IOS)

Cisco Internetworking Operating System (IOS) software runs on Cisco internetworking platforms and allows communication among a variety of network protocols. The IOS software provides a command language for controlling network elements, enabling configuration, fault management, performance management, provisioning and so on over the network.

Cisco IOS software supports users and applications throughout the enterprise and provides security and data integrity for the internetwork. Cisco IOS software manages resources in a cost effective manner by controlling and unifying complex, distributed network information. It also functions as a flexible vehicle for adding new services, features, and applications to the internetwork.

Cisco IOS software includes the following primary features:

- Scalability
- Adaptive Routing
- Remote Access and Protocol Translation Functionality
- WAN Optimization
- Management and Security

Adaptive Routing and Scalability   Cisco IOS software provides adaptive routing using scalable routing protocols to avoid congestion, overcome inherent protocol limitations, and bypass many of the obstacles that result from the complex scope and geographical dispersion of an internetwork. Features include route filtering, protocol termination and translation, smart broadcasts, and helper address services.

Route filtering and route redistribution save network resources by preventing data from being unnecessarily broadcast to nodes that do not need it. Priority output queuing and custom queuing grant priority to important sessions when network bandwidth is scarce. Load balancing uses every available path across the internetwork to preserve bandwidth and improve network performance. Cisco IOS software also provides scaling for network applications that require transparent or source-route bridging algorithms.

By distributing routing intelligence and switching functions to create "virtual LANs," CiscoFusion multilayer switching capabilities increase bandwidth while simplifying moves, additions, and changes across the enterprise. CiscoFusion extends the Cisco IOS software to include ATM and LAN switches.

Remote Access and Protocol Translation Functionality Depending on the product, a Cisco device connects terminals, modems, microcomputers, and networks over serial lines to LANs or WANs. Cisco products provide network access to terminals, printers, workstations, and other networks. On LANs, terminal services support TCP/IP on UNIX machines with Telnet and rlogin connections, IBM machines with TN3270 connections, an Digital machines with LAT connections. Customers can use the router or access server's protocol translation services to make connections between hosts and resources running different protocols including router and access server connections to X.25 machines using X.25 PAD.

Access servers provide remote configuration through Telnet and Digital Equipment Corporation's Maintenance Operation Protocol (MOP) connections to virtual ports.

Cisco IOS software supports the following types of server operation:

- Remote node services — Connect devices over a telephone network using AppleTalk Remote Access (ARA), Serial Line Internet Protocol (SLIP), compressed SLIP (CSLIP), Point-lD-Point Protocol (PPP), and Xremote
- Terminal services — Connect asynchronous devices to a LAN or WAN through network and terminal-emulation software including Telnet, rlogin, Digital's Local Area Transport (LAT) protocol, and IBM TN3270.
- Protocol translation services — Convert one virtual terminal protocol into another protocol.
- Asynchronous remote access routing — Enables full-featured Internet Protocol (IP), Novell Internet Packet Exchange (IPX), and AppleTalk routing over asynchronous interfaces.

WAN Optimization Cisco IOS software accommodates circuit-switched WAN services such as Integrated Services Digital Network (ISDN), switched T1, and dial-up telephone lines. Cisco IOS software features such as dial-on-demand access and dial backup capabilities provide alternatives to point-to-point switched leased lines. Support for advanced, packet-switched services such as X.25, Frame Relay, Switched Multimegabit Data Service (SMDS), and ATM extends the internetwork across the broad range of WAN interface alternatives now available.

In addition to remote node WAN connectivity with ARA, SLIP, PPP, or Xremote, other WAN services include dial-on-demand routing (DDR) of IP and IPX, X.25, Frame Relay, and SMDS.

Management and Security Cisco IOS provides several management features that are built into the Cisco routers and access servers. These management features include configuration services, as well as monitoring and diagnostic services.

Cisco IOS software includes a tool kit for partitioning resources and prohibiting access to sensitive or confidential information and processes. Multidimensional filters prevent users from knowing that other users or resources are even on the network. Encrypted passwords, dial-in authentication, multilevel configuration permissions, and accounting and logging features provide protection from and information about unauthorized access attempts.

24.2 COMPOSITION OF THE RULESET

The Element Management System (EMS) consists of standard NetExpert runtime software modules and specialized rules operating within the runtime modules. The rules are specific to each application and govern the behavior of the runtime modules for that application. The term "rules" generally applies to all implementation work required to create an application. The rules included as part of this package consist of the following:

- **Object Model**   The Object Model is a series of classes and attributes used to describe the managed network and other entities related to that network. The object model for this ruleset is based on the Cisco MIB.
- **Identification and Parsing Rules**   These rules interpret ASCII data from a device or network element and tell the Generic Gateway how to translate that data into events.
- **Analysis Rules**   These rules, performed upon receipt of an event, are processed within IDEAS and perform the analysis and correlation within an application.
- **Alert Definitions**   When an alert is displayed to an operator, the alert definition describes the attributes and visual appearance of the alert.
- **Dialogs**   Dialogs are processed by the Generic Gateway. They include a state tree of commands and anticipated responses, and are used to send commands to a network element or other management systems.
- **VisualAgent Screens**   These screens are the bitmapped representations of the managed network, and the management data to be displayed to the operator. They also contain the dynamic behaviors to be executed when alert or object data is received by the VisualAgent Server.
- **VisualAgent Map Files**   These files map between VisualAgent screens and the NetExpert Object Model. They describe what attributes and alerts should be displayed.
- **DataArchiver Configuration Files**   These files map the event data received from the Generic Gateway to DataArchiver database tables.

Three primary element management functions are provided by this OSI ruleset: (1) Configuration Management, (2) Performance Management, and (3) Fault Management.

Figure 24.1 and the following pages show an overview of these functions.

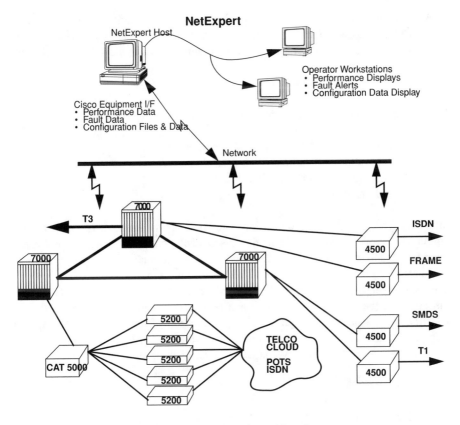

FIGURE 24.1   Overview of functions.

Configuration Management   The following configuration functions are supported:

- Provisioning
- Configuration setting
- Status reporting
- Change control and reporting

The configuration information is stored in a database in the NetExpert MIB. Users can download the stored configuration to selected network elements and can commit those to NVRAM.

This provides a central repository for the configuration of network elements, and a single point from which the configuration of multiple network elements can be controlled.

The class hierarchy for configuration management consists of the following classes:

- CONFIG_EVENTS

Performance Management   Real-time performance monitoring is provided via a monitor panel that shows current CPU, memory buffer, interface, and device errors. The class hierarchy for performance management consists of the following classes:

- PERFORM_EVENTS

Fault Management   The following fault management functions are supported:

- Surveillance via SNMP traps and syslog messages
- Fault location and isolation
- Alarm filtering and correlation
- Trouble management

Surveillance consists of the collection, display, and management of network and equipment errors as they occur. In the Cisco EMS, fault information is collected from two locations: SNMP traps received from agents that reside on the network elements and the messages reported by IOS through the UNIX syslog utility. All alarms received from the elements are directly translated to NetExpert alerts and are displayed on the operator's console. Each fault is one alarm line in the NetExpert Alert Display windows. The graphical representation of the network element will also change color to indicate the severity of the alarm.

The class hierarchy for surveillance consists of the following classes

- SNMP_COMMUNITY
- SYSLOG_CLASS
- MONITOR_EVENTS

Fault Location and Isolation   Fault management is the detection of a problem, isolation and correction to normal operation. The NetExpert system polls the managed objects in search of error conditions and then illustrates the problem in either a graphic format or an alert message. In NetExpert, the faults that are detected, the isolation and correction routines, and the display are determined by the rules written using the NetExpert development tools.

The fault management rules support MIB-II (rfc-1213), the Cisco Enterprise MIB (v11.1 and prior) traps and object definitions, and Cisco IOS-generated error condition messages that are forwarded to syslog.

Correlation   Within the EMS, it is possible to perform correlation within a single network element. This will take the form of consolidating duplicate alarms, thresholding alarms that must occur many times to be considered relevant, and suppressing alarms which are symptomatic of an existing fault.

Trouble Management   Alarms that are generated will be forwarded for attention to network engineers through two mechanisms, trouble tickets and paging. The specific trouble ticket package that will be used may be different for each customer.

Therefore, within the EMS, trouble tickets will be modeled as objects and the BSS interface will be responsible for translating those objects to the appropriate messages to be sent to the trouble ticket package.

Alarms can also result in pages sent to technicians. Pages can be sent automatically when an alarm is generated, or can be manually triggered by the operator. Pages include the description of the alarm that initiated the page and the name of the object in alarm state.

24.3 THE CISCO MIBS

The OSI ruleset for Cisco products is built upon the foundation established by certain Cisco management information base (MIB) files. These files establish object IDs (OIDs) which are used in the OSI ruleset to call particular events relating to the MIB objects. The MIB files used as the basis for the ruleset are listed below:

- CISCO-PRODUCI S-MIB.my
- OLD-CISCO-SYSMIB.my
- OLD-CISCO-INTERFACES-MIB.my
- OLD-CISCO-IP-MIB.my
- OLD-CISCO-TCP-MIB.my
- OLD-CISCO-TS-MIB.my
- OLD-CISCO-FLASH-MIB.my
- OLD-CISCO-CHASSIS-MIB.my
- CISCO-TCP-MIB-VlSMI.my
- CISCO-ENVMON-MIB-VlSMI.my
- CISCO-PING-MIB-VlSMI.my
- CISCO-SNAPSHOT-MIB-VlSMI.my
- CISCO-CDP-MIB-VlSMI.my
- CISCO-STUN-MIB-VlSMI.my
- CISCO-ISDN-MIB-VlSMI.my
- CISCO-QUEUE-MIB-VlSMI.my
- CISCO-CONFIGMAN-MIB-V1SMI.my
- IF-MIB-VlSMI.my
- RS-232-MIB-VlSMI.my
- RFC1253-MIB.my
- RMON-MIB.my
- ETHERLIKE-MIB.my
- RFC1406-MIB.my
- BGP4-MIB-VlSMI.my
- SNMP-REPEATER-MIB.my
- CISCO-ICSUDSU-MIB-VlSMI.my
- CISCO-IMAGE-MIB-VlSMI.my
- BRIDGE-MIB.my

Table 24.1 shows a segment of the public MIB including RMON-PDUs; Table 24.2 shows one segment of the private part of the MIB, including Cisco Products-PDUs.

TABLE 24.1
Example of RMON-PDUs

```
rmon ................................. 1.3.6.1.2.1.16
statistics .......................... 1.3.6.1.2.1.16.1
etherStatsTable ................... 1.3.6.1.2.1.16.1.1
etherStatsEntry ................... 1.3.6.1.2.1.16.1.1.1
etherStatsIndex .................. 1.3.6.1.2.1.16.1.1.1.1
etherStatsDataSource ............ 1.3.6.1.2.1.16.1.1.1.2
etherStatsDropEvents ............ 1.3.6.1.2.1.16.1.1.1.3
etherStatsOctets ................ 1.3.6.1.2.1.16.1.1.1.4
etherStatsPkts ................... 1.3.6.1.2.1.16.1.1.1.5
etherStatsBroadcastPkts .......... 1.3.6.1.2.1.16.1.1.1.6
etherStatsMulticastPkts .......... 1.3.6.1.2.1.16.1.1.1.7
etherStatsCRCAlignErrors ......... 1.3.6.1.2.1.16.1.1.1.8
etherStatsUndersizePkts .......... 1.3.6.1.2.1.16.1.1.1.9
etherStatsOversizePkts ........... 1.3.6.1.2.1.16.1.1.1.10
etherStatsFragments .............. 1.3.6.1.2.1.16.1.1.1.11
etherStatsJabbers ................ 1.3.6.1.2.1.16.1.1.1.12
etherStatsCollisions ............. 1.3.6.1.2.1.16.1.1.1.13
etherStatsPkts64Octets ........... 1.3.6.1.2.1.16.1.1.1.14
etherStatsPkts65to127Octets ...... 1.3.6.1.2.1.16.1.1.1.15
etherStatsPkts128to255Octets ..... 1.3.6.1.2.1.16.1.1.1.16
etherStatsPkts256to511Octets ..... 1.3.6.1.2.1.16.1.1.1.17
etherStatsPkts512to1023Octets .... 1.3.6.1.2.1.16.1.1.1.18
etherStatsPkts1024to1518Octet .... 1.3.6.1.2.1.16.1.1.1.19
etherStatsOwner .................. 1.3.6.1.2.1.16.1.1.1.20
etherStatsStatus ................. 1.3.6.1.2.1.16.1.1.1.21
history .......................... 1.3.6.1.2.1.16.2
historyControlTable .............. 1.3.6.1.2.1.16.2.1
historyControlEntry .............. 1.3.6.1.2.1.16.2.1.1
historyControlIndex .............. 1.3.6.1.2.1.16.2.1.1.1
historyControlDataSource ......... 1.3.6.1.2.1.16.2.1.1.2
historyControlBucketsReq ......... 1.3.6.1.2.1.16.2.1.1.3
historyControlBucketsGranted ..... 1.3.6.1.2.1.16.2.1.1.4
historyControlInterval ........... 1.3.6.1.2.1.16.2.1.1.5
historyControlOwner .............. 1.3.6.1.2.1.16.2.1.1.6
historyControlStatus ............. 1.3.6.1.2.1.16.2.1.1.7
etherHistoryTable ................ 1.3.6.1.2.1.16.2.2
etherHistoryEntry ................ 1.3.6.1.2.1.16.2.2.1
etherHistoryIndex ................ 1.3.6.1.2.1.16.2.2.1.1
etherHistorySampleIndex .......... 1.3.6.1.2.1.16.2.2.1.2
etherHistoryIntervalStart ........ 1.3.6.1.2.1.16.2.2.1.3
```

TABLE 24.1 (continued)
Example of RMON-PDUs

etherHistoryDropEvents 1.3.6.1.2.1.16.2.2.1.4

etherHistoryOctets 1.3.6.1.2.1.16.2.2.1.5

etherHistoryPkts 1.3.6.1.2.1.16.2.2.1.6

etherHistoryBroadcastPkts 1.3.6.1.2.1.16.2.2.1.7

etherHistoryMulticastPkts 1.3.6.1.2.1.16.2.2.1.8

etherHistoryCRCAlignErrors 1.3.6.1.2.1.16.2.2.1.9

etherHistoryUndersizePkts 1.3.6.1.2.1.16.2.2.1.10

etherHistoryOversizePkts 1.3.6.1.2.1.16.2.2.1.11

etherHistoryFragments 1.3.6.1.2.1.16.2.2.1.12

etherHistoryJabbers 1.3.6.1.2.1.16.2.2.1.13

etherHistoryCollisions 1.3.6.1.2.1.16.2.2.1.14

etherHistoryUtilization 1.3.6.1.2.1.16.2.2.1.15

alarm 1.3.6.1.2.1.16.3

alarmTable 1.3.6.1.2.1.16.3.1

alarmEntry 1.3.6.1.2.1.16.3.1.1

alarmIndex 1.3.6.1.2.1.16.3.1.1.1

alarmInterval 1.3.6.1.2.1.16.3.1.1.2

alarmVariable 1.3.6.1.2.1.16.3.1.1.3

alarmSampleType 1.3.6.1.2.1.16.3.1.1.4

alarmValue 1.3.6.1.2.1.16.3.1.1.5

alarmStartupAlarm 1.3.6.1.2.1.16.3.1.1.6

alarmRisingThreshold 1.3.6.1.2.1.16.3.1.1.7

alarmFallingThreshold 1.3.6.1.2.1.16.3.1.1.8

alarmRisingEventIndex 1.3.6.1.2.1.16.3.1.1.9

alarmFallingEventIndex 1.3.6.1.2.1.16.3.1.1.10

alarmOwner 1.3.6.1.2.1.16.3.1.1.11

alarmStatus 1.3.6.1.2.1.16.3.1.1.12

hosts 1.3.6.1.2.1.16.4

hostControlTable 1.3.6.1.2.1.16.4.1

hostControlEntry 1.3.6.1.2.1.16.4.1.1

hostControlIndex 1.3.6.1.2.1.16.4.1.1.1

hostControlDataSource 1.3.6.1.2.1.16.4.1.1.2

hostControlTableSize 1.3.6.1.2.1.16.4.1.1.3

hostControlLastDeleteTime 1.3.6.1.2.1.16.4.1.1.4

hostControlOwner 1.3.6.1.2.1.16.4.1.1.5

hostControlStatus 1.3.6.1.2.1.16.4.1.1.6

hostTable 1.3.6.1.2.1.16.4.2

hostEntry 1.3.6.1.2.1.16.4.2.1

hostAddress 1.3.6.1.2.1.16.4.2.1.1

hostCreationOrder 1.3.6.1.2.1.16.4.2.1.2

hostIndex 1.3.6.1.2.1.16.4.2.1.3

hostInPkts 1.3.6.1.2.1.16.4.2.1.4

hostOutPkts 1.3.6.1.2.1.16.4.2.1.5

hostInOctets 1.3.6.1.2.1.16.4.2.1.6

TABLE 24.1 (continued)
Example of RMON-PDUs

hostOutOctets 1.3.6.1.2.1.16.4.2.1.7
hostOutErrors 1.3.6.1.2.1.16.4.2.1.8
hostOutBroadcastPkts 1.3.6.1.2.1.16.4.2.1.9
hostOutMulticastPkts 1.3.6.1.2.1.16.4.2.1.10
hostTimeTable 1.3.6.1.2.1.16.4.3
hostTimeEntry 1.3.6.1.2.1.16.4.3.1
hostTimeAddress 1.3.6.1.2.1.16.4.3.1.1
hostTimeCreationOrder 1.3.6.1.2.1.16.4.3.1.2
hostTimeIndex 1.3.6.1.2.1.16.4.3.1.3
hostTimeInPkts 1.3.6.1.2.1.16.4.3.1.4
hostTimeOutPkts 1.3.6.1.2.1.16.4.3.1.5
hostTimeInOctets 1.3.6.1.2.1.16.4.3.1.6
hostTimeOutOctets 1.3.6.1.2.1.16.4.3.1.7
hostTimeOutErrors 1.3.6.1.2.1.16.4.3.1.8
hostTimeOutBroadcastPkts 1.3.6.1.2.1.16.4.3.1.9
hostTimeOutMulticastPkts 1.3.6.1.2.1.16.4.3.1.10
hostTopN 1.3.6.1.2.1.16.5
hostTopNControlTable 1.3.6.1.2.1.16.5.1
hostTopNControlEntry 1.3.6.1.2.1.16.5.1.1
hostTopNControlIndex 1.3.6.1.2.1.16.5.1.1.1
hostTopNHostIndex 1.3.6.1.2.1.16.5.1.1.2
hostTopNRateBase 1.3.6.1.2.1.16.5.1.1.3
hostTopNTimeRemaining 1.3.6.1.2.1.16.5.1.1.4
hostTopNDuration 1.3.6.1.2.1.16.5.1.1.5
hostTopNRequestedSize 1.3.6.1.2.1.16.5.1.1.6
hostTopNGrantedSize 1.3.6.1.2.1.16.5.1.1.7
hostTopNStartTime 1.3.6.1.2.1.16.5.1.1.8
hostTopNOwner 1.3.6.1.2.1.16.5.1.1.9
hostTopNStatus 1.3.6.1.2.1.16.5.1.1.10
hostTopNTable 1.3.6.1.2.1.16.5.2
hostTopNEntry 1.3.6.1.2.1.16.5.2.1
hostTopNReport 1.3.6.1.2.1.16.5.2.1.1
hostTopNIndex 1.3.6.1.2.1.16.5.2.1.2
hostTopNAddress 1.3.6.1.2.1.16.5.2.1.3
hostTopNRate 1.3.6.1.2.1.16.5.2.1.4
matrix 1.3.6.1.2.1.16.6
matrixControlTable 1.3.6.1.2.1.16.6.1
matrixControlEntry 1.3.6.1.2.1.16.6.1.1
matrixControlIndex 1.3.6.1.2.1.16.6.1.1.1
matrixControlDataSource 1.3.6.1.2.1.16.6.1.1.2
matrixControlTableSize 1.3.6.1.2.1.16.6.1.1.3
matrixControlLastDeleteTime 1.3.6.1.2.1.16.6.1.1.4
matrixControlOwner 1.3.6.1.2.1.16.6.1.1.5
matrixControlStatus 1.3.6.1.2.1.16.6.1.1.6

TABLE 24.1 (continued)
Example of RMON-PDUs

matrixSDTable 1.3.6.1.2.1.16.6.2
matrixSDEntry 1.3.6.1.2.1.16.6.2.1
matrixSDSourceAddress 1.3.6.1.2.1.16.6.2.1.1
matrixSDDestAddress 1.3.6.1.2.1.16.6.2.1.2
matrixSDIndex 1.3.6.1.2.1.16.6.2.1.3
matrixSDPkts 1.3.6.1.2.1.16.6.2.1.4
matrixSDOctets 1.3.6.1.2.1.16.6.2.1.5
matrixSDErrors 1.3.6.1.2.1.16.6.2.1.6
matrixDSTable 1.3.6.1.2.1.16.6.3
matrixDSEntry 1.3.6.1.2.1.16.6.3.1
matrixDSSourceAddress 1.3.6.1.2.1.16.6.3.1.1
matrixDSDestAddress 1.3.6.1.2.1.16.6.3.1.2
matrixDSIndex 1.3.6.1.2.1.16.6.3.1.3
matrixDSPkts 1.3.6.1.2.1.16.6.3.1.4
matrixDSOctets 1.3.6.1.2.1.16.6.3.1.5
matrixDSErrors 1.3.6.1.2.1.16.6.3.1.6
filter 1.3.6.1.2.1.16.7
filterTable 1.3.6.1.2.1.16.7.1
filterEntry 1.3.6.1.2.1.16.7.1.1
filterIndex 1.3.6.1.2.1.16.7.1.1.1
filterChannelIndex 1.3.6.1.2.1.16.7.1.1.2
filterPktDataOffset 1.3.6.1.2.1.16.7.1.1.3
filterPktData 1.3.6.1.2.1.16.7.1.1.4
filterPktDataMask 1.3.6.1.2.1.16.7.1.1.5
filterPktDataNotMask 1.3.6.1.2.1.16.7.1.1.6
filterPktStatus 1.3.6.1.2.1.16.7.1.1.7
filterPktStatusMask 1.3.6.1.2.1.16.7.1.1.8
filterPktStatusNotMask 1.3.6.1.2.1.16.7.1.1.9
filterOwner 1.3.6.1.2.1.16.7.1.1.10
filterStatus 1.3.6.1.2.1.16.7.1.1.11
channelTable 1.3.6.1.2.1.16.7.2
channelEntry 1.3.6.1.2.1.16.7.2.1
channelIndex 1.3.6.1.2.1.16.7.2.1.1
channelIfIndex 1.3.6.1.2.1.16.7.2.1.2
channelAcceptType 1.3.6.1.2.1.16.7.2.1.3
channelDataControl 1.3.6.1.2.1.16.7.2.1.4
channelTurnOnEventIndex 1.3.6.1.2.1.16.7.2.1.5
channelTurnOffEventIndex 1.3.6.1.2.1.16.7.2.1.6
channelEventIndex 1.3.6.1.2.1.16.7.2.1.7
channelEventStatus 1.3.6.1.2.1.16.7.2.1.8
channelMatches 1.3.6.1.2.1.16.7.2.1.9
channelDescription 1.3.6.1.2.1.16.7.2.1.10
channelOwner 1.3.6.1.2.1.16.7.2.1.11
channelStatus 1.3.6.1.2.1.16.7.2.1.12

TABLE 24.1 (continued)
Example of RMON-PDUs

```
capture ............................ 1.3.6.1.2.1.16.8
bufferControlTable ................. 1.3.6.1.2.1.16.8.1
bufferControlEntry ................. 1.3.6.1.2.1.16.8.1.1
bufferControlIndex ................. 1.3.6.1.2.1.16.8.1.1.1
bufferControlChannelIndex ........ 1.3.6.1.2.1.16.8.1.1.2
bufferControlFullStatus ........... 1.3.6.1.2.1.16.8.1.1.3
bufferControlFullAction ........... 1.3.6.1.2.1.16.8.1.1.4
bufferControlCaptureSliceSize .... 1.3.6.1.2.1.16.8.1.1.5
bufferControlDownloadSlice ....... 1.3.6.1.2.1.16.8.1.1.6
bufferControlDownloadOffset ...... 1.3.6.1.2.1.16.8.1.1.7
bufferControlMaxOctetsReq ........ 1.3.6.1.2.1.16.8.1.1.8
bufferControlMaxOctetsGranted .... 1.3.6.1.2.1.16.8.1.1.9
bufferControlCapturedPackets ..... 1.3.6.1.2.1.16.8.1.1.10
bufferControlTurnOnTime .......... 1.3.6.1.2.1.16.8.1.1.11
bufferControlOwner ............... 1.3.6.1.2.1.16.8.1.1.12
bufferControlStatus .............. 1.3.6.1.2.1.16.8.1.1.13
captureBufferTable ................ 1.3.6.1.2.1.16.8.2
captureBufferEntry ................ 1.3.6.1.2.1.16.8.2.1
captureBufferControlIndex ........ 1.3.6.1.2.1.16.8.2.1.1
captureBufferIndex ............... 1.3.6.1.2.1.16.8.2.1.2
captureBufferPacketID ............ 1.3.6.1.2.1.16.8.2.1.3
captureBufferPacketData .......... 1.3.6.1.2.1.16.8.2.1.4
captureBufferPacketLength ........ 1.3.6.1.2.1.16.8.2.1.5
captureBufferPacketTime .......... 1.3.6.1.2.1.16.8.2.1.6
captureBufferPacketStatus ........ 1.3.6.1.2.1.16.8.2.1.7
event ............................. 1.3.6.1.2.1.16.9
eventTable ........................ 1.3.6.1.2.1.16.9.1
eventEntry ........................ 1.3.6.1.2.1.16.9.1.1
eventIndex ........................ 1.3.6.1.2.1.16.9.1.1.1
eventDescription ................. 1.3.6.1.2.1.16.9.1.1.2
eventType ........................ 1.3.6.1.2.1.16.9.1.1.3
eventCommunity ................... 1.3.6.1.2.1.16.9.1.1.4
eventLastTimeSent ................ 1.3.6.1.2.1.16.9.1.1.5
eventOwner ....................... 1.3.6.1.2.1.16.9.1.1.6
eventStatus ...................... 1.3.6.1.2.1.16.9.1.1.7
logTable ......................... 1.3.6.1.2.1.16.9.2
logEntry ......................... 1.3.6.1.2.1.16.9.2.1
logEventIndex .................... 1.3.6.1.2.1.16.9.2.1.1
logIndex ......................... 1.3.6.1.2.1.16.9.2.1.2
logTime .......................... 1.3.6.1.2.1.16.9.2.1.3
logDescription ................... 1.3.6.1.2.1.16.9.2.1.4
```

TABLE 24.2
Example of Cisco Products PDUs

experiment 1.3.6.1.3
private 1.3.6.1.4
enterprises 1.3.6.1.4.1
cisco 1.3.6.1.4.1.9
ciscoProducts 1.3.6.1.4.1.9.1
ciscoGatewayServer 1.3.6.1.4.1.9.1.1
ciscoTerminalServer 1.3.6.1.4.1.9.1.2
ciscoTrouter 1.3.6.1.4.1.9.1.3
ciscoProtocolTranslator 1.3.6.1.4.1.9.1.4
ciscoIGS 1.3.6.1.4.1.9.1.5
cisco3000 .... 1.3.6.1.4.1.9.1.6
cisco4000 .... 1.3.6.1.4.1.9.1.7
cisco7000 .... 1.3.6.1.4.1.9.1.8
ciscoCS500 .... 1.3.6.1.4.1.9.1.9
cisco2000 .... 1.3.6.1.4.1.9.1.10
ciscoAGSplus 1.3.6.1.4.1.9.1.11
cisco7010 .... 1.3.6.1.4.1.9.1.12
cisco2500 .... 1.3.6.1.4.1.9.1.13
cisco4500 .... 1.3.6.1.4.1.9.1.14
cisco2102 .... 1.3.6.1.4.1.9.1.15
cisco2202 .... 1.3.6.1.4.1.9.1.16
cisco2501 .... 1.3.6.1.4.1.9.1.17
cisco2502 .... 1.3.6.1.4.1.9.1.18
cisco2503 .... 1.3.6.1.4.1.9.1.19
cisco2504 .... 1.3.6.1.4.1.9.1.20
cisco2505 .... 1.3.6.1.4.1.9.1.21
cisco2506 .... 1.3.6.1.4.1.9.1.22
cisco2507 .... 1.3.6.1.4.1.9.1.23
cisco2508 .... 1.3.6.1.4.1.9.1.24
cisco2509 .... 1.3.6.1.4.1.9.1.25
cisco2510 .... 1.3.6.1.4.1.9.1.26
cisco2511 .... 1.3.6.1.4.1.9.1.27
cisco2512 .... 1.3.6.1.4.1.9.1.28
cisco2513 .... 1.3.6.1.4.1.9.1.29
cisco2514 .... 1.3.6.1.4.1.9.1.30
cisco2515 .... 1.3.6.1.4.1.9.1.31
cisco3101 .... 1.3.6.1.4.1.9.1.32
cisco3102 .... 1.3.6.1.4.1.9.1.33
cisco3103 .... 1.3.6.1.4.1.9.1.34
cisco3104 .... 1.3.6.1.4.1.9.1.35
cisco3202 .... 1.3.6.1.4.1.9.1.36
cisco3204 .... 1.3.6.1.4.1.9.1.37
ciscoAccessProRC 1.3.6.1.4.1.9.1.38
ciscoAccessProEC 1.3.6.1.4.1.9.1.39
cisco1000 .... 1.3.6.1.4.1.9.1.40
cisco1003 .... 1.3.6.1.4.1.9.1.41

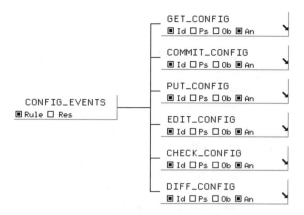

FIGURE 24.2   Class hierarchy for the configuration ruleset.

24.4 DESCRIPTION OF THE RULESET

This NetExpert application concentrates on three principal management areas:

- Configuration management
- Performance management
- Fault management

24.4.1 CONFIGURATION MANAGEMENT RULES

The Cisco configuration management ruleset is described on the following pages. The class hierarchy is presented first, followed by event definitions and descriptions of relationships.

The purpose of the configuration ruleset is to allow remote configuration of Cisco equipment using NetExpert. Figure 24.2 shows the Class Hierarchy for the Cisco EMS Configuration ruleset.

Table 24.3 shows the Configuration Event Class Definions.

TABLE 24.3
Configuration Event Class Definitions

Event	Function
CONFIG_EVENTS	Manager class for configuration events listed below
GET _CONFIG	Gets router config data and writes to a file named "routername-config"
COMMIT_CONFIG	Commits previously-sent configuration file to non-volatile RAM. Generates an alert to confirm success or failure of commitment.
PUT_CONFIG	Instructs router to read config file from NetExpert server to active RAM for holding until committed to non-volatile RAM
EDIT_CONFIG	Opens config file for editing
CHECK_CONFIG	Runs NetSys Checker to check for configuration errors
DIFF_CONFIG	Gets router configuration and writes it to tftp server as a file. Then does a diff comparison of current router configuration to what is on tftpboot server directory.

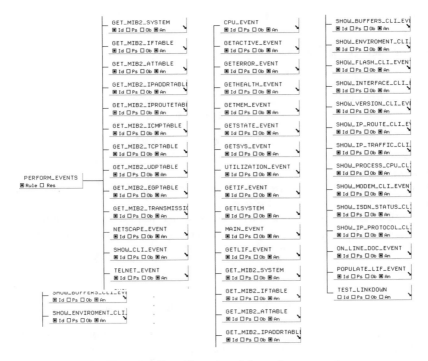

FIGURE 24.3   Class hierarchy of the performance ruleset.

24.4.2 PERFORMANCE MANAGEMENT RULES

The Cisco performance management ruleset is described in some depth. The class hierarchy is presented first, followed by event definitions and description of relationships. The purpose of the performance ruleset is to work interactively with Visual Agent graphics to provide active monitoring of Cisco equipment performance.

Figure 24.3 shows the Class Hierarchy for the Cisco EMS Performance ruleset. Table 24.4 shows the Performance Event Class Definions.

24.4.3 FAULT MANAGEMENT RULES

The Cisco fault management ruleset is described on the following pages. The class hierarchy is presented first, followed by event definitions and descriptions of relationships.

The purpose of the fault ruleset is to monitor SNMP trap and syslog messages sent out by the Cisco equipment indicating faults in the Cisco system and to generate appropriate NetExpert operator alerts indicating the faults.

The fault management class hierarchy consists of three primary classes:

1. SNMP_COMMUNM
2. SYSLOG_CLASS
3. MONITOR_EVENTS

TABLE 24.4
Performance Event Class Definitions

Event Class	Function
PERFORM_EVENT	Manager class for the performance events below
CPU_EVENT	Gets current CPU load statistics
GETACTIVE_EVENT	Gets terminal server line characteristics
GETERROR_EVENT	Gets list of interface errors, such as framing, CRC, runt and grunt errors
GETHEALTH_EVENT	Gets current memory error data
GETMEM_EVENT	Gets memory utilization data, such as memory free, buffer size, and buffer failures
GETSTATE_EVENT	Gets data about each interface (port), such as address and netmask
GETSYS_EVENT	Gets router configuration information, such as config file name and processor RAM available
UTILIZATION_EVENT	Gets current utilization percentage for each interface (port)
GETIF_EVENT	Gets router interface (port) table of values to allow pre-populating database
GETLSYSTEM	Gets information about the local system hardware, such as RAM available and firmware version
GET_MIB2_SYSTEM	Gets system name, location, contact, and other system information
GET_MIB2_IFTABLE	Gets MIBII interface information
GET_MIB2_IPADDRTABLE	Gets MIBII IP Address Table
GET_MIB2_IPROUTETABLE	Gets MIBII IP Route Table
GET_MIB2_ICMPTABLE	Gets MIBII ICMP Table

Fault SNMP Trap Class Hierarchy　　The top-level of the SNMP Trap class hierarchy is the SNMP Community class. This class consists of events triggered by Cisco trap messages received through the SNMP gateway and trap daemon. These messages may convey information about faults within the Cisco hardware or network, or they may include responses to SNMP Get requests.

Data Flow within the SNMP Community Class Hierarchy　　The data flow for an SNMP Community event occurs as follows (Figure 24.4). A Cisco device sends a trap (fault message) to the NetExpert SNMP trap daemon. The trap daemon decodes the trap message and maps it (using the Object IdentifierWlD) to a corresponding event within NetExpert IDEAS. IDEAS processes the message, passing it through the analysis rules in the class hierarchy, resulting in the generation of a specific event.

The class hierarchy for the Cisco ruleset consists of the SNMP_COMMUNIB class, which serves as the manager class for the event called SNMP_EVENT. Additional events that inherit from SNMP_COMMUNITY and SNMP_EVENT include standard SNMP events (cold Start warmStart, etc.) and the SNMP_enterpriseSpecific event, which contains the rules specific to Cisco products. The analysis rules within the SNMP_enterpriseSpecific event generate the event called Cisco_Traps, which includes analysis rules for generating specific fault Cap events named CISCO_)CXX_TRAP.

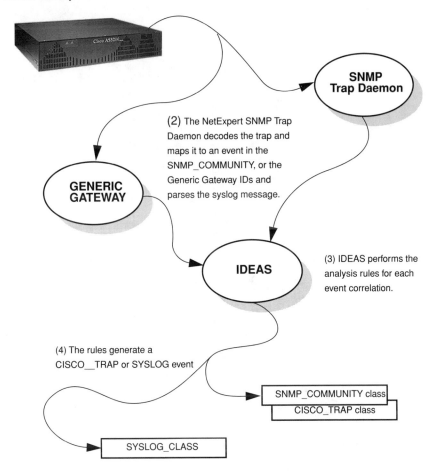

(1) A Cisco device sends a trap to the SNMP Gateway or a syslog message to the Generic Gateway

SNMP Trap Daemon

(2) The NetExpert SNMP Trap Daemon decodes the trap and maps it to an event in the SNMP_COMMUNITY, or the Generic Gateway IDs and parses the syslog message.

GENERIC GATEWAY

IDEAS

(3) IDEAS performs the analysis rules for each event correlation.

(4) The rules generate a CISCO__TRAP or SYSLOG event

SNMP_COMMUNITY class
CISCO_TRAP class

SYSLOG_CLASS

FIGURE 24.4   Data flow of an SNMP community event.

Figure 24.5 shows the SNMP Trap Fault Management Classes Hierarchy ruleset, while Table 24.5 shows the Fault SNMP Trap Class Definions.

Fault Monitor Class Hierarchy   The Cisco fault monitor ruleset is described in some detail on the following pages. The class hierarchy is presented first, followed by parent/child, manager/managed, and other important relationships. Monitor events watch certain performance characteristics of the Cisco equipment and report any faults or problems with the monitored items.

Figure 24.6 shows the Monitor Class Hierarchy for the Fault Monitor ruleset, while Table 24.6 shows the Fault Monitor Event Class Definions.

FIGURE 24.5   SNMP trap fault management hierarchy ruleset.

24.5   PRESENTATION OF RESULTS

It is extremely important to offer easy-to-understand graphics to assist the work of operators. This NetExpert application offers a number of graphics addressing various levels of details. A few examples are given for a number of various operational areas.

Graphic windows are opened from the Client Manager menu. The first view is a world, national, or regional view. From this view, menu options are provided to access local network and element views, followed by various graphical representations of element data.

VisualAgent drives the graphic displays and provides pop-up and cascading menus for executing element management commands. VisualAgent uses "template" models which are dynamically driven by attributes within the events in the NetExpert

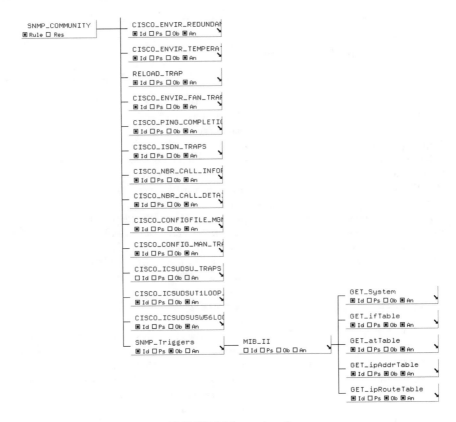

FIGURE 24.5 (continued)

Management Information Base (MIB). As events are triggered, data from the specified element in the event updates the event attributes.

Most windows display a legend at the bottom of the screen describing the severity colors that appear in the graphics.

The graphical hierarchy (Figure 24.7) shows all the graphics available in the Cisco ruleset.

The market view (Figure 24.8) depicts an overview map of the market region with colored buttons representing the central locations of management centers. The color indicates the type of alarm for that particular location. By clicking on any of those location buttons, a new window is presented with a detailed view of the location.

The view of the management centers (Figure 24.9) shows and lists all the elements located on the network at a particular geographical area or office. Clicking on the element causes a new window to appear, showing the physical layout of the selected element.

Equipment view is presented next (Figure 24.10). It shows the back of the selected device cabinet, where certain components and buttons are highlighted. The

TABLE 24.5
Fault SNMP Trap Class Definitions

Event Class	Function
SNMP_EVENT	Performs analysis to see if node is known. Forwards SNMP events to Cisco_Traps event
SNMP_coldStart	Generates a cold start alert
SNMP_warmStart	Generates a warm start alert
SNMP_linkDown	Generates a link down alert for the given node
SNMP_linkUp	Generates a link up alert
SNMP_authentication Failure	Generates an authorization failure alert
SNMP_egpNeighborLoss	Generates an SNMP EGP Neighbor Loss alert
SNMP_enterpriseSpecific	Generates a CISCO_TRAPS event
SNMP_unknownObjectID	Generates an unknown object alert
CISCO_PING_TRAPS	Forwards ping traps to the appropriate ping trap events
TCPCONNECTION_ CLOSE_TRAP	Generates a TCP connection closed alert
CISCO_TRAPS	Forwards Cisco traps to appropriate Cisco trap event. Used to logically group traps
CISCO_FLASH_TRAPS	Forwards Cisco flash traps to the appropriate flash trap event
CISCO_FLASH_MISCOP_ COMP_TRAP	Generates a FLASH_MISCOP alert
CISCO_ENVIRONMENT_TRAPS	Forwards Cisco environment traps to the appropriate Cisco environment trap event
CISCO_FLASH_COPY_ COMP_TRAP	Generates a flash copy complete alert
CISCO_FLASH_DEVICE_CHANGE_TR AP	Generates a flash device change alert
CISCO_FLASH_ PARTITION_COMP_TRAP	Generates a flash partition comp alert
CISCO_STANDARD_ TRAPS	Forwards Cisco standard traps to appropriate Cisco trap event
CISCO_ENVIR_SHUT DOWN_TRAP	Generates an environment shutdown alert
CISCO_ENVIR_ VOLTAGE_TRAP	Generates an environment voltage alert
CISCO_ENVIR_REDUN- DANT_SUPPLY_ TRAP	Generates an environment redundant supply alert
CISCO_ENVIR_TEMPERATURE_TRAP	Generates an environment temperature alert
RELOAD_TRAP	Generates a reload alert
CISCO_ENVIR_ FAN_TRAP	Generates an environment fan alert
CISCO_PING_ COMPLETION_TRAP	Generates a ping completion alert
CISCO_ISDN_TRAPS	Forwards NBR traps to appropriate NBR trap events

TABLE 24.5 (continued)
Fault SNMP Trap Class Definitions

Event Class	Function
CISCO_NBR_CALL_ INFORMATION_TRAPS	Generates NBR call information alert
CISCO_NBR_CALL_ DETAIL_TRAP	Generates NBR call detail alert
CISCO_CONFIGFILE_ MGMT_TRAPS	Forwards configuration file management traps to appropriate configuration file management events
CISCO_CONFIG_ MAN_TRAP	Generates configuration management alert
CISCO_ICSUDSU_TRAPS	Forwards ICSU/DSU traps to the appropriate events
CISCO_ICSUDSUT1LOOP_STATUS_TR APS	Forwards ICSU/DSU T1 loop status traps to the appropriate ICSU/DSU status events
CISCO_ICSUDSUSW56 LOOP_STATUS_TRAPS	Generates a switched 56 loop status alert
SNMP_TRIGGERS	Forwards SNMP triggers to the appropriate SNMP events

operator can click on these components or on the buttons to view data relating to that aspect of the equipment.

If the operator clicks on one of the port components on the back of the equipment, port data is displayed as shown in Figure 24.11.

If the operator clicks on the Router Check button on the back of the router display, a view of router data appears as in the example of Figure 24.12.

Various performance panels can be opened from the control panel. With this panel, (Figure 24.13) the operator can view equipment performance characteristics, such as buffer utilization levels.

In order to configure systems, the configuration panel is very helpful. With this panel, (Figure 24.14) the operator can select a number of system configuration functions, such as getting the current system configuration file and making changes to it.

In order to display current configuration and performance data, the show commands panel is helpful (Figure 24.15). For example, the operator can view the selected IP protocol, the software version, or the size of the memory buffers.

24.6 SUMMARY

This NetExpert application helps to manage a large number of different routers from Cisco. CiscoWorks, a well-known application from Cisco, is a successful element management system, supporting fault, performance and configuration management for different types of Cisco routers. NetExpert can be considered as the umbrella platform to integrate the various element management systems from Cisco and also from other vendors.

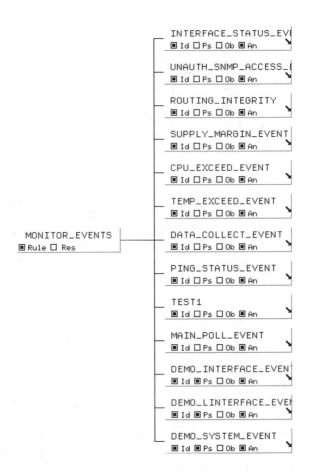

FIGURE 24.6 Monitor class hierarchy for the fault monitor ruleset.

TABLE 24.6
Fault Monitor Class Definitions

Event Class	Function
MONITOR_EVENTS	Manager class for the surveillance events below
INTERFACE_STATUS	Walks through interfaces to check administrative and operational status
UNAUTH_SNMP_ACCESS	Generates an alert if it detects unauthorized community strings
ROUTING_INTEGRITY	Checks router number and types of routing error messages being generated. Sends an alert if buffer full errors occur too frequently
DATA_COLLECT	Checks for correct access to various buffers
CPU_EXCEED	Checks current CPU utilization percentage. Generates an alert if above desired value
TEMP_EXCEED	Checks air intake and exhaust temperatures. Generates an alert if temperature exceeds desired value or if router is close to temperature-caused shutdown

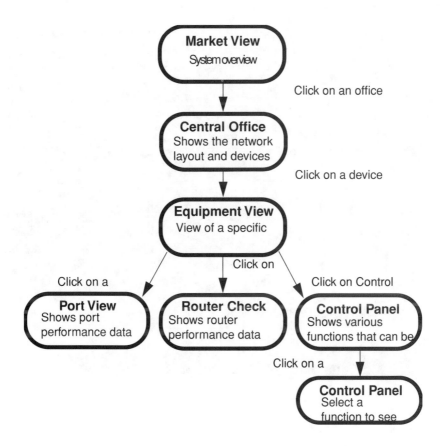

FIGURE 24.7 Graphics overview.

FIGURE 24.8 Overview map.

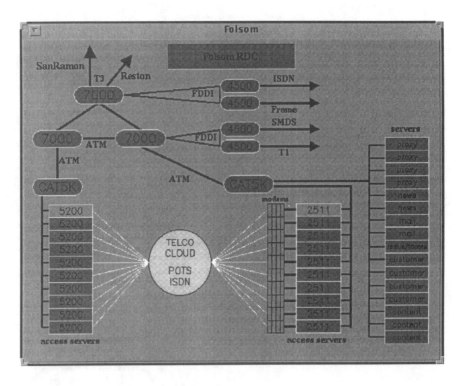

FIGURE 24.9 Overview of management centers.

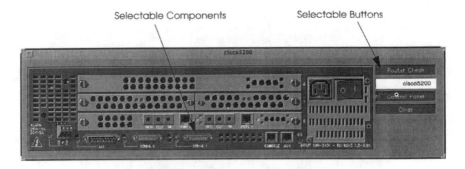

FIGURE 24.10 Equipment view.

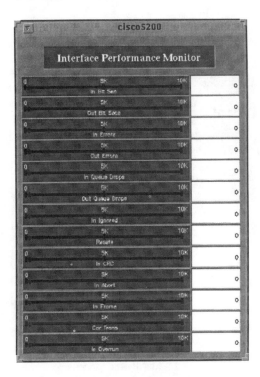

FIGURE 24.11   Display of equipment ports.

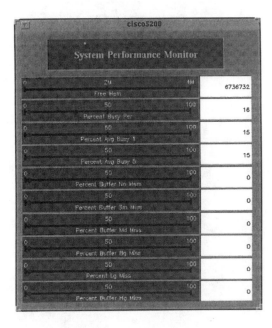

FIGURE 24.12   Viewing router data.

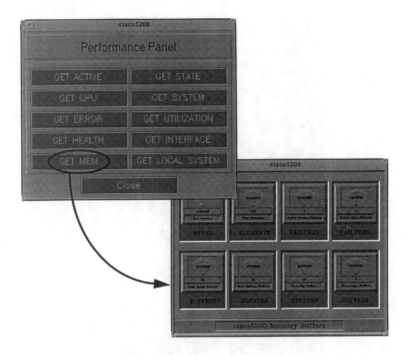

FIGURE 24.13   Performance panels.

FIGURE 24.14   Configuration panel.

FIGURE 24.15 Display of the show command panel.

25 Integrated Network Management Case Studies

CONTENTS

FIGURES

TABLES

Introduction   Due to the significantly increasing demand for Internet access and for implementing Web technology in enterprises, the role of Internet Service Providers (ISPs) is increasing as well. However, this market is crowded. ISPs are looking to differentiate their services from those of the competition. The expectations are high, and in most cases include the following list:

- Good performance
- Bandwidth available on demand
- Easy change management
- Predictable billing
- Roaming to service mobile users
- Reasonable security
- Selective user access

Powerful and integrated management solutions are necessary to meet all of these requirements. This chapter focuses on three innovative companies that are using NetExpert as their core network management product. In all three cases, service

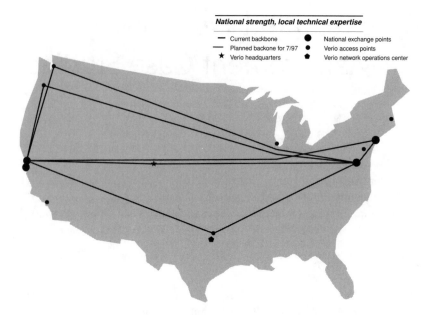

FIGURE 25.1   Verio's national network.

creation, rapid deployment, powerful change management, and efficient trouble shooting were the deciding factors for choosing the NetExpert framework.

25.1   VERIO INC. CASE STUDY

Verio Inc. is the nation's first Internet Service Provider (ISP) to combine the power of a high-speed national network with local, home-town providers. Dozens of well-established, locally managed Verio Internet affiliates in major markets across the country are helping businesses of all sizes — from small to mid-sized companies, to larger institutions — take advantage of the many benefits of doing business on the Internet. (See Figure 25.1.)

At its founding in 1996, the Verio network was backed by $80 million in private funding from several big players in the capital investment and telecommunications industries. Financing is being used to purchase or invest in the cream of the crop among the best-managed and most respected ISPs in the country, many of whom were among the original pioneers in the Internet industry. Affiliation with the Verio group gives these providers access to a sophisticated network operations center, expert engineering support, 24-hour customer care, and an advanced billing system. In addition, affiliates can participate in volume purchase agreements with some of the biggest names in Internet technology, including Cisco Systems, from whom they can purchase routers of all sizes — from 7513 routers for high-capacity to 2514 routers for small businesses, and 1005 routers for people working out of their homes.

ISPs in the Verio group tie into a high-capacity national backbone through nodes in major metropolitan areas. At the core of this complex network is Objective System Integrator's NetExpert framework, which integrates all of Verio's technology, applications and systems into one seamless network management system. Because of NetExpert's flexibility, Verio's business customers can access the Internet at whatever speed works best for them, from low-speed access for the small or home office to T1 and T3 connectivity needed by large corporations. They also benefit from a full range of Internet access tools, including audio and video features, web page design and web hosting, the potential to conduct business 24-hours a day over the Internet, and the ability to handle secure financial transactions.

Verio chose OSI's NetExpert framework because it is the one management system that can handle all the disparate functions of the network from one basic framework. At the point when Verio was gathering its system elements, NetExpert's framework had already proven itself in the management of telco and wireless feeds. Verio needed something that could handle every one of its system's 40,000 different data tables. Verio picked OSI's framework to manage its complex wireless systems because it could integrate all the system elements and also handle the telco feeds.

Verio needed to get its network up and running quickly. There were some good network element managers in the existing tools on the market, but they could not integrate all the elements Verio used, and they did not allow Verio to integrate them across other systems like those that handled billing and customer care. Without OSI's framework solution, Verio would have had to buy a lot of products and integrate them themselves, or ask their operators to be constantly scanning several different screens.

Verio has nearly twenty affiliates in its family of ISPs and will continually add to them. Among these affiliates are multi-sized networks using equipment from different manufacturers with varying configurations. It is important that affiliates be able to integrate seamlessly and that all have access to Verio's enterprise systems. NetExpert is the transaction manager that allows Verio to maintain continuity of operations every time it adds a new customer or a new element to the network. As Verio continues to grow, this ability to glue all its disparate systems together so they can function efficiently around the world will be essential to its continued success.

Verio also appreciates the framework's ability to integrate tools from other manufacturers into their system. This keeps Verio from having to develop everything from scratch.

Today NetExpert primarily manages the backbone of the Verio network. In the future it will be expanded to allow local Verio ISPs to monitor their networks all the way out to the customer premises.

25.2   NetExpert MANAGES @HOME NETWORK

As the World Wide Web quickly becomes the World Wide Wait, consumers and businesses alike are looking for ways to reduce delays by increasing Internet access bandwidth.

@Home Network of Redwood City, CA, is one of the pioneers in providing high speed Internet access through the cable television infrastructure. Using a hybrid fiber/coax network, @Home Network can provide access speeds of up to 27 Mbps, far higher than even the 128 Kbps of ISDN lines. @Home Network's @Work division offers additional secure network communications and telecommuting services.

@Home has business agreements with cable companies such as Comcast, Cox, interMedia Partners, Marcus Cable, Rogers, Shaw, and TCI.

The good news is that @Home got access to that many more homes. However, this does require a larger network infrastructure to manage.

So when @Home selected a network management system, it needed one that was highly scalable to handle the growing number of subscribers and the burgeoning network infrastructure necessary to support them.

@Home also wanted the system to allow proactive management of its subscriber base and infrastructure. The cable television plant is typically not managed. By actively monitoring and managing the cable modem devices, @Home can be more proactive and differentiate themselves with level of service. @Home wants its people to be able to respond to calls by saying, 'it looks like the outage in your area was caused by X and it will be fixed by Y'.

@Home chose NetExpert to manage its infrastructure because it is very powerful and flexible. @Home is currently engaged in a pilot project, using NetExpert in one regional data center and the central NOC (network operations center). The company plans to extend NetExpert to manage a total of 50 regional data centers.

So far, @Home has been pleased with NetExpert's ability to meet its requirements in the pilot project. NetExpert's architecture provides the necessary scalability by allowing @Home to distribute network polling and trap processing across the network, thereby distributing server CPU workload and minimizing network traffic.

NetExpert enables @Home to proactively manage its network through distributed error detection and troubleshooting capabilities. @Home will use NetExpert components such as the Intelligent SNMP Gateway Threshold Crossing Agent to make it easy for each of the 50 regional data centers to identify subscriber's service problems without having to poll each subscriber modem.

When network errors occur, trap management capabilities and NetExpert rulesets will automatically diagnose errors. The system can even recommend a course of corrective action to the human network administrators, who are all located in a single, centralized operational support center, or automatically initiate corrective measures.

The distribution and automation of its network management system enables @Home to troubleshoot and correct problems more quickly. At the same time, it also reduces the number of operators necessary to manage the system, thereby cutting costs.

While the system has yet to be deployed to the regional centers, @Home is encouraged with early results. They are pleased with the threshold crossing agent, and the connectivity agent can do what they want with a reasonable delay. They can poll their entire infrastructure in seconds rather than hours, so they feel that when they scale up to the full 50 regional centers, they will be able to handle that easily.

In future phases of the project, they hope to use NetExpert to manage services like email, news, chat, and the web. They also hope to use it to document service levels for the corporate networks being put in place by the @Work division.

NetExpert is helping @Home and @Work help end the World Wide Wait.

TABLE 25.1
Rulesets Used by @Home

Rule Set	Interface
Cisco 7200/7500	SNMP
Catalyst 5000	SNMP
Xylogics terminal server	SNMP
LANCity modem (LCP)	SNMP

The first phase of the implementation deals with fault management which includes developing and delivering fault management rulesets for Cisco routers, Catalyst switches, Xylogic terminals, and LANCity cable modems. Also, object-modeling and overall graphics development are included.

The rulesets for @Home are listed in Table 25.1. Each ruleset includes an interface to the device using SNMP, device level graphics and generic events and functions for providing fault management.

The fault management rule packages cited above provide a set of functions to detect and isolate abnormal operations of the network and its environment. For fault management on @Home's devices, NetExpert will receive unsolicited SNMP traps (fault messages) to NetExpert SNMP trap daemon. The trap daemon decodes the trap message and maps it (using the Object Identifier-OID) to a corresponding event within NetExpert IDEAS. IDEAS processes the message, passing it through the analysis rules in the class hierarchy, resulting in the generation of alerts which affects a graphical object(s). See Figure 25.2.

For proactive maintenance, polling and thresholds will be used to ensure the health of the device. To provide @Home with the functionality desired, several OSI tools will be used including NetExpert rulesets, NetExpert SNMP GUI, and intelligent agents.

NetExpert rules will be developed to process MIB-2, vendor-specific and all TCA-generated SNMP traps. Each trap processed by the SNMP gateway will generate the appropriate alert and cause the appropriate graphical objects to change state.

The NetExpert SNMP GUI is a suite of tools to augment the standard NetExpert framework. These tools will provide quick, standard access to network variables which are available from the various network elements. The SNMP GUI tools that will be delivered in Phase I are: 2DGraph tool and MIB browser. The 2DGraph tool will allow graphical representation of network variables. These variables can be polled or obtained via an "on demand" method. OSI will provide configuration files for various objects mentioned in the requirements document. These files are easy to maintain and can be modified to suit the needs of @Home. The MIB browser will allow the "Get"ing and "Set"ing of network variables on an "on demand" basis. Appropriately compiled MIBs will be provided to @Home to be used with the browser.

The SNMP GUI tools are being provided to aid in the trouble shooting and on demand access to information stated in the requirements. They will allow a more flexible and detailed approach when trouble shooting problems with the various network elements occur.

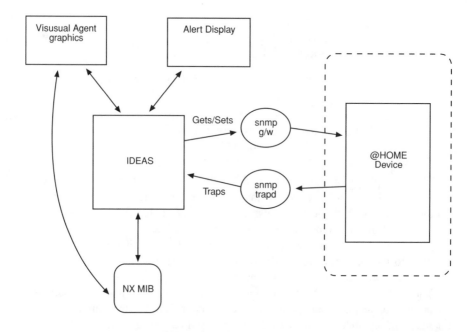

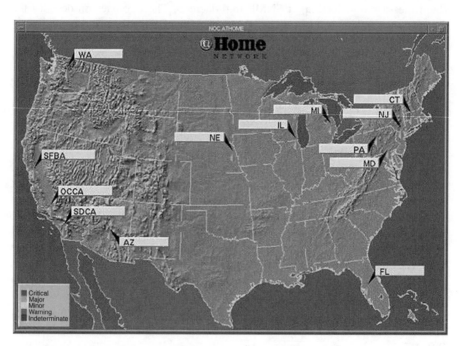

FIGURE 25.2 Overview of @Home Device (Cisco/Xylogic/LanCity/Catalyst) fault management system.

One of the intelligent agents is a tool that will be used to proactively monitor counter type objects of a device. It is driven via a configuration file. This file is currently maintained by editing an ASCII file. The configuration file contains the following information:

- How often a device should be polled
- Which devices to poll
- What variables to poll for a device
- Which trap to send when a threshold is crossed
- The variables to be sent in the trap when a threshold is crossed

It also can be used as a data collector. If configured correctly, it will poll the devices described in the configuration file, and/or output the information to a flat file.

A new enhancement has been added to allow the sending of information by using various RMON objects. These objects are:

- alarmValue (either the actual counter value or delta — determined by alarmSampleType)
- alarmVariable (the object name that caused the alarm)
- alarmSampleType [either absolute Value(1) or deltaValue(2)]
- alarmRisingThreshold (the MAX threshold value from the configuration file)
- alarmFallingThreshold (the MAX threshold value from the configuration file)

The other intelligent agent will provide a user-configurable tool to check for the connectivity of a device. This tool will be driven by an ASCII configuration file that will contain the following:

- Polling interval for retrieving sysUpTime and interface information
- number of retries for SNMP requests
- time out for SNMP requests and ICMP requests
- list of devices to poll
- trap numbers for the various traps generated by each

It will be executing on a server in the RDC. Each Remote Diagnostics Center (RDC) will have its own set of configuration files set up specifically for the devices it will be monitoring for the region.

25.3 ACSI'S INTEGRATED SONET/ATM NETWORK OPENS NEW MARKETS

Possibly one of the most ambitious and aggressive undertakings in the communications industry in 1996 was accomplished by American Communications Services, Inc. (ACSI).

In just under 60 days, this fast-growing alternative access provider deployed a national, integrated data, transport, and switched services ATM network providing 22 cities with ATM, Frame Relay, and Internet services. The company has 36 city networks SONET-ready and 40 city networks connected to the ATM backbone. Plans are to offer services in 50 markets by mid-1998. ASCI's SONET/ATM digital network provides dedicated services, local-exchange services, voice-messaging products, and data-transport and networking solutions to businesses and government organizations. This network-construction feat also makes ACSI one of the first companies in the U.S. to offer competitive switched dial tone over its own network instead of reselling RBOC services.

Using state-of-the-art digital voice switching, ACSI's local fiber optic rings connect seamlessly to its coast-to-coast ATM broadband data network. The rings, based on self-healing SONET technology, assure high quality and reliability.

NetExpert was installed to manage five transport networks. The job called for a network management system that could interface with multiple communication protocols and elements and provide traffic management. At that time, OSI's framework approach was chosen for its flexibility and potential to keep up with network growth. Today, ACSI's business plan supports 50 cities with SONET communications, 80 cities with data communications, and 28 cities with switching services — a 15-fold increase in the number of deployed network elements. NetExpert has proven to be very scalable, not only vertically, but also horizontally. It handled the sudden shift to data networks with new equipment. The company was using NetExpert to manage the network's SONET equipment fault tolerance features and decided to add NetExpert applications for managing performance, provisioning, facility security, and accounting. Foresight in choosing a network management framework resulted in ACSI getting the best option in a consolidated solution that could integrate both fault management and provisioning across the board for switching, transport, and data networks.

The NetExpert iSAC module, combined with the implementation of a shadow database, is one of the most succesful components in OSI's product suite for provisioning at ACSI. OSI's integrated Service Activation Controller provides intelligent service activation and configuration management across hybrid network domains. iSAC maintains the relationship between the "customer service view" and "network view" for reference, activation coordination, and rollback. iSAC uses a shadow database, which is a mirror representation of switch data used for provisioning, audits, modeling, and backups. For ACSI, integrated provisioning means truly dynamic service delivery. iSAC handles the provisioning distribution center. ACSI is testing provisioning rulesets for the POTS, DID/DOD, and DACS IV/II, as well as data fault management and switch billing systems. The fault management system for switch, SONET, and transport equipment is in full production. Integrated fault management brings a lower mean time to repair because alarms are correlated between data/switch devices and the SONET transport equipment. To ACSI, service-based management, as opposed to element-based management, also means a lean network management workforce and lower operating costs. Besides handling the burgeoning number of network elements, NetExpert also accommodates the addition of ATM, Frame Relay, and switch equipment. For ACSI, NetExpert integrates the management of two other element managers and approximately 20 different types of elements from five vendors.

NetExpert also addresses ACSI's need to correlate data. The product allows ACSI to consolidate all of their data, switch, and transport management on a central workstation. Local technicians also benefit from the ability to manage elements within their regions rather than having to dispatch to the element location in a different city or state. An ACSI national network administrator can diagnose and isolate faulty network elements, rearrange bandwidth to accommodate traffic patterns, and monitor security access points from one workstation at ACSI's state-of-the-art Network Management Center at ACSI headquarters in Annapolis Junction, Maryland.

Network integration is essential for ACSI to leverage its data, SONET, and switched services. Service levels for existing products can be kept high and new products and services can be more easily introduced. Integration makes the network efficient, increases reliability, and makes it possible for ACSI to maintain the highest operating standards and lower operating costs. By integrating products, services, and marketing efforts within the Communications Services Division, ACSI expects to gain significant efficiencies and market leverage in cross-selling into small, medium, and large business and government accounts. ACSI plans to be a one-stop shop for services that are high-quality, reasonably priced, and reliable. ACSI can deliver this guarantee by controlling costs through integrated management and automated provisioning.

An integrated network infrastructure has allowed not only ACSI to effectively compete, but its business customers as well. ACSI believes its new network brings business customers an essential competitive edge through better service. ACSI will challenge its competitors with faster network restoral times and shorter provisioning intervals. For colleagues and customers, the systems present an opportunity to dynamically link local transport, switch, and national ATM facilities and services with customer's enterprise networks and the long distance carriers' network for a total, end-to-end view of all facilities. The next step in ACSI's vision is to provide customers with the means to economically control their own portion of the ACSI network through customer network management.

25.4 SUMMARY

ISPs with the best service/price ratio will survive. ISPs are expected to serve a wide variety of consumers, including enterprise users, small businesses, telecommuters, and also mobile users. The service to these different users requires a wide range of access technologies, such as dedicated lines, dial-up-circuits, asynchronous subscriber lines, cable, and wireless connections. The operations support system is expected to provide means of rapidly defining, provisioning, and deploying differentiated new services for which premium prices may be charged. The majority of these services will often be based on emerging IP capabilities such as multicasting, RSVP, priority, and precedence and security features which will be provided by device manufacturers and management framework providers.

Part VII

Cost/Benefit Ratio of Management Frameworks

The development and operation of an Operations Support System (OSS) represents a significant investment. Deploying OSSs always requires a business decision. Decision makers usually work with hard facts, expenditures on one side and savings on the other. Financial analysis includes three principal techniques:

- Return on investment (ROI)
- Cash flow analysis
- Quantifying the payback period of OSS investments.

Before deciding on a particular product, the decision whether to build or buy an OSS should be made.

This part of the book includes an in-depth analysis of the cost of internal development of an OSS. The analysis displays and contrasts the two basic development models, called the Waterfall Model and the Spiral Model. Many telecommunications providers make the mistake of custom design of operation support systems. *Chapter 26* proves the superiority of the spiral model based on a management framework with rapid prototyping capabilities.

In all cases, time-to-market is very important, and should be weighted high. Other evaluation criteria may include:

- scalability
- rapid change management
- separation of technical inventory from operations
- use of object technology

- support of all relevant management processes
- rapid deployment of new services
- better first diagnosis of problems
- less referrals for allocating troubles
- less undetected problems in networks and in their components

An operations support system can be the best investment a service provider ever makes. The right flexible OSS can help a service provider generate greater revenue while prviding high-quality service to its customers. *Chapter 27* summarizes the benefits of an innovative OSS technology.

26 Internal Development: A Look At Costs*

CONTENTS

FIGURES

Introduction A primary goal of any service provider is to ensure that its networks are managed by the most capable Operations Support Systems (OSSs). The multitude of possible services and the variety of business models available creates a need to deploy OSSs that are customized for the provider's environment. Such deployments usually demand in-house or outsourced custom development, forcing the service provider to enter the world of software development.

Several issues can be anticipated with custom OSS software development, whether using high-level, fourth-generation frameworks or lower-level APIs and platforms. Issues discussed include the impact internal development ventures have on the focus and core expertise of the provider, clarification of the expected and hidden costs of taking on such development projects, and recommendations about how to avoid the "legacy development trap." In addition, a comparison is made between two major software development methodologies: the Waterfall Model and the Spiral Model.

Intellectual Investment In the process of operating any business, a company must invest its time and resources in two major areas. First is the "core" area, which is

---

* This chapter is based on the work of Mo Nikain.

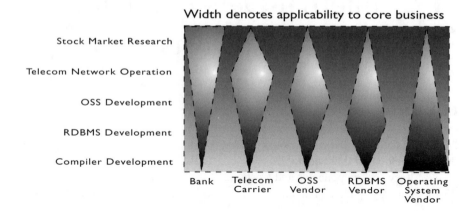

FIGURE 26.1   Intellectual investment comparison.

tied directly to the products and services the company offers to its customers. The second is the "support" area which, directly or indirectly, supports the activities of the core area.

Figure 26.1 shows a sample set of functional areas (on the vertical bar) and a few types of businesses (on the horizontal bar). The graph shows the applicability of each functional area to a core business. Where the graph is wide, the applicability is strong, and where it is narrow, the area supports core business. The idea illustrated is simple and straightforward — possibly deceptively so. A company's ability to apply this concept can be a key factor in its success.

Intellectual investment can be defined as the way a company focuses its expertise and where it chooses to expand such expertise. It governs, among other things, the learning capacity of the organization, including decisions regarding training, technology transfer, research, and on-the-job experiences. As with any investment, it involves the process of spending limited resources to attain maximum returns. In the context of intellectual investment, the "limited resource" is employee expertise or the capacity to build it. This expertise is "spent" on several different areas by focusing employees on these areas.

The question, of course, is where to focus or "invest" this expertise. For example, consider two companies that develop banking applications. Company A purchases off-the-shelf, fourth-generation components to build applications. As such, its intellectual investment is focused on banking applications and how they can be improved. In contrast, Company B decides to develop its applications from scratch. In this case, the intellectual investment is spread over software development techniques and debugging in addition to banking application features. Assuming equal technical staffs, it is clear that A's expertise in banking application features will grow significantly faster than B's. However, B will have added expertise in building development tools and will have better software developers.

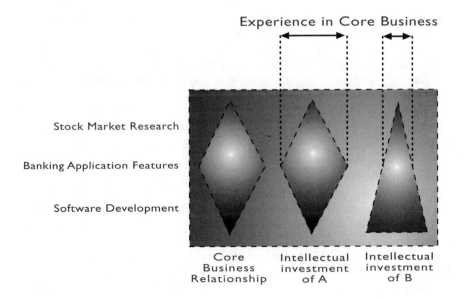

FIGURE 26.2   Maximizing expertise.

In general, the goal is to maximize expertise in core areas. Consider the extension of the table presented previously as it applies to Companies A and B. (See Figure 26.2.)

Note that the area of the intellectual investment graphs should be the same for A and B, assuming that they have similar resources. However, in the case of B, investment does not follow the core business relationship, hence B is building expertise in support areas and is left with less expertise in the core area. Maximizing expertise in core areas will not only give a company a competitive advantage, but it will significantly increase job security for the employees by ensuring that they will directly contribute to the business, since the likelihood of shedding or outsourcing core business components is very narrow.

26.1   LEGACY OSSs: HOW TO AVOID CREATING THEM

Legacy OSSs exert a major impact on OSS development in deploying projects. In fact, a major issue in the evolution of large telecommunications providers is the upgrading or replacing of legacy OSSs and the need to avoid producing another legacy OSS. Legacy OSSs are the main stumbling blocks in offering new and innovative services because they are expensive to maintain and to modify. Many firms are planning a transition to more modern and flexible systems. However, most legacy OSSs, deeply entrenched in a company's operations, perform quite complex functions, making their replacement a major, and sometimes costly, process. Though it presents a seemingly overwhelming undertaking, most companies that want to guarantee a competitive future and escape slow and costly legacy modifications "bite the bullet" and pursue replacing legacy systems.

The term *legacy OSS* refers to a class of large OSS in use by telecommunications carriers and service providers. While what constitutes a "legacy OSS" varies from company to company, it usually refers to a larger system that is custom made, inflexible and expensive to modify.

The danger companies face in incurring the cost of developing new OSSs and grinding through the replacement process is that, when the dust settles, they may find that they have created yet another legacy OSS. It is easy to underestimate the likelihood of this happening. However, consider that the developers of current legacy OSSs did not intend to make legacy OSSs either. In fact, most OSSs that now carry the "legacy" label were state-of-the-art at the time of implementation. They were designed and developed by the most advanced and knowledgeable people available. The conditions that made earlier products "legacy OSSs" may be the same conditions current developers face.

So what causes an OSS to become legacy and how can this be avoided? While there is no definitive answer, there are certainly danger signals in the design and development of systems that can warn that a legacy OSS may result. Some of these signs are:

- Long development cycles with large development teams
- Use of low-level programming environments
- Highly customized and optimized systems
- No exit strategy

The first two items can be somewhat anticipated. If cycles are long for development, then cycles will be long for modification. Sometimes the development plan is extended to build flexibility into the product. However, in reality, building true flexibility is a massive task, and the usual in-house development can barely crack the surface. Fourth-generation tool vendors spend thousands of man-years on their products to achieve a certain level of flexibility that can be provided more economically by spreading the cost over many customers. Expecting the same level of flexibility, or even a reasonable subset, in a "one-shot" or customized development process is probably too optimistic.

So while some systems end up being more flexible than others, the hopes of being able to reuse code and make rapid modifications may not be reality after custom development efforts are complete. In fact, most of the flexibility available in the resulting systems is usually a feature of the platform or framework used, not the application. It can be said, with reasonable confidence, that the flexibility of a custom application is no greater than the flexibility of its underlying tools and framework.

To avoid creating a "highly customized and optimized system" appears almost contradictory to any development process. But if it were, there would be fewer legacy OSSs. The fact is that increased customization and optimization result in a larger deviation from available off-the-shelf tools. A company that forces itself to use more off-the-shelf equipment at the cost of customization and optimization can take advantage of an array of software industry activity that has caused applications to evolve. On the other hand, a highly optimized and customized system relies on an in-house team (usually very small compared with the rest of the industry) for its evolution. Clearly, the latter will eventually fall behind regardless of the team's competence. Service providers can strive for the best of both worlds by using off-the-shelf tools that are also highly customizable.

The last issue, no exit strategy, may be the most misunderstood. Usually, when a development project is being planned, the long-term replacement of the product is not considered. While flexibility and scalability can be built into most products, no system will last from now until the "end of time." (Although it might last from now until the end of the business!) At some point, every system must be replaced. A development plan that takes this into account will consider the finite lifetime of the product and produce a system that yields itself to evolution as more frameworks and tools become available.

Successful framework vendors provide an upgrade path and, to stay competitive, rollover product lines every few years. The massive costs of these roll-overs are offset by a large customer base and potential new customers. For a custom application development organization, the cost of such a roll-over is the sum of the cost of the framework roll-over and the application (customization) roll-over. It is important to consider that the roll-over of the framework is relatively inexpensive because it has been distributed over many customers and may even be covered by maintenance contracts. But the roll-over of the application is expensive because it involves customization. Therefore, the cost of the roll-over is essentially determined by the amount of customization and framework upgrade paths. For systems that use low-level tools and a higher degree of development, the cost of roll-over can be prohibitive. But without roll-overs, the system will eventually degenerate into a legacy OSS.

The term *rollover* refers to a major upgrade of the core functions, sometimes involving a nearly complete replacement of code.

With consideration of the warning signs presented above, the benefits of using a high-level framework against using a low-level tool or platform will be compared. But first, development methodology, an important related issue, will be discussed.

26.2   DEVELOPMENT METHODOLOGIES

There are two general methods for software/system development: the Waterfall Model and the Spiral Model. The Waterfall Model is a traditional method in which complete requirements are developed in detail and passed, as a waterfall, to a development group that produces the final system. The Spiral Model, on the other hand, is a more innovative method that breaks development into cycles, each cycle consisting of a more refined set of requirements and resulting in a more complete version of the code. While both methods seem straightforward, the requirements for each and their advantages and disadvantages are not always apparent. The following sections describe each method in more detail.

The Waterfall Model   Organizations using this traditional method usually have two groups: requirements and development. The requirements group is focused on the service aspects of the product and the market but does not have much information about development tools and methodologies. The development group, on the other hand, is usually not as informed about the product or markets, but has expertise in software development issues, tools, and environments.

Differences in the knowledge, capabilities, and culture of the two groups generally require that, to avoid ambiguity, the details be communicated through a formal, detailed requirements document. Further, because of the typical lack of understanding of available development tools and environments, the requirements group may not take significant advantage of off-the-shelf utilities. The result can be a highly customized solution with a long development cycle. The effort needed to develop such detailed documents, along with the costs of a long development cycle, reduces the potential to exchange ideas and results in a basically one-way flow of information from the requirements group to the development group. Hence, the term "waterfall."

This method, while suitable for producing a highly customized solution in a static environment, is not favorable in a fluid, competitive environment. The rapidly evolving telecommunications environment can render requirements obsolete before the lengthy development cycle has been completed. This is prevalent, as witnessed by the increasing number of projects that are abandoned after thousands of dollars of resources have been spent on requirements and their partial development. Further, restricted communication between the development and requirements groups can result in impractical solutions.

The Spiral Model   The Spiral Model involves a number of cycles, each including requirements, development, testing, and feedback. In this method, the organization can still consist of two groups (requirements and development), but the groups will have an increased need for contact and thus will work much more closely together. In many organizations, the groups actually merge and provide a more "project-oriented" approach to the development process. This project-oriented approach designates a group to be responsible for all aspects of a system's deployment (including requirements, development, and testing). Another group will be responsible for another system, and so on.

In spiral development, the project is completed in phases or cycles. At the end of each cycle, a more refined prototype is produced and tested, and the results are fed back to the requirements team. A shorter development cycle, which greatly increases the ability to receive feedback and correct problems, significantly reduces the amount of work and detail necessary for the requirements team. With a spiral methodology, the requirements writer can afford to give more freedom and flexibility to the development team to leverage their ingenuity and knowledge of the tools. This is possible because, if the resulting prototype is unacceptable, the requirements writer has ample opportunity to make corrections in subsequent phases. In short, rapid development cycles reduce risk, yet promote creativity and the exploration of many possibilities. Another major advantage of spiral development is that significant market changes that occur during development can be met by design modifications in the next or subsequent cycles.

The increased communication inherent in the Spiral Model brings the added benefit of bridging the cultural differences between requirements and development

Development Aspect	Spiral Model	Waterfall Model
Time to Market	Fast	Slow
Organizational Structure	Closely Coupled Groups of Project teams	Separate Requirements & Development teams
Incorporation of Feedback	High	Low
Flexibility & Market Alignment	High/Long-term	Low/Short-term
Tendency to Develop Legacy OSSs	Low	High
Required Development Tools	Fourth Generation/ Rapid Development	Lower Level tools (C/C++APIs)
Cost of Development Environment	High	Low
Cost of Development	Low	High

FIGURE 26.3 Comparison of development methodologies.

groups, and triggers the well-documented advantages of a high-performance team. Another important edge is reduced time-to-market. A shorter development cycle provides a subset of functionality that can be capitalized on early in the process.

Achieving success with the spiral technique, however, generally requires higher-level, fourth-generation tools and rapid prototyping environments. Lower-level tools demand a much longer minimum amount of time for each development cycle. This makes it impossible to create an early prototype that behaves like a complete system, which, even though functionality is reduced, is needed for meaningful feedback. Using the wrong tools reduces the effectiveness of the spiral methodology.

Figure 26.3 provides a quick comparison between the two models.

26.3 LIFECYCLE COMPONENTS AND COSTS

Deployment and maintenance of an OSS, and for that matter any major system, has a number of cost components. However, in planning and starting such projects, most companies focus on a smaller subset of more "celebrated" components and trivialize or ignore the rest. This is not because of a lack of knowledge of these other components. Its main cause is that one group lacks knowledge about some components. The phenomena is similar to what happens with the proverbial "blind men and the elephant," where several blind men feel various parts of an elephant and deduce seemingly conflicting views of what an elephant looks like. For example, higher management might consider the initial cost of development but ignore support issues. Development might consider development costs but not worry too much about time-to-market and lost opportunity costs. Operations center staff might only consider things such as the cost of retraining. These problems are not unique to organizations that are new to "competitive" in-house systems development. They are experienced by even the most seasoned software development organizations.

To understand these issues better, the costs associated with the lifecycle of a software system will be discussed. While the focus is OSS deployment, what is presented applies to most large software systems. Several aspects are described, some more obvious than others. Also noted is the relationship to core business and intellectual investment based on the assumption that the software is not the main

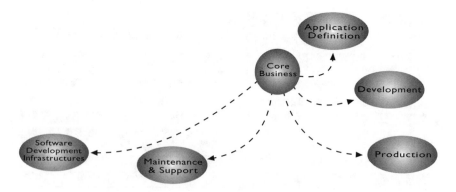

FIGURE 26.4   Magnitude of divergence from core business.

product of the business but is being developed to support key business functions, such as an OSS development project for a telecommunications services provider. The relationship of core business, where intellectual investment should be focused, to areas required in software development is represented in Figure 26.4 by how far each area is from the core business sphere. Hence, the graph shows to what degree intellectual investment is actually applied to core, rather than support, areas.

Application Definition   The main component of application definition is the specification of requirements for the development process. Requirements define what features are needed in the product. Requirements work demands excellent understanding of the application and may involve passive or active research on operational modes and necessary features. Creating requirements involves two major components: high-level functional description and detailed requirements. The major difference, other than the level of detail, is the "language" of the requirements: functional description generally covers the application area; detailed requirements involve technical issues such as software architecture, interfaces, and protocols.

The cost of detailed requirements can easily overshadow the cost of other requirements activities. Add this to the fact that detailed requirements are farther from core business than all the other requirements activities, therefore making the least contribution to core expertise. Fortunately, this area can be minimized the most by optimizing the development approach. The extent of detailed requirements depends heavily on software development methodology. At one extreme is spiral development using fourth-generation tools, which calls for the least effort to be spent on detailed requirements, and at the other, the Waterfall Model using low-level tools, which demands the greatest focus on creating detailed requirements.

The main reason for this difference in effort is that fourth-generation tools have already addressed many of the detailed technical specifications called for in the requirements. Also, since only widely used frameworks should be considered, it is safe to assume that these tools have been seasoned by many implementations and have incorporated a wider and more mature set of specifications than can be gained without such experience. When fourth-generation frameworks are coupled with

Waterfall Model & Traditional Tools   Spiral Model & 4GL Tools

Lighter-shaded regions represent activities that increase expertise in core areas.

FIGURE 26.5   Magnitude of necessary requirements development.

spiral development methodology, there is less need to rely on the judgment of the development staff, a risk that is the basis for creating detailed requirements in the first place. With the Waterfall Method, however, there is a high degree of dependency from the outset on the developers being correct down to the smallest detail. Accurate details are vital when using the Waterfall Model because, with a longer development cycle, any misinterpretation of the requirements will most likely not be detected until late in the process. Even worse, the impact of a misinterpretation may be too entrenched in the code to easily alter. Remedying such issues presents a formidable task. Figure 26.5 summarizes this issue.

Development costs include the cost of writing and debugging code and the cost associated with acquiring the specific tools and framework necessary to complete the task. This area also includes costs associated with the study of the requirements and tools. Development costs are a major component of the overall system cost, and are typically given the most consideration. In fact, in many cases, this is the only cost considered when deciding whether to develop a system or subsystem in-house or to purchase it from a vendor.

It is important to realize, however, that the cost of development is heavily dependent on the type of tools used. With fourth-generation tools, the cost is shifted from head-count/duration to framework. Figure 26.6 illustrates this.

The difference between the type of tool and its associated development cost is clear. As shown above, fourth-generation tools generally cost more initially but reduce development time, require less testing and debugging, and involve shorter integration times. Often, however, the impact the type of tool used has on later development phases and real, overall cost is not considered.

The test and debug phase is reduced as a result of two things: (1) the code being developed is smaller, so the potential number of bugs drops accordingly; (2) higher-level environments are less prone to bugs because the programs and functionality are more closely related and better understood by the team.

The integration and final modification phases are shorter with fourth-generation tools for the same reason that development time is shorter, but with an added advantage: developers clearly understand the modifications since closer communication with the requirements team has increased their application knowledge, and they can, in many cases, implement changes on the spot.

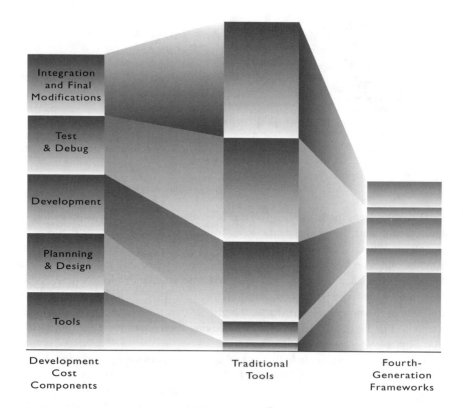

FIGURE 26.6   Tool types and associated development costs.

Productization   With most in-house development efforts, productization costs are heavily underestimated. These costs cover necessary manuals, additional testing and quality assurance, installation/extraction utilities, upgrade utilities, extensive online help, and other enhancements that allow the software to be operated without the developers at hand. This cost is not fully considered because, with in-house development, the development group is usually at hand. However, money saved by less-formal productization is usually spent on additional support. Productization may seem a small component of overall costs, but in the case of in-house development it can become quite large. In fact, productization costs usually exceed development costs.

On average, the ratio between the cost of *productization* and cost of developing a fully functional "lab" system is about 70 to 30 (the cost of productization is more that twice the cost of development). This document discusses a development step to produce a product somewhat more mature that a lab system, but not as productized as an off-the-shelf system.

Significant reductions in productization costs, as well as improved productization results, can be achieved with fourth-generation tools. The fourth-generation framework already has been productized and includes many of the enhancements listed earlier. The customized portions can easily reference and leverage framework and vendor resources. Further, superior fourth-generation tools also provide support for application productization.

Support and maintenance can be the largest and most troublesome components of product cost. The main culprit is that there is no clear end to support activity, and modifying custom code can become quite expensive.

Despite object-oriented methodology and emphasis on ease of modification and code reuse, cost and time-frame constraints force most in-house development to focus on current features and make extensibility of code secondary. Consider that even larger companies developing fourth-generation tools have problems building extensibility into their products and that is their major goal. Smaller in-house development organizations can expect major problems with extensibility as a secondary goal.

The need to support a product exists long after the excitement of developing a new product fades. Further, some support activities, especially the need for new features, are not easily predicted. In many cases, the psychological effects of shifting from a high-profile, organized development effort to routine, bland support activity significantly impacts the productivity of the development staff. Support activities include patches and code fixes, help desk, enhancements and new features, and porting and scaling.

Given that providing support is problematic, one may wonder how software companies survive. Actually, many do not. But for those that do, the key is the number of customers who will underwrite their product. Both the high cost of support and the psychological effects of having to provide it can be remedied with a large customer base. First, costs are distributed over the customer base. A permanent support organization can be formed and staffed with support, rather than development, personnel. In-house development efforts for in-house applications generally have only one customer. Hence, support costs cannot be distributed, nor can a special support organization be formed, again because of cost.

Support costs and related issues are the primary reason that more and more companies are abandoning systems (or systems development) and opting for outsourcing or seeking high-level frameworks. Support is not only an area of significant cost, but also it is one where cost becomes progressively more difficult to justify. Usually, development cost is expected to decrease after development is complete. Surprisingly, in reality, decreases are smaller than expected, particularly in the quickly evolving telecommunications industry. Figure 26.7 illustrates this.

If you add the cost of modifications, the end of the graph might actually continue to rise rather than fall or flatten out. Why? Because retrofitting features and modifying custom code is generally expensive. This is confirmed, in a reasonable number of instances, by development groups choosing to develop new systems rather than retrofit the old.

Infrastructure costs are additional costs associated with making development a long-term activity. These include costs associated with maintaining a state-of-the-art development environment (upgrading hardware and software tools) and keeping an informed and capable development team to do training and research on development methodologies, tools, and environments. Most of these efforts build expertise in non-core areas and are not the company's best intellectual investment.

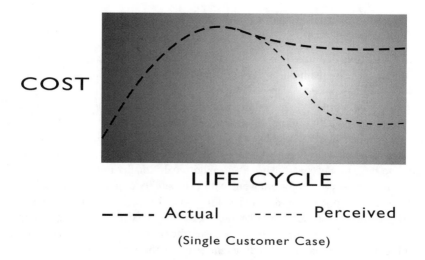

LIFE CYCLE

- - - - Actual   - - - - - Perceived

(Single Customer Case)

FIGURE 26.7   Actual and perceived cost/cycle comparison.

If the goal is traditional software development, including maintenance aspects, then the cost will be a quickly outdated, feature-poor product. This scenario ultimately increases development costs and damages or eliminates competitive advantage, particularly considering that savvy competitors may be advancing the features and functions of their system using spiral methods and fourth-generation frameworks.

Figure 26.8 presents a summary of the cumulative costs of development.

26.4   SUMMARY

Customized OSS development is considered a necessary evil by some and a competitive advantage by others. Either way, such efforts represent a major investment for most telecommunications service providers. It pays to compare two approaches for the development of custom OSSs: the traditional Waterfall Method, which uses low-level tools (C/C++ APIs, libraries, etc.), and the preferred Spiral Method that takes advantage of high-level, fourth-generation tools that provide programmerless expert engines for rapid development.

In the case of traditional tools, the attractive smaller startup cost is rapidly overtaken by an avalanche of hidden costs, lost opportunities, and divergent focus. The main factors are support and modification costs that linger long after the initial glamour and excitement of a "new project" are gone. These costs alone can force an organization to abandon low-level development for higher-level tools and applications. In addition, development efforts using fourth-generation frameworks focus the expertise of an organization on its core business, increase competitive advantage, and help maintain staff productivity.

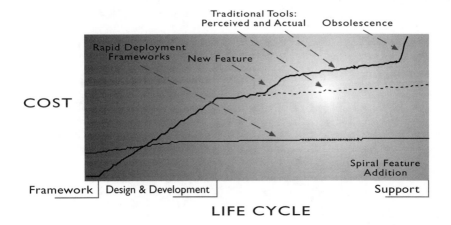

FIGURE 26.8 Cumulative costs of development.

27 Summary and Benefits

An operations support system (OSS) can be the best investment a service provider ever makes. Besides reaping optimal performance benefits from network equipment, the right OSS can help a service provider generate greater revenue while providing the lowest cost, highest quality service to its customers.

Some attributes of an excellent OSS are database independence, inherent flexibility, scalability, and multivendor equipment support. An OSS should make it possible for business processes to evolve rather than revolve by giving the service provider control over them, plus the ability to reflect them in the processes of the OSS. When integrated at the service and business operations levels, an automated operations support system delivers high quality service, reduces operational costs, and streamlines information processing, enabling businesses to successfully compete in today's crowded telecommunications marketplace.

Yet many chief operating officers view the OSS as just another expenditure or necessary evil, when in reality an intelligent OSS can be a cost saver that drives business decisions to fruition. Today's intelligent operations support system greatly reduces the labor necessary to support processes in engineering, maintenance, records keeping, and provisioning. The provisioning labor cost reductions alone result in savings that can be used to purchase the system. Businesses can quickly realize their return on the OSS investment through equipment and labor cost savings.

The same cannot be said for older, unintelligent operations support systems. The intelligence in the older systems was based on switches so that copper, telephones, and so on, had to come to them for their functionality. Prior to TR-303 and SONET, even the most intelligent system was only able to report whether or not the elements were working. They could not deliver complete status reports and an overlay alarm network served as the only reporting system.

Today's OSS is an essential part of the entire network rather than just an external transport system for carrying limited data back to switch locations. Consequently, service providers can access detailed information from intelligent network elements. Most intelligent network elements have nonvolatile memory that retains the cross-connect information regarding not only itself but also that of the adjacent network elements.

The OSS can access intelligent network elements to retrieve all essential data they have been designed to store and report, rather than having to take the data back to a switch location. In this way, the new OSS is able to make the most use of equipment functionality, and pass on this ability to network administrators. Armed with network element information and an integrated OSS, network administrators can remotely provision, remotely diagnose for service conditions, proactively self-report element conditions, and automatically retain element status records.

Increased intelligence built into SONET and TR-303 based network elements (that is, digital switches) has significantly changed the provisioning processes. It is in the provisioning process where network administrators are realizing substantial cost savings and benefits affecting both network management and business operations.

27.1   REMOTE PROVISIONING AND PROACTIVE MANAGEMENT

The distribution of intelligent network elements throughout the network lets network administrators monitor the equipment and effectively use all network assets. This distributed intelligence is remotely accessible by administrators. Administrators can determine where, when, and how equipment has been used in the business. Intelligent switches, for instance, can handle their own rerouting and congestion management by reporting their actions to the gateway.

Gateways located close to the devices convert the devices' messages into a common format. The gateways can filter out events, or route data, and consolidate information, such as status reports and alerts, before delivering it to the OSS. Multiple fault indications are delivered in a single alarm, enabling network administrators to react quickly to network events. Immediate intervention of a network crisis minimizes the cost of lost data. And because an intelligent OSS facilitates communication with devices, administrators can "direct" them to automatically perform specific actions, such as provisioning and testing.

In this way, the OSS allows the network administrator to remotely provision circuits in shelves via a network operations center (NOC) rather than sending field technicians out to a remote site and turning up circuits using a laptop computer. Element management layer (EML) systems, provided by the network equipment vendors, are avilable, but they typically can be applied to only that vendor's equipment. This makes it difficult, if not impossible, to have the flow-through provisioning just described.

Proactive maintenance also is possible by setting the quality of service thresholds via the OSS and monitoring the service performance. Consolidating labor processes and remotely handling them through an OSS quickly can result in process savings and reduce the number of required NOCs. The savings realized from field technician travel alone warrants investing in an intelligent OSS to handle remote provisioning.

Similar savings are possible from the proactive maintenance capabilities afforded by the right OSS.

Example: Network Operations Labor Cost Savings   Eight to nine technicians staff a network operations center open year round, 24-hours a day, seven days a week. If the typical technician costs a company $54,080 a year ($20 an hour plus another 30% for benefits), a company can save at least $432,640 for each NOC it does not have to staff.

27.2   AFFORDABLE BANDWIDTH-ON-DEMAND PERFORMANCE

Some network technologies such as SONET support bandwidth management that, if used properly, can result in savings of up to 40 percent in hardware costs. Extra hardware (that is, DS1s) to increase bandwidth is unnecessary if bandwidth is rearranged between existing customer locations. But the more obvious benefit is consistent high-performance network service. By rearranging bandwidth from slow areas to high-traffic areas, a service provider can meet traffic demands head on by making the most out of the bandwidth. This is possible with a good bandwidth utilization plan and an OSS that supports bandwidth allocation.

One reason why many administrators do not take advantage of bandwidth management features is the perception that bandwidth record keeping is complex. It is true that traditional record keeping processes make it difficult to track bandwidth use. However, network administrators can track bandwidth in real time by using the intelligence built into the network elements and the elements' electronic record keeping systems.

Example: Remote Provisioning Savings Benefits   The industry average labor rate costs (wages, trucks, and tools) of a field technician is estimated at $100 an hour. Let's say travel and testing time is conservatively two hours, then the labor cost per circuit is $200. On the other hand, the OSS provisioning process takes about 30 minutes, resulting in a cost savings per circuit of $150. Now add the cost of DS1 services priced at from $100 to $200 per month. At this rate, it would take at least an additional month's revenues to cover the cost of manual field installation. At 100 circuits per day for 260 business days, the savings from OSS provisioning is $3,900,000.

27.3   ELECTRONIC RECORD KEEPING IN REAL TIME

Record keeping is simple and cost effective when gathering and storing information through the OSS. OSS record keeping is fulfilled in real time using the intelligent network element's databases. In the same way the network elements electronically deliver alert information, they also store element configuration data. The OSS can look through all network elements, regardless of vendor, and provide the same level of information as a traditional record keeping system. Technicians are no longer needed to physically check every piece of equipment to discover what is used, where it is used, and how it is being used.

With inventory information readily available, businesses can then make well-informed decisions about network operations. Network administrators can efficiently deploy resources to reduce or eliminate under-used equipment. Purchasing decisions can be based on accurate, updated information about the use of existing equipment.

Electronic record keeping via the OSS eliminates the cost associated with manual record keeping. The annual cost of maintaining and updating a records database are reduced which means reduced payments to a third party for software updates.

27.4   CUSTOMER CARE

Intelligent network management and elements can assist a service provider in implementing customer service process improvements that result in savings benefits, and at the same time, address the customer's service requirements.

Here is an example of how this might work:

> A customer service representative enters a customer's requested dates, service types, and service attributes into a PC. Transparent to the representative, the order initiates an OSS-based inquiry to a facilities database that assesses whether or not the customer's expectations can be met. Then, if the customer agrees to the feedback from the inquiry, the appropriate network facilities are initialized and service is turned on in a matter of minutes with little human intervention.

> The customer request information is electronically forwarded to the appropriate departments so fewer employees are needed to handle the order, or manage the process. The only exception when a request is taken offline to the engineering department for problem resolution. A platform-specific element management system, in this case, might not be able to respond to customer requests as quickly as can be accomplished with flow-through OSS processing.

Other departments, such as billing and engineering, also benefit from electronic order processing. Electronic interfacing between service representatives, engineering, and billing ensures the accuracy and quality of information since the data is not altered by rekeying it into the system.

Problems are more easily resolved when the representatives can access call termination information commonly stored in customer call history. Quick and convenient access to the customer call records lets service representatives proactively resolve disputes by referring to the call terminal codes or service activity records. Costly billing errors can be reduced by accessing billing records at the switches.

The cost benefits for electronic customer order processing are enormous. In addition, paperwork and record keeping is reduced since the customer's request is now forwarded electronically.

Example: Customer Order Processing Savings   Assume that good bandwidth engineering and subsequent equipment implementation would enable 95 percent of all orders to be flowed through in a good OSS. If a company processes 100 orders a day with an OSS-based labor cost of $50 an order, then the flow-through costs would be approximately $4,750 for the 95 flow-through orders. If a vendor-specific

EML process which entails engineering involvement is used, installation commitments, records updating, and service representative rehandling, each order that could not flow through would cost $100 per order. Using the OSS process, the cost for the 100 orders would be approximately $5,250 when automatically processing 95 of every 100 orders. Without the OSS flow-through ability, 100 orders per day processed via the EML system would cost a company $10,000 per day in the service order process. The EML cost penalty is $4,750 a day. Multiply this number by 260 business days, and the annual cost savings would be $1,235,000.

27.5 INTEGRATING DATA SOURCES FOR BETTER MOBILE EFFICIENCY

By combining different data sources, it becomes possible to achieve a deeper understanding of network performance and subscriber behavior. This is a capability that has been difficult to achieve using stand-alone point element management solutions that have no intrinsic integration. OSS by definition allows otherwise disparate data sources to be compared and analyzed.

Opportunities to combine these data sources are limited only by the imagination of workers. Several examples are given below to illustrate some of the potential of OSS. The examples arise from the data sources as defined by the TMN model: fault, configuration and performance management.

Integrating Fault and Performance Data   Alarm data is used primarily to indicate network health and to trigger reactions to catastrophic network failures, such as a base station outage. Performance data records usage and service quality indicators. The creative use of the two together can yield significant operational improvements to the network.

When a BTS fails or otherwise becomes unavailable all voice channels deny service to all incoming (hand-in) calls. Origination as well as hand-overs is denied until the BTS is restored to service. The impact on the network is to increase the number of dropped calls in two ways.

First, at the time of the BTS failure, all calls on the failing BTS are instantly dropped. This increases the dropped call count by as much as several hundred, one for every call in progress on the BTS. Only by preventing the failure in the first place can these drops be prevented.

Second, all of the neighboring BTS are affected. In normal operation, calls are passed between neighboring BTS. When a BTS fails, calls from neighboring BTS, called a hand-over source BTS, which normally would be handed into the failed BTS (called a hand-over target BTS) before call quality deteriorated to an unacceptable level, are held longer on the source BTS. This has the impact of reducing call quality during the time that those calls are held longer than normal and these calls eventually drop when the signal becomes weak enough.

OSS can be used to reduce the second impact in several ways. By detecting the failure more rapidly, the period of outage can be dramatically reduced. OSS also can be set to dynamically reprogram the neighbor BTS to extend the time that they hold calls before dropping.

OSS also could be used to introduce a self-modifying feedback loop for use in this example. Consider this scenario in the AMPS network: OSS rules detect consistent and chronic hand-over failure messages from two neighboring BTS. A command-and-response dialogue is initiated by OSS to verify that the neighbor lists are symmetric and to determine the current power settings for the failing channels. If the lists are asymmetric, OSS could automatically input the commands, repair them, or notify an Engineer to make the change. If the hand-over lists are not symmetric, OSS can log the errors and can input the necessary commands to fix them.

If the hand-over lists are symmetric, OSS could input commands to modify the hand-over parameters in the 2 BTS, such as increment the power level of one of the channels and decrement the power of the other, while logging the before-and-after settings. The hand-over failure count for these channels would then be reset in OSS and monitored for subsequent hand-over failures. Improvements would be e-mailed to the Engineer responsible for the affected BTSs. Additional failures could stimulate a rollback of the change or a further modification.

Integrating Configuration and Performance Data   Configuration parameters are used by cellular switches and BTS to set the relationship between switches and BTS, and to determine the behavioral characteristics of the network.

Cellular operators and engineers continually optimize the operation of the network. One area where this is done is balancing the traffic and call-processing load across the various network elements. Periodically network operators will move or "re-home" BTS from one switch to another to even the distribution of traffic load among the switches and BSC. Load-balancing is often a manual effort by RF and Traffic Engineers poring over volumes of performance and traffic data, testing various sample configurations trying to find one that is workable. Many constraints are imposed by the RF environment, by switch configuration and by the load itself.

OSS can reduce the amount of manual effort required to determine a workable network configuration, and can produce configurations which are more accurate. OSS will automatically collate and summarize performance data from the OSS performance database to gain an accurate measure of present traffic load per BTS using configuration parameters also stored in the OSS database, to test one-by-one, different combinations of BTS and switch. There are hundreds and possibly thousands of BTS and switch combinations that are possible. OSS evaluates each combination in turn and logs the results. When all combinations are tested, the results are prioritized based upon how balanced the resulting traffic load would be. Engineers then must simply review the resulting list for compliance with the myriad constraints and select the best candidate.

The benefit to the carrier is that much of the process is automated, which will save many engineers weeks of analysis and guesswork. Because of this automation, more combinations can be tried, and each combination can be more accurately analyzed. The whole process becomes not only less time-consuming, but also much more accurate, too. The creation of command files to input rehoming commands to the BTS once the best network configuration is selected also can be automated.

By automating the process it can be done more frequently, thus reducing the deviation from a balance condition in the first place. Due to the amount of work involved in the manual re-home process, this is presently impractical.

Integrating Fault and Configuration Data   Configuration data also can be integrated with fault data for a further analysis of the relationship between the network configuration and network failures.

Many cellular switches and base station controllers provide alarm or warning messages when RF interference on voice channels is detected. OSS collects all alarms, warnings and status messages and stores them in the database. Interference messages can be organized to isolate the interfering frequency. OSS would then access the configuration database to identify other channels using the same frequency, and to orient these other channels relative to the channel reporting the interference. This data also can be displayed graphically to make the geographic relationship between potential interference clear. OSS can then identify the most likely source of the interference. This process can be automated completely, saving RF Engineers considerable time and reducing the duration that such interference impacts Quality of Service.

Cross-Domain Integration   The previous examples all focus on the advantages of integrating different data types within a single wireless technology. If a carrier has two cellular technologies, say, AMPS and CDMA, there also is the possibility to integrate data across differing technologies. This must be done cautiously to avoid comparing data having fundamentally dissimilar meaning. Nonetheless, it will be reasonable for the carrier to do this. At some point, it will become uneconomical for the carrier to continue operating the AMPS network. The decision to discontinue AMPS service will be based upon many factors, among them will be the dropped call rate and the effect it has on quality of state.

Dropped calls are an important measurement in both AMPS and CDMA technologies. The dropped call rate is often calculated from several measurements reported by the switch and radio subsystems such as Hand-Over Failures and Lost Signal. Engineers and analysts might compare the Dropped Call rate between AMPS and CDMA to help guide the decision and timing to discontinue AMPS service.

A similar exercise will be useful when the PCS network is built and operated. Financial planners will compare the performance of the different technologies to help guide additional investment and advertising decisions.

Another example is the Service Network. Presently voice bandwidth for carrying calls between cellular switches or between cellular switches and the CGS is managed separately. This is less efficient that pooling all voice bandwidth regardless of whether it is AMPS, CDMA or PCS. Combining all these separate components will yield economies of scale in usage and maintenance. OSS will be used to determine how to best allocate trunks within that bandwidth. This is a fundamental property of telecommunications which states that a single pool (say, a trunk group) of many servers (many trunks) offers more capacity than multiple pools having the same total number of servers. In short, breaking bandwidth into separate groups for AMPS,

CDMA and PCS is less efficient, and more total bandwidth, and thus higher costs, will be required in such an arrangement.

This is yet another way OSS delivers more means of a better and deeper understanding of the service.

27.6 SUMMARY

Service providers must be able to compete at cost, quality, and service levels in today's competitive telecommunications environment. A sophisticated buyer understands excess costs shouldered by a service provider will ultimately be passed on to them as price increases. An intelligent network management system lets service providers take advantage of the advanced intelligence residing in their networks and control internal costs at the same time. A flexible network management system can even accomodate the customer's service needs, or shift to support a service provider's business strategy. It stands to reason then that the company that uses the network functionality to either lower the cost of sale, or enhance the services delivered to customers, will likely gain the greatest market share.

28 Future Trends in the Industry

Telecommunications providers are in the process of repackaging and reenginering their existing operations and business support systems. The reason to do it is the Telecommunications Management Network (TMN) model. TMN has not yet been finalized, but the present status of the documentation is sufficient to indicate the following:

- Segregate or integrate management teams to clarify responsibilities
- Streamline management instruments and applications
- Facilitate decision making for frameworks and instruments
- Renegotiate service contracts with customers
- Define and deploy interfaces between internal and external instruments and management applications
- Set priorities for future implementation and roll out plans

The TMN message has arrived at management framework suppliers as well. They try to regroup their offers in accordance with TMN layers. The suppliers represent various groups, such as those depicted below:

- Framework and platform providers (for example, Bellcore, Clear, Cross-keys, GTE, OSI, Saville, TCSI, and Teradyne)
- Consultants, integrators, and outsourcers (for example, AMS, Anderson, Cap Gemini, and EDS)
- Hardware platform vendors (for example, Digital, Hewlett-Packard, IBM, and Sun)
- Network equipment manufacturers (for example, Alcatel, Cabletron, Lucent Technologies, Nokia, Nortel, and Tellabs)

There is not a single provider who can support all TMN layers with the required depth and performance. Alliances are very likely. They could be just projects-based or strategic. In these alliances, NetExpert could play the role of an umbrella manager in both network management and service management layers supported by other point products for element management and by applications for business management.

The liberalization and competition will drive telecommunications suppliers to publish their interfaces towards their customers and define APIs for their customers. Most likely, the web technology will become very helpful to find the common denominator between suppliers and customer/subscribers. The web interface will support both directions:

- Customers inquire about service level indicators, trouble tickets, trouble resolution progress, and billing information using a universal browser on their own premises
- Using the universal browser, customers can initiate various processes such as service provisioning, trouble shooting, fraud investigation, and bill processing.

It is expected that more processes will be automated as a result of reenginering of provisioning, service creation, service assurance, and billing for resource use. It is also expected that independent software vendors (ISV) will provide specific applications, such as customer care, network design, network optimization, trending, data mining, and statistics that may be integrated into existing management frameworks, such as NetExpert.

Telecommunications service providers most likely will operate multiple support systems in the future. Full integration is very difficult, but collaboration is possible. However, no one can design, implement, and maintain specific bilateral interfaces between support systems. The solution is the use of a mediation device. TMN gives the necessary answers by defining interfaces, but now the rolled out products come in different flavors. Mediation devices, either CORBA or DCOM-based, will play short — and medium term — a very important role. Again, ISVs are expected to contribute and provide solution alternatives.

Telecommunication providers will leverage their service offers at comparable and competitive prices. But the differentiator will remain the management frameworks and management applications. This will ensure leaner operations for the providers. The winners are the customers in gaining the benefits of getting good or premium service at reasonable prices.

References and Bibliography

ABER97   Aber, R., "xDSL Supercharges Copper," *Data Communications*, March 1997, pp. 99–105.

ADAM96   Adams, E. K. and Willets, K. J., *The Lean Communications Provider — Surviving the Shakeout through Service Management Excellence*, NMF and McGraw-Hill, New York, 1996.

AIDA98   Aidarous, S. and Plevyak, T., *Telecommunication Networks Management*, IEEE Series on Network Management, IEEE Press, New York, 1998.

ALEX95   Alexander, P. and Carpenter, K., "ATM Net Management — A Status Report," *Data Communications*, McGraw-Hill, September 1995, pp. 110–116.

BALL94   Ball, L. L., *Network Management with Smart Systems*, McGraw-Hill Series on Computer Communications, New York, 1994.

BATE95   Bates, B. and Gregory, D.. *Voice and Data Communications Handbook*, McGraw-Hill, New York, 1995.

BLAC94   Black, U., *Emerging Communications Technologies*, Prentice Hall Series in Advanced Communication Technologies, Englewood Cliffs, NJ, 1994.

DAMO94   Damodaram, R. & Co., "Network Management for the NCIH," *IEEE Network*, November–December 1994, pp. 48–54.

DESA95   Desai, V., "The ATM Management Roller Coaster," *Internetwork*, August 1995, pp. 41–44.

DORF93   Dorf. C. R., *Handbook — Electrical Engineering*, CRC Press, Boca Raton, FL, 1993.

DOWN96 Downey, T., "Tag switching promises scalability, high performance for internetworks," *Network World*, Oct. 7, 1996, p. 57.

FORU92   *Network Management Forum*, Statement of User Requirements for Management of Networked Information Systems, Morristown, NJ, 1992.

FOWL95   Fowler, H., "TMN-based Broadband ATM Network Management," *IEEE Communications Magazine*, March 1995, pp. 74–79.

GARE95   Gareis, R. and Heywood, P., "Tomorrow's Networks Today," *Data Communications*, September 1995, pp. 55–65, McGraw-Hill, New York, 1995.

GHET97   Ghetie, I. G., *Networks and Systems Managment — Platforms, Analysis and Evaluations*, Kluwer Academic, Norwell, MA, 1997.

GIBS96   Gibson, J. D., *The Mobile Communications Handbook*, CRC Press, Boca Raton, FL, 1996.

GROW94	Grovenstein, L. W. & Co., "NCIH Services, Architecture, and Implementation," *IEEE Network*, November–December 1994, pp. 18–22.

HOLL95	Holliman, G. and Cook, N., "Get ready for real Customer Network Management," *Data Communications*, McGraw-Hill, September 1995, pp. 67–72.

NMF95	*Network Management Forum: Discovering OMNIPoint 1 and OMNIPoint 2 — A Common Approach to the Integrated Management of Networked Information Systems*, Prentice Hall, Englewood Cliffs, NJ, 1995.

PATH94	Pathak, G. & Co., "Integrated Network and Service Management for the NCIH," *IEEE Network*, November–December 1994, pp. 58–63.

ROGE95	Rogers, C., "Telcos versus Cable TV: The Global View," *Data Communications*, September 1995, pp. 75–80.

STAL96	Stalling, W., *SNMP, SNMP2 and RMON — The practical guide to network management standards*, Addison-Wesley Publishing Co., Reading, MA, 1996.

TERP96	Terplan, K., *Benchmarking for Effective Network Management*, McGraw-Hill, New York, 1995.

TERP92	Terplan, K., *Communication Networks Management*, Second Edition, Prentice Hall, Englewood Cliffs, NJ, 1992.

TOWL95	Towle, T. T., "TMN as Applied to the GSM Network," *IEEE Communications Magazine*, March 1995, pp. 68–73.

YAMA95	Yamagishi, K. & Co.., "An Implementation of a TMN-Based SDH Management System in Japan," *IEEE Communications Magazine*, March 1995, pp. 80-8788.

TMN STANDARDS BIBLIOGRAPHY

The following list identifies a set of common TMN and TMN-related recommendations:

1. M.3000	*Overview of TMN Recommendations*
2. M.3010	*Principles for a Telecommunications Management Network*
3. M.3020	*TMN Interface Specification Methodology*
4. M.3100	*Generic Network Information Model*
5. M.3101	*Conformance Statement Proformas for Recommendation*
   M.3100	*Generic Network Information Model*
6. M.3180	*Catalogue of TMN Management Information*
7. M.3200	*TMN Management Services: Overview*
8. M.3300	*TMN Management Capabilities Presented at F Interface*
9. M.3400	*TMN Management Functions*
10. M.3640	*Management of the D-Channel Data Link and Network Layer*
11. M.3641	*Management Information Model for the Management of the Data Link and Network Layer of the ISDN D-Channel*

12. G.773   *Protocol Suite for Q-Interface for Management for Transmission Systems*
13. G.774   *SDH Management Information Model for the Network Element View*
14. G.803   *Architecture of Transport Network Based on the SDH Description*
15. Q.811   *Lower Layer Protocol Profiles for the Q3-Interface*
16. Q.812   *Upper Layer Protocol Profiles for the Q3-Interface*
17. Q.821   *Stage 2 and Stage 3 Description for the Q3-Interface-Alarm Surveillance*
18. Q.822   *Stage 1, Stage 2, and Stage 3 Descriptions for the Q3-Interface-Performance Management*
19. X.700   *Management Framework for Open Systems Interconnection for CCITT Application*
20. X.701   *Information Technology-Open Systems Interconnection- Systems Management Overview*
21. X.710   *Common Management Information Service Definition for CCITT Applications*
22. X.711   *Common Management Information Protocol Specification*
23. X.720   *Information Technology-Open Systems Interconnection-Structure of Management Information: Management Information Model*
24. X.720 Series   This series of recommendations define common aspects of *Open Systems Interconnection Management* for the understanding and specification of management information.
25. X.730/740 Series   This series of recommendations defines *Open Systems Interconnection Management Services*, which are used by TMN System Management Services.

Acronyms

4GL 4th (Fourth) Generation Language

A

AAL ATM Adaption Layer

ABR Available Bit Rate

ACI Administrative and Control Interface

ACSE Association Control Service Element

ACSI American Communications Services, Inc.

ADPCM Adaptive Differential Pulse Code Modulation

ADM Add/Drop Multiplexer

ADSL Asynchronous Digital Subscriber Line

AF Auxiliary carry Flag

AIN Advanced Intelligent Network

AIS Alarm Indication Signal

AMA 1. Accounting Management Application, 2. Automatic Message Accounting

AMO Affected Managed Object

AMATS Automatic Message Accounting Teleprocessing System

AMPS Advanced Mobile Phone Service

AMX Administrator and Maintenance interface

API Application Programming Interface

ARA Apple Remote Access

ARM Asynchronous Response Mode

ARP Address Resolution Protocol

ASCII American Standard Code for Information Interchange

ASN.1 Abstract Syntax Notation.1

ASR 1. Automated Speech Recognition, 2. Address Space Register, 3. Automated Send/Receive

ATM Asynchronous Transfer Mode

AuC Authentication Center

B

B-NT2 Network Termination 2 (used in fig. 2.20)

B-NT1 Network Termination 1 (used in fig. 2.20)

B-ISDN Broadband Integrated Services Digital Network

BECN Backward Explicit Congestion Notification

BER Bit Error Rate

BH Busy Hour

BLSR Bidirectional Line Switch Ring

BML Business Management Level

BOSS Broadband Operations Support System

BRI Basic Rate Interface

BRLC Broadband Remote Line Concentrator

BSC Base Station Controller

BSMS Broadband Service Management System

BSS Base Station System

BTS Base Transceiver Station

C

CAD Computer Aided Design

CAD/CAM Computer Aided Design/ Computer Aided Manufacturing

CAN Customer Access Network

CAP 1. Competitive Access Carrier, 2. Computer Aided Publishing, 3. Carrierless Amplitude Phase

CARS Command and Response System

CATV Community Antenna TeleVision

CBR Constant Bit Rate

CCITT Consultative Committee on International Telegraphy and Telephony

CDDI Cable Distributed Data Interface/ Copper Data Distribution Interface

CDMA   Code (or Call) Division Multiple Access

CDPD   Cellular Digital Packet Data

CDR   Call Detail Records

CDR/AMA   Call Detail Records/ Automatic Message Accounting

CEPT   European Conference of Posts and Telecommunications

CES   Circuit Emulation Switching

CFP   CNM Feature Packages

CI   Critical action Indicator

CIR   Committed Information Rate

CLEC   Competitive Local Access Carrier

CLP   Cell Loss Priority

CLR   Circuit Layout Records

CMIP   Common Management Information Protocol

CMIS   Common Management Information Services

CMISE   Common Management Information Service Element

CMOL   CMIP Over Logical link control

CMOT   CMIP Over TCP/IP

CO   Central Office

CORBA   Common Object Request Broker Architecture

CNM   Customer Network Management

CPE   Customer Premises Equipment

CPU   Central Processing Unit

CSLIP   Compressed Serial Line Interface Protocol

CSMA/CD   Carrier Sense Multiple Access/with Collision Detection

CSU   Channel Service Unit

CT2   Cordless Telephony Generation 2

CTI   Computer Telephone Integration

CU   Control Unit

DACS   Digital Access Cross-connect System

D

DCE   Data Communication Equipment

DCN   Data Communication Network

DCS   1. Digital Cross-Connect, 2. Distributed Communication System

DCSC   Data Customer Support Center

DE   Discrete Element

DLC   1. Data Link Connection, 2. Digital Loop Carrier

DLCI   Data Link Connection Identifier

DLS   Data Link Switching

DME   Distributed Management Environment

DMH   Data Message Handler

DMI   Desktop Management Interface

DMT   Discrete Multitone

DMTF   Desktop Management Task Force

DNS   Domain Name System

DPE   Distributed Processing Environment

DQDB   Dual Queue Dual Bus

DS1   T1 1.544 Mbps (digital signal)

DS2   T2 6.312 Mbps (digital signal)

DS3   T3 44.736 Mbps (digital signal)

DS4   T4 274.176 Mbps (digital signal)

DSn   Digital Speed Number Elements

DSL   Digital Subscriber Line

DSU   Data Service Unit

DTE   Data Terminal Equipment

E

EA   1. Extension Address, 2. Emergency Action

EA/CD   Emergency Action/Control Display Channel

EB   Electronic Bonding

EDI   Electronic Data Interface

EFD   Event Forwarding Discriminators

EGP   Exterior Gateway Protocol

EIR   1. Equipment Identity Register, 2. Equipment Identity Requester

EM   Element Manager

EML   Element Management Layer

EMS   Element Management System

ESN   Emergency Service Number

ESMR   Enhanced Specialized Mobile Radio

Euro-ISDN   European-Integrated Services Digital Networks

F

FCC   Federal Communications Commission

FDDI   Fiber Distributed Data Interface (data)

FDDI II   Fiber Distributed Data Interface (voice and data)

FDM   Frequency Division Multiplexing

FEBE   Far End Block Error

FECN   Forward Explicit Congestion Notification

FERF   Far End Reporting Failure

FHR   Fixed Hierarchical Routing

FIFO   First In, First Out

FITL   Fiber In The Loop

FOTS   Fiber Optic Terminating Systems

FRADs   Frame Relay Access Devices

FR   Frame Relay

FRS   Frame Relay Switch

FTTC   Fiber To The Curb

FTMP   File Transfer Management Protocol

FTP   File Transfer Protocol

G

GDMO   Guidelines for the Definition of Managed Objects

GIS   Graphical Information System

GNE   Gateway Network Element

GOS   Grade Of Service

GSM   Global System for Mobile communication

GSM   Groupe Speciale Mobile

GUI   Graphical User Interface

H

HDSL   High-bit-rate Digital Subscriber Line

HDT   Host Digital Terminals

HDTV   High Definition Television

HEC   Header End Control

HFC   Hybrid Fiber/Coax

HFW   Hybrid Fiber/Wireless

HLR   Home location register

HMMP   HyperMedia Management Protocol

HMMS   HyperMedia Management Schema

HMOM   HyperMedia Object Manager

HTML   HyperText Markup Language

HTTP   HyperText Transfer Protocol

HU   High Usage

I

ICI   Interexchange Carrier Interface

ICMP   Internet Control MessageProtocol

IDEAS   Intelligent Dynamic Event Analysis Subsystem

IDL   interface definition language

IETF   Internet Engineering Task Force

IEX   InterEXchange carrier

IH   Intermediate High usage

ILMI   Interim Local Management Interface

ILMI MIB   Interim Local Management Interface Management Information Base

IMEI   International Mobile Equipment Identity

IN   1. Intelligent Network, 2. Intelligent Node

IOS   Internetwork Operating System

IP   Internet Protocol

IPMH   Interprocess Message Handler

IPMHD   Interprocess Messaging Handler Daemon

IPX   Internet Package Exchange

IPX/SPX   Internet Package Exchange/ Sequenced Package Exchange

iSAC   integrated Service Activation Controller

iSAC/CA   integrated Service Activation Controller client application

iSAC/GUI   integrated Service Activation Controller Graphical User Interface

iSAC-NSAM   integrated Service Activation Controller Network Service Activation Manager

iSAC/PDT   integrated Service Activation Controller process definition tool

iSAC/TDT integrated Service Activation Controller/Task Definition Tool

iSAC/WE integrated Service Activation Controller/Workflow Engine

ISCP intelligent Service Control Point

ISDN integrated Services Digital Network

ISDN PRI Integrated Services Digital Network Primary Rate

ISO International Standards Organization

ISP Internet Service Provider

ISSI InterSwitching Interface

ISV Independent Software Vendor

ITU International Telecommunications Union

ITU-T Telecommunications Standardization Sector of the International Telecommunications Union

IXC Interexchange Varrier

IU International Union

J

JaMAPI Java Management Application Programming Interface

K

Kbps Kilo bits per second (Thousand bits/second)

L

LAN Local Area Network

LAPD Link Access Procedure D

LAT Local Area Transport

LATA Local Access and Transport Area

LCN Logical Channel Number

LEC Local Exchange Carrier

LLC Logical Link Control

LLC1 Logical Link Control 1

LOS Loss of Service

LT Logical Terminal

LTE Line Terminating Equipment

M

M2M Manager-to-Manager

MAC Media Access Control

MAN Metropolitan Area Network

Mbps Megabits per second (Million bits/second)

MCU Multipoint Control Unit

MF Mediation Function

MI Management Information

MIB Management Information Base

MIF Management Information Format

MIM Management Information Model

MO Managed Object

MOCS Managed Objects Configuration System

MOM Manager Of Managers

MOP Maintenance Operations Protocol

MR Modified Read

MS Mobile Station

MSC Mobile Switching Center

MTSO mobile telephone switching office

MUX Multiplexor

N

NCIH North Carolina Information Highway

NE Network Element

NEF 1. Network Element Function, 2. Network Element Facility

NIU Network Interface Units

NMA Network Monitoring and Analysis

NMF Network Management Forum

NML Network Management Layer

NML/SML Network Management Layer/ Service Management Layer

NMS Network Management System

NNI 1. Network Node Interface, 2. Network to Network Interface

NOC Network Operations Center

NSAP Network Service Access Point

NSDB Network and Service DataBase

NT Network Termination

NT1 Network Termination type 1

NT2 Network Termination type 2

NVRAM Non-Volatile Rapid Access Memory

NX NetExpert (Please w. o.: Search and Replace NX w/ NetExpert)

NXCMIP NetExpert Common Management Information Protocol

NXV NetExpert Vectors

O

OAM Operation, Administration, and Maintenance

OAM&P Operation, Administration, Maintenance, and Provisioning

OCE Operation Creation Environment

ODP Open Distributed Processing

OID Object Interface Definition

OMA 1. Object Management Architecture, 2. Open Management Architecture

OMG Object Management Group

OMG CORBA Object Management Group CORBA

OMNIPoint Open Management Interoperability Point

OPX Operator interface

ORB Object request broker

OS Operating system

OS/NE Operating system/Network Element

OSF Operations Systems Functions

OSI Open System Interconnection

OSS Operations Support System

P

PAD Packet Assembler/Disassembler

PBX Private Branch Exchange

PC Personal Computer

PCM Pulse Code Modulation

PCMCIA Personal Computer Memory Card International Association

PCS Personal Communication Service

PDU Protocol Data Unit

PHY Physical Layer

PICS Plug-in Inventory Control System

PIN Personal Identification Number

PLMN Public Land Mobile Network

PMD Physical Medium Dependent

POP Point Of Presence

POTS Plain Old Telephone Service

PPP Point to Point Protocol

PRI Primary Rate Interface

PS Packet Switching

PSTN Public Switched Telephone Network

PTT Postal Telephone and Telegraph

PVC Permanent Virtual Circuit

Q

QAF 1. Quality Assurance Function, 2. Q Adaptor Function

QoS Quality of Service

R

RAM Rapid Access Memory

RBOC Regional Bell Operating Company

RADSL Rate-Adaptive Digital Subscriber Line

RDC Remote Diagnostics Center

RDT Recall Dial Tone

RF Radio Frequency

RMON Remote Monitoring Specification

ROI Return On Investment

ROSE Remote Operations Service Element

S

SA Service Adapter

SCMP Simple Connection Management Protocol

SCP Service Control Point System Control Program

SDH Synchronous Data Hierarchy

SDSL Single-line Digital Subscriber Line

SEFS Severely Erred Framing Seconds

SGML Standardized Generalized Markup Language

SIM SIMulator

SIP SMDS Interface Protocol

SIR Sustained Information Rate

SLIP Serial Line Internet Protocol

SMAE System Management Application Entities

SMASE System Management Application Service Element

SMDS 1. Switched Megabit Data Service, 2. Switched Multimegabit Data Service

SMF 1. Systems Management Function, 2. Station Management Function

SMI Structure of Management Information

SML Service Management Layer

SMS Service Management System

SNA Systems Network Architecture

SNI System network Interface

SNMP Simple Network Management Protocol

SO Switching Office

SOCS/SOAC Service Order Control System and Service Order Analysis and Control

SONET Synchronous Optical Network

SPINA Subscriber PersonalIdentification Number Access

SPINI Subscriber Personal Identification Number Intercept

SPIRIT Service Providers Integrated Requirements for Information Technology

SPVC Semi-Permanent Virtual Circuit

SQL Structured Query Language

SRES Signed Response

SS7 Signaling System number 7

SSP Service Switching Point

STP Signaling Transfer Point

STDM Statistical Time Division Multiplexing

STP Signaling Transfer Point

SVC 1. Switched Virtual Call, 2. Switched Virtual Circuit, 3. Switched Virtual Connection

SW Switch

SXF AccessCNM Core

T

T1/E1 High capacity networks designed for the digital transmission of voice, data, and video.

TARP Target Address Resolution Protocol

TCA Threshold Crossing Alarms

TCP/IP Transmission Control Protocol/Internet Protocol

TDM Time Division Multiplexing

TDMA Time Division Multiple Access

TE1 Terminal Type 1

TE2 Terminal Type 2

Telco Telephone Companies

TFTP Telnet File Transfer Protocol, sometimes referred to as Trivial

TL1 Transaction Language 1

TINA Telecommunications Information Networking Architecture

TIRKS Trunk Integrated Record Keeping System

TM Terminal Multiplexer

TMN Telecommunications Network Management

TONICS TelOps Network Integrated Control System

V

VC Virtual Circuit

UBR Unspecified Bit Rate

UDP User Datagram Protocol

UDP/IP User Datagram Protocol/Internet Protocol

UME UNI Management Entry

UNI User Network Interface

UPSR Unidirectional Path Switch Ring

URC 1. Uniform Resource Citations, 2. Uniform Resource Characteristics

URL Uniform Resource Locator

URN Universal Resource Names/Numbers

V

VA 1. VisualAgent, 2. Virtual Address

VBR Variable Bit Rate

VC   Virtual Connection
VCC   Virtual Channel Connection
VCI   Virtual Circuit Identifier
VCL   Virtual Channel Link
VDSL   1. Very high-rate Digital Subscriber Line, 2. Virtual Digital Subscriber Line
VLR   1. Visitor Location Register, 2. Visible Location Requester
VPC   Virtual Path Connection
VPI   Virtual Path Identifier
VPN   Virtual Private Network
VT   Vertical Tab

W

WAN   Wide Area Network

WBEM   Web-Based Enterprise Management
WDM   Wave Division Multiplexing
WFA   Work Force Administration
WSF   Work Station Function

X

xDSL   X Digital Subscriber Line (where x is generic)
XMP   X/open Management Protocol
XOM   X/open OSI-abstract data Manipulation
XOT   X.25 Over TCP/IP

Index

Application redundancy, 15
Application rules, NetExpert, 18–23
Apply Engineering Data, 196
Architecture
　amaMANAGER, 232–233
　IPMH, 132–133
　IPMHD, 132
　mobileMASTER, 278
　NCIH, 307–308
　NetExpert framework, 122, 124–147,
　　161–163
　　gateways, 126–130
　　operator workstation, 139–147
　　servers, 130–139
　open, growth potential, 254
　TINA, 38
　TMN NetExpert support, 167–170
Archive daily data process, 197
ASN.1 protocol, 172
Associate customer, service creation, 189–190
Associate Data, service creation, 190, 192–193
Associate Service, service creation, 191
Association Control Service Element (ACSE),
　94, 96
Asymmetric digital subscriber line
　(ADSL), 72
Asynchronous communication gateways, 127
Asynchronous Digital Subscriber (ADSL)
　systems, 331–332
ATM, 16, 59–63
　ASCI case study, 415–417
　challenges and status of network management,
　　300–302
　layers supported by, 77
　NCIH equipment, 308–310, see North Carolina
　　Information Highway
　standards, evolution of, 303–306
　tag switching, 76
　TONICS, 314–319
　transportMASTER and, 292
ATM Adaptation Layer (AAL), ATM standards,
　303, 304
Attributes, Identification Rules, 155
Audit trails, DataArchiver and, 237
AuthAgent features/properties, 135–136
Authentication, see Security management; User
　authentication
Authentication center (AC), GSM, 242–243, 249
Authorization Editor, 147, 159
AutoCad, 150
Automatic Message Accounting (AMA)
　performance monitoring and management, 257
　amaMANAGER, see amaMANAGER
Automatic service provisioning, switchMASTER
　features, 213

Automation
　configuration management, competition
　　in marketplace, 255–257
　mobileMASTER, 272
Availability, 15–16

B

Back hauling, SONET/SDH, 63
Backup
　increasing availability, 15
　mobileMASTER, 272
Backup interfaces, increasing
　availability, 16
Backward Explicit Congestion Notification
　(BECN), 50
Bandwidth
　cable LAN, 332
　customer network management services,
　　356–357
　emerging technologies, see specific emerging
　　technologies
　SONET management of, 288–289
　T1 subrates, 35
Bandwith-on-demand performance, 437
Base Station Controller (BSC)
　GSM, 244, 245, 251
　NetExpert framework interoperability, 264
Base Station Subsystem (BSS)
　functional architecture, 242–243
　operations, 250–251
Base Transceiver Station (BTS), 242–244, 251
Baud rates, service creation, 188
Bellcore ISDN technology, 45
BellSouth BOSS, 311–314
Bidirectional line switch ring (BLSR), 284
Billing
　accounting management, 260, see also
　　Accounting management
　amaGATEWAY, 232
　business model of operations, 6–7
　DataArchiver and, 235
　layers of management, 5
　NetExpert capabilities, 268
　performance monitoring and management, 257
　TONICS, 315, 316
Bit error rate (BER)
　DataArchiver and, 234–235
　SONET, 289
Black-listed status, GSM security, 249
BOSS (Broadband Operations Support System),
　311–314
Broadband services
　ISDN, SONET/SDH, 67
　market trends, 16–17